Structure/Performance Relationships in Surfactants

ACS SYMPOSIUM SERIES 253

Structure/Performance Relationships in Surfactants

Milton J. Rosen, EDITOR

*Brooklyn College of the
City University of New York*

Based on a symposium sponsored by
the Division of Colloid and Surface Chemistry
at the 186th Meeting
of the American Chemical Society,
Washington, D.C.,
August 28–September 2, 1983

American Chemical Society, Washington, D.C. 1984

Library of Congress Cataloging in Publication Data

Structure/performance relationships in surfactants.
  (ACS symposium series, ISSN 0097–6156; 253)

  "Based on a symposium sponsored by the Division of
Colloid and Surface Chemistry at the 186th Meeting of
the American Chemical Society, Washington, D.C.,
August 28–September 2, 1983."

  Includes bibliographies and indexes.

  1. Surface active agents—Congresses. 2. Surface
chemistry—Congresses.

  I. Rosen, Milton J. II. American Chemical Society.
Division of Colloid and Surface Chemistry.
III. American Chemical Society. Meeting (186th: 1983:
Washington, D.C.) IV. Title. V. Series.

TP994.S77   1984   668'.1   84–6384
ISBN 0–8412–0839–5

ACS Symposium Series

M. Joan Comstock, *Series Editor*

Advisory Board

FOREWORD

The ACS SYMPOSIUM SERIES was founded in 1974 to provide a medium for publishing symposia quickly in book form. The format of the Series parallels that of the continuing ADVANCES IN CHEMISTRY SERIES except that in order to save time the papers are not typeset but are reproduced as they are submitted by the authors in camera-ready form. Papers are reviewed under the supervision of the Editors with the assistance of the Series Advisory Board and are selected to maintain the integrity of the symposia; however, verbatim reproductions of previously published papers are not accepted. Both reviews and reports of research are acceptable since symposia may embrace both types of presentation.

CONTENTS

EFFECT OF STRUCTURE ON PERFORMANCE
IN VARIOUS APPLICATIONS

PREFACE

Worldwide consumption of surfactants, even excluding soaps, is now measured in billions of pounds. In the United States alone, surfactants are a multibillion dollar industry, with hundreds of different types used in industrial and consumer products for a multitude of different purposes.

Although a vast literature exists on the properties of surfactants, the amount of data from which reliable chemical structure/property relationships can be drawn is surprisingly meagre. A major reason for this paucity of data is the marked effect that very small quantities of highly surface-active impurities can have on the properties of surfactants in solution, coupled with the difficulty of removing some of these impurities when they are present. Moreover, many investigations into the properties of surfactants have dealt with commercial materials that are mixtures of surface-active compounds, and structure/property correlations so based are often of questionable validity because of insufficient characterization of the materials investigated. On the other hand, the effect of surface-active impurities has been known for about 40 years now, and careful investigators have compiled a body of data that, although limited in scope, has made available some information on chemical structure/property relationships.

Recent years have seen a revival of interest in the study of surfactants and their properties, in part due to their potentialities for use in enhanced oil recovery. In addition, greater awareness of the effects of impurities, the availability of a variety of high-purity surfactants from a number of commercial sources, and improved methods for characterizing and purifying materials have resulted in an increased number of investigations containing data on surfactant properties from which reliable conclusions can be drawn.

The symposium on which this book is based assembled the results of worldwide research on surfactant structure/performance relationships by most of the active groups throughout the world. Most of the papers included here embody recent research results; a few are invited overview papers.

The use of the facilities of the Department of Chemistry of Brooklyn College, City University of New York, greatly expedited the editing of the manuscripts and is gratefully acknowledged. I should like to thank Theresa Rudd and Barbara Fudge of the secretarial staff of the department for typing revised portions of the manuscripts and the necessary correspondence. My thanks also to the unnamed referees for their conscientious examination of the manuscripts, and to the authors for their interest in the symposium and their cooperation in making this volume possible.

MILTON J. ROSEN
Great Neck, New York

February 1984

EFFECT OF STRUCTURE ON SURFACE AND MICELLAR PROPERTIES OF INDIVIDUAL SURFACTANTS

Interfacial and Performance Properties of Sulfated Polyoxyethylenated Alcohols

M. J. SCHWUGER

Henkel KGaA, (4000) Düsseldorf, Box 1100, Germany

Alkyl ether sulfates are, after alkyl benzene sulfonates (LAS), the group of technically important anionic surfactants with the largest production volume and product value. They have in comparison with other anionic surfactants special properties which are based on the particular structure of the molecule. These are expressed, for example, in the general adsorption properties at different interfaces, and in the Krafft-Point. Alkyl ether sulfates may be used under conditions, at which the utilization of other surfactant classes is very limited. They possess particularly favorable interfacial and application properties in mixtures with other surfactants. The paper gives a review of all important mechanisms of action and properties of interest for application.

Alkyl ether sulfates with chain lengths ranging from C_{12} to C_{14} are quantitatively the most important products currently based on fatty alcohols. It is estimated, that about 20 % of all surfactant alcohols - about 40 % of all fatty alcohols in the coconut range (C_{12}-C_{14}) - are used in the form of alkyl ether sulfates (1). Alkyl ether sulfates are the most important group of anionic surfactants after linear alkyl-benzenesulfonate (LAS) (2).

In 1980, the total consumption of alkyl ether sulfates in Western Europe equaled that of all other anionic surfactants with the exception of alkylbenzene-sulfonates (Fig. 1). Since alkyl ether sulfates are the most expensive group of anionic surfactants, which are produced in larger quantities, their importance on a

0097–6156/84/0253–0003$06.00/0

value basis is even more pronounced. The main areas of application of fatty ether sulfates in Europe are cosmetic rinse-off preparations (shampoos, bubble baths, shower baths) and manual liquid dishwashing detergents.

In addition, in the USA ether sulfates are used on a large scale in laundry detergents. For that reason, their market share in the USA is even larger than in Europe (3). At present, alkyl ether sulfates are primarily used in the form of their Na salts. In the past, however, cosmetic preparations containing ammonium and Mg salts have also been quite common (4).

The great technical and economic importance of this product group was reached despite its higher price only because of its special properties. Due to the ionic sulfate group and the adjacent ether groups, ether sulfates combine the classical elements of ionic and nonionic surfactants in one molecule. This provides a number of properties, one of which, the Krafft-Point, is of special importance for the technical application of these compounds.

Krafft Points

The Krafft Point may be defined as the temperature above which the solubility of a surfactant increases steeply. At this temperature, the solubility of the surfactant becomes equal to the critical micelle concentration (c_M) of the surfactant. Therefore, surfactant micelles only exist at temperatures above the Krafft Point. This point is a triple point at which the surfactant coexists in the monomeric, the micellar, and the hydrated solid state (5, 6).

The temperature dependence of the solubility is demonstrated in Fig. 2 for Na dodecyl sulfate (7). Below the Krafft Point, the surfactant dissolves in a molecularly dispersed manner until the saturation concentration is reached. At higher concentrations, a hydrated solid is in equilibrium with individual molecules. Above the Krafft Point, the hydrated solid is in equilibrium with micelles and individual molecules.

Therefore, the physical meaning of the solubility curve of a surfactant is different from that of ordinary substances. Above the critical micelle concentration the thermodynamic functions, for example, the partial molar free energy, the activity, the enthalpy, remain more or less constant. For that reason, micelle formation can be considered as the formation of a new phase. Therefore, the Krafft Point depends on a complicated three phase equilibrium.

With increasing length of the n-alkyl chain an

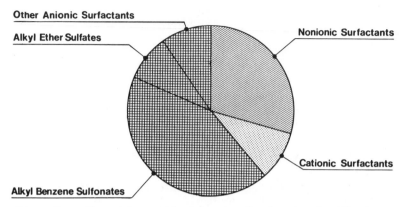

Fig. 1   Use of synthetic surfactants in Western
          Europe (1980)

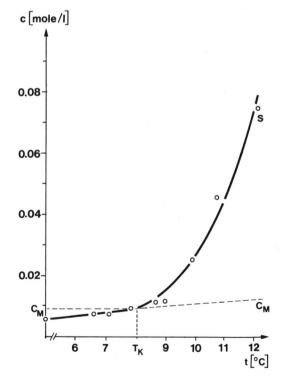

Fig. 2   Solubility of sodium dodecyl sulfate
          (purity 99.8%)

increase of the Krafft Point is observed. The Krafft
Points of many common surfactants, such as alkyl sulfates, alkane sulfonates, p-n-alkyl benzene sulfonates,
lie above room temperature (7, 8). Since the solubility
of the surfactants depends both on the aqueous and on
the crystalline phase, the heat of formation of
crystals must be increased in order to effect a depression of the Krafft Point. This can be achieved by
branching in the hydrophobic portion of the surfactant
or by using surfactant mixtures of different chain
length. With Na salts, these measures lead to a sufficiently large depression of the Krafft Point. An
application of the respective surfactants in Ca salt
form is not possible, because the Krafft Point is
mostly too high.   For example, for Na-tetradecyl
sulfate, the Krafft Point is 21 ^oC, whereas the Ca salt
has a value of 67 ^oC (9). Therefore, in a number of
large fields of application, such as detergents, many
surfactants may only be used in combination with
complexing agents and/or ion exchangers.
　　An especially effective reduction of the Krafft
Point results from the insertion of ether groups into
the molecule of the anionic surfactant. In table I this
is examplified with Na dodecyl sulfate and Na-tetradecyl sulfate in comparison to various n-alkyl ether
sulfates of the same chain length (10). As a measure of
the Krafft Point, a temperature is defined at which a
1 % solution dissolves clearly. By the incorporation of
oxyalkylene groups into the molecule, the Krafft Point
and the melting point are greatly depressed. This
depression is especially effective if there is
branching in the oxyalkylene groups.
　　The depression of Krafft Points of Ca salts (11)
is of special importance from an application point of
view. As shown in table II, Ca dodecyl sulfate has a
Krafft Point of 50 ^oC. The introduction of one oxyethylene group into the molecule results in a 35 ^oC
reduction. The reduction is more strongly pronounced
than with the corresponding Na salts. These peculi-
arities of alkyl ether sulfates are of great importance for the selection of these surfactants in a
number of fields of application. The Krafft Points are
shifted to higher temperatures with increasing length
of the hydrocarbon chain and, within a particular group
of the Periodic System, with increasing atomic weight of
the cation, as well as, within a period of the Periodic
System, with increasing valency (12).

TABLE I

Temperature for the Solubility of 1 % Solutions
and Melting Points of R $(OCH_2CHR')_mOSO_3Na$

R	$(OCH_2CHR')_m$	Melting Point ($^\circ$C)	Temperature ($^\circ$C)
$C_{12}H_{25}$	None	190 - 95	16
$C_{12}H_{25}$	OC_2H_4	143 - 146	11
$C_{12}H_{25}$	$(OC_2H_4)_2$	126 - 136	< 0
$C_{12}H_{25}$	$OCH_2CH(CH_3)$	137 - 142	< 0
$C_{12}H_{25}$	$[OCH_2CH(CH_3)]_2$	87 - 93	< 0
$C_{12}H_{25}$	$OCH_2CH(C_2H_5)$	77 - 82	< 0
$C_{12}H_{25}$	$[OCH_2CH(C_2H_5)]_2$	< 25	< 0
$C_{14}H_{29}$	None	182 - 183	30
$C_{14}H_{29}$	OC_2H_4	146 - 150	25
$C_{14}H_{29}$	$(OC_2H_4)_2$	130 - 134	< 0
$C_{14}H_{29}$	$OCH_2CH(CH_3)$	139 - 140	14
$C_{14}H_{29}$	$[OCH_2CH(CH_3)]_2$	82 - 87	< 0
$C_{14}H_{29}$	$OCH_2CH(C_2H_5)$	74 - 76	13
$C_{14}H_{29}$	$[OCH_2CH(C_2H_5)]_2$	< 25	< 0

TABLE II

Krafft Points ($^\circ$C) of n-Dodecyl Ether Sulfates

Surfactant anion	Na salt	Ca salt	Sr salt	Ba salt
$C_{12}H_{25}OSO_3^-$	9	50	64	105
$C_{12}H_{25}OCH_2CH_2OSO_3^-$	5	15	32	62
$C_{12}H_{25}(OCH_2CH_2)_2OSO_3^-$	- 1	< 0	--	35
$C_{12}H_{25}(OCH_2CH_2)_3OSO_3^-$	< 0	< 0	< 0	12

The Special Character of Oxyethylene Groups in Alkyl Ether Sulfates

The lower Krafft Points resulting from the incorporation of oxyethylene groups into the surfactant molecule is an essential, but not sufficient, property for the utilization of alkyl ether sulfates.

Several variations in chemical constitution, which lead to a depression of the Krafft-Point (for example, branching of the hydrophobic part of the molecule), frequently result in diminished hydrophobicity of the molecule. At constant molecular weight, the critical micelle concentration (c_M) is shifted with increased branching to higher concentrations, the surface activity diminishes, the tendency to adsorb at hydrophobic interfaces decreases, etc. (13, 14, 15). Therefore, the nature of the oxyethylene groups in alkyl ether sulfates is of major importance.

From a physical chemical viewpoint, oxyethylene groups located adjacent to the ionic sulfate group may be considered at the first glance as additional hydrophilic groups, in analogy to nonionic ethylene oxide (EO) adducts.

For pure nonionic EO adducts, increase in the number of oxyethylene groups in the molecule results in a decrease in the tendency to form micelles and an increase in the surface tension of the solution at the critical micelle concentration (16) (17). This change in surface activity is due to the greater surface area of the molecules in the adsorption layer and at the micellar surface as a result of the presence there of the highly hydrated polyoxyethylene chain. The reduction in the tendency to form micelles is due to the increase in the free energy of micelle formation as a result of partial dehydration of the polyoxyethylene chain during incorporation into the micelle (16) (17).

In the case of alkyl ether sulfates, an increase in the number of oxyethylene groups produces an opposite result (19 - 22). Fig. 3 shows a decrease of the critical micelle concentration (c_M) and the concentration for a given surface tension (for example, 50 mN/m). Only the same increase in the surface tension for solutions at the critical micelle concentration is observed. Therefore, EO incorporation into the surfactant molecule seems to increase the hydrophobic nature of the molecule. This is apparently in contradiction to known correlations between solubility and hydrophobicity.

Nonionic EO adducts always have much smaller c_M values than ionic surfactants with the same hydrophobic group. For this reason it was proposed that the introduction of oxyethylene groups into an alkyl sulfate ion weakened its ionic character. This should result in an increased approximation of the properties to those of nonionic surfactants (19, 23, 24). This conception was, however, always very questionable because the sulfate group remain ionic.

The dissociation of the counter-ions of micelles is a characteristic feature of ionic surfactants. For alkyl ether sulfates the degree of dissociation α of the counter-ions of the micelles has been determined by several authors by various methods (21, 25, 26). With increasing number of oxyethylene groups in the molecule the degree of dissociation increases. These data are summarized in Fig. 4. Although the agreement of the absolute values between individual authors is not quite satisfying, they all show an increase of dissociation. This is just contrary to what should be expected with a decrease in ionic character.

Moreover, it is characteristic of ionic surfactants that c_M is greatly reduced by electrolyte addition and that, simultaneously, the surface tension curves in the nonmicellar region are shifted significantly to smaller concentrations (Fig. 3). With nonionic EO adducts the influence of electrolyte is barely present. The curves of alkyl ether sulfates are shifted in the same way as simple alkyl sulfate by the addition of salt. This shift is even larger than with Na dodecyl sulfate. An additional indication of the ionic nature of alkyl ether sulfates is derived from static light scattering. The micelle aggregation numbers are presented in Table III. As can be expected for ionic surfactants (27), they increase in the presence of NaCl.

Therefore, the hypothesis of an increasing nonionic character of alkyl ether sulfates with increasing number of oxyethylene groups is not tenable. Some time ago (30), it was suggested that a certain hydrophobic nature can be attributed to the polyoxyethylene chain of alkyl ether sulfates. At first, this appears to be in contradiction to the decidedly hydrophilic character of the polyoxyethylene chain for nonionic surfactants. However, the possibility of EO group hydration impairment by the sulfate group cannot be excluded.

Table III shows some data regarding the possible hydrophobic nature of ether sulfates. From several investigations, it is known that, for nonionic surfactants with identical hydrophobic groups, an increase in the hydrophilic part of the molecule causes a decrease in the aggregation number (28). This is caused by the

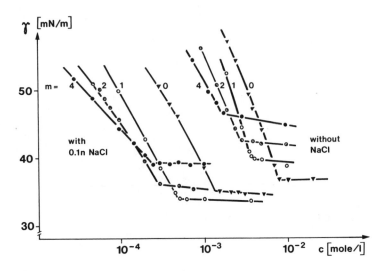

Fig. 3   Surface tension of aqueous solutions of $C_{12}H_{25}(-O-CH_2-CH_2-)_mOSO_3Na$ at 25°C (purity 98-99.5%)

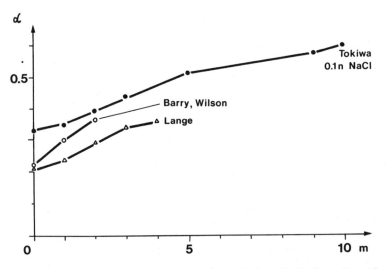

Fig. 4   Degree of dissociation (α) of dodecyl ether sulfate micelles
m = number of oxyethylene groups in the molecule

increased affinity of the polyethylene chain for water.
Comparing the change of the micelle aggregation number
at the transition from dodecyl sulfate to dodecyl monoglycol ether sulfate, an increase can be observed with
and without salt. This corresponds to the known increase of the micelle aggregation number with increasing length of the n-alkyl chain (29). However,
this increase is less pronounced in comparison with
dodecyl and tetradecyl sulfate.

TABLE III

Micelle Aggregation Numbers at 25 °C

Substance	without salt	with 0.1 n NaCl
$C_{12}H_{25}OSO_3Na$	60	91
$C_{12}H_{25}OCH_2CH_2OSO_3Na$	72	96
$C_{14}H_{29}OSO_3Na$	80	--

In addition, there may be still another possible
reason for the observed effects. The hydration of the
polyglycol portion of the micelle increases with
increasing number of oxyethylene groups (31). The area
demand of a molecule adsorbed at the surface, and,
therefore, also the distances between the charged
groups will become larger. Since, in addition, the
chains are more or less stretched because of the
electrical repulsion of the terminal sulfate groups,
the distance between sulfate groups also increases.
For this reason, the work to overcome the electrical
repulsion is diminished, the c_M reduced, and the
aggregation number increased. This model was checked
experimentally with the adsorption at the water/air
interface (21, 22).
    Fig. 5, curve 1, shows - for dodecyl ether
sulfates - the areas occupied by a molecule at c_M as a
function of the number, m, of EO groups. The values were
calculated by the Gibbs' equation from the surface
tension measurements. For $0 \le m \le 2$ there is an area
increase with increasing m, however, considerably

weaker than for m > 2. For comparison the area values for dodecyl polyglycol ethers are also given. The two different curves were taken from different sources (16, 18). With alkyl ether sulfates the increase of area per molecule and, therefore, also of the distance between the terminal groups in the adsorption layer is smaller, at m = 0 to m = 2, and larger, at m > 2, than for the corresponding nonionic compounds.

Parallel results were found with micelles (31). It was concluded from measurements of sedimentation, diffusion, and viscosity, that the hydration of the micelles of the dodecyl ether sulfates at m = 0 - 2 shows only a little increase, whereas a strong one was observed at m > 2. A similar trend should also exist with the distance of the terminal groups on the surface of the micelles.

For compounds with one and two oxyethylene groups in their molecule, the increase of the micelle formation tendency can be explained by a contribution of these groups to the hydrophobic part of the molecule. With a longer polyoxyethylene chain in the molecule, however, the increased tendency to form micelles is primarily caused by the increased distance between the charged groups due to increased hydration of the ether groups.

Fig. 3 shows that increase in the number of oxyethylene units in the dodecyl ether sulfate molecule results in increased adsorption at the aqueous solution/air interface at concentrations below c_M. It was interesting to determine whether an analogous effect is shown for adsorption onto solid surfaces. Measurements of the adsorption of Na dodecyl sulfate and dodecyl ether sulfates onto activated carbon at the constant and very low surfactant concentration of $1 \cdot 10^{-4}$ mol/l, which is well below c_M (22, 23), show that the incorporation of oxyethylene groups leads to an increase of adsorption (Table IV). The effect, however of one group is far weaker than the influence of an extension of the hydrocarbon chain by two CH_2 groups. Contrary to this (Fig. 6), the saturation values of adsorption onto graphon above the critical micelle concentration decrease with increasing length of the polyoxyethylene chain, as the surface area per adsorbed molecule increases. This observation corresponds with the result of studies on adsorption at the aqueous solution/air interface.

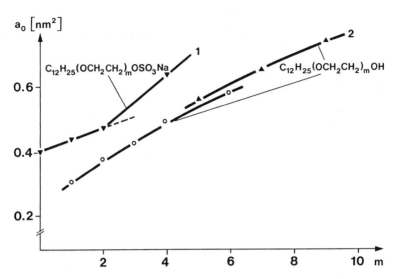

Fig. 5   Areas per molecule at the aqueous solution/
air interface as a function of the number of
oxyethylene groups

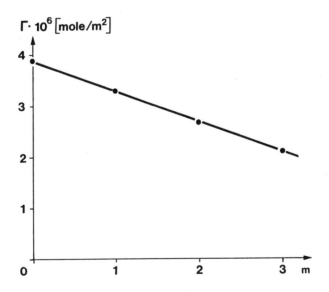

Fig. 6   Adsorption of alkyl ether sulfates on graphon
(plateau values) at 25°C
(product purity 98-99.5%)

TABLE IV

Adsorption of Surfactants on Active Charcoal

Substance	$Q \cdot 10^5$ (mole/g)
$C_{12}H_{25}OSO_3Na$	5.40
$C_{12}H_{25}(OCH_2CH_2)_1OSO_3Na$	7.29
$C_{14}H_{29}OSO_3Na$	8.12
$C_{12}H_{25}(OCH_2CH_2)_2OSO_3Na$	8.82
$C_{16}H_{33}OSO_3Na$	12.67

Practical Results

Washing and Cleaning Action. The properties of alkyl ether sulfates, due to the good solubility and the special hydrophilic/hydrophobic properties of the molecule, are of particular practical interest. From the investigations described in sections 2 and 3, it can be concluded that, in addition to the decrease in the Krafft Point, favorable properties for practical applications can be expected as a result of the inclusion of the oxyethylene groups into the hydrophobic part of the molecule. As is true for other anionic surfactants, the electrical double layer will be compressed by the addition of multivalent cations. By this means, the adsorption at the interface is increased, the surface activity is raised, and, furthermore, the critical micelle concentration decreased. In the case of the alkyl ether sulfates, however these effects can be obtained without encountering undesirable salting out effects.

In Fig. 7, this is exemplified with surface tension concentration curves for Na n-tetradecyl diethyleneglycol ether sulfates (33). Less soluble surfactants would produce with increasing water hardness increased formation of sparingly soluble Ca salts. Therefore, the critical micelle concentration would be shifted toward much larger concentrations.

However, in the case of Na tetradecyl dioxyethylene sulfate, the surface tension and the critical micelle concentration will be reduced in the presence of water hardness. If a complexing agent is added, the effect is weakened because of the complexing of the

multivalent alkaline earth ions. Thus alkyl ether sulfates have special advantages in hard water and in
products containing no complexing agents.

In Fig. 8, the soil removal from woolens by
various technically important surfactants are presented
as a function of the water hardness (34). The favorable
properties of alkyl ether sulfates in comparison to
other anionic surfactants at higher water hardnesses
are clearly evident.

If sufficient amounts of complexing agents or of
ion exchangers are added, the soil removal of water
hardness-sensitive surfactants, for example LAS,
becomes independent of the Ca ion concentration. However, in the case of alkyl ether sulfates an increase
in the washing effect can still be observed with
increasing water hardness (33). In this case, the
stronger influence of Ca and Mg ions than that of
Na ions can be attributed to their stronger compression
of the electrical double layer at the interface. Only
in the lower range of hardness does the addition of
sodium sulfate bring a certain advantage. In analogy to
the surface tension measurements, there is even a small
deterioration of the soil removal as consequence of
complexing by Na triphosphate present in solution
(Fig. 9).

The results indicate that in detergent formulations without complexing agents or ionic exchangers,
alkyl ether sulfates may have definite advantages.
Detailed investigations of washing conditions in the
United States showed best results with tetradecyl 3 EO
sulfate (35). It may be used preferably in liquid
heavy duty detergents without phosphate (36).

TABLE V

Foam Volume in cm^3 Determined with the Beating Method
Using a Perforated Disk after 30 sec. at $40^{\circ}C$, 1 g/l

Substances (techn.)	Water hardness $10^{\circ}d$	$17^{\circ}d$	Foam Decrease %
C_{12} Sulfate (Na salt)	780	170	78
C_{12-18} Sulfate (Na salt)	870	145	83
C_{12-16} 4EO Sulfate (Na Mg salt)	840	560	33
C_{12} 2EO Sulfate (Na salt)	830	675	19

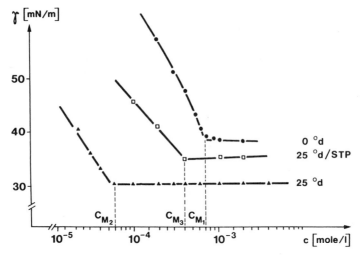

Fig. 7   Surface tension of aqueous solutions of
          sodium tetradecyl diglycol ether sulfate at
          25°C (technical product)

Fig. 8 Soil removal from wool by different anionic surfactants. Test conditions: temperature, 30 °C; time, 15 min; concentration, surfactant 015 g/L + Na_2SO_4 1.5 g/L; soil, sebum/pigment mixture; surfactant (1) C_{12-14} alcohol 2EO sulfate, (2) C_{15-18} olefinic sulfonate, (3) C_{16-18} sulfo fatty ester, (4) C_{12-18} fatty alcohol sulfate, (5) C_{13-18} alkyl sulfonate, (6) C_{10-13} alkyl benzene sulfonate.

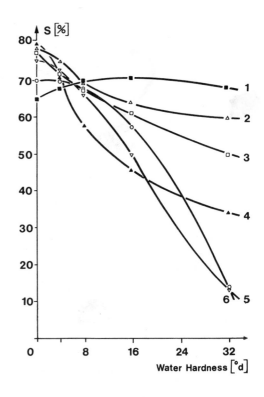

Cosmetic Rinse off Preparations. In certain cosmetic products, for example hair shampoos, it is not possible to use complexing agents because of the irritation of mucous membranes. Here a low sensitivity of surfactants to water hardness is a precondition for their application. Foam formation is generally considered to be a measure of shampoo quality (Table V). With increase in water hardness the foam volume of alkyl sulfates decreases very much, whereas with the corresponding alkyl ether sulfates this decrease is relatively small. For cosmetic applications, the good skin compatibility and low irritation to mucous membranes of alkyl ether sulfates is of high importance (37).

Bulk Properties. Because of viscosity and stability requirements for product manufacture and processing, the appearance of mesomorphous phases in mixtures with water is very important. In the case of dodecyl monoglycol ether sulfate and dodecyl sulfate, a highly viscous middle phase is observed up to a concentration of 80 %. For the corresponding diglycol ether compound, however, the middle phase is present only in a concentration range up to about 65 % (38). Above this concentration, a lamellar neat phase exists.

The middle phase is much more viscous than the neat phase, though the concentration of the latter is higher. If a concentrated system of an alkyl ether sulfate and water existing initially as a neat phase is stepwise diluted, the range of the middle phase will be reached. This is accompanied by a steep increase in viscosity. The formation of the neat phase allows the manufacture and handling of highly concentrated fluid preparations. This property, however, can lead to processing problems upon dilution.

Table VI shows the results of polarized light microscopic observations. Sometimes isotropic regions and the middle phase exist simultaneously. The region of the middle phase is marked by heavy lines. The range of the especially viscous middle phase narrows with transition from two to three oxyethylene groups in the surfactant molecule. Up to 27 %, the system appears optically isotropic. In this concentration range the viscosity can be increased strongly by addition of NaCl, as shown in table VII.

In the case of sodium dodecyl sulfate, there is no corresponding effect. This thickening is at any rate not associated with the formation of the middle phase since the products remain isotropic. However, x-ray diffraction measurements indicate the presence of a crystalline, randomly oriented phase (39). It is still unsettled, whether or not thickening is caused by the formation of a gel system, as described in the literature (40). Various salt additions (Hoffmeister Series) have little effect on the range of existence of the middle phase. The possibility to change the viscosity by means of surfactant concentration or by electrolyte addition has advantages for the manufacture of fluid and gel products containing alkyl ether sulfates.

TABLE VI

Phases of Concentrated Solutions of n-Dodecyl Ether Sulfates

Weight %	Number of EO-Groups			
	0	1	2	3
25		–	–	
30	–	– (M)	– (M)	–
35	M	M	M (–)	–
40	M	M	M	M
50	M	M	M	M
60	M	M	M	M
65			M (N)	N
70	M	M	N (M)	N
80		M		N

– optically isotropic, M middle phase, N neat phase

TABLE VII

Influence of Sodium Chloride on the Viscosity (mPa·s) of Surfactants

Substance (techn.)	Sodium chloride addition %		
	3	5	7
C_{12-14} Sulfate	< 100	< 100	< 100
C_{12-14} 2EO Sulfate	< 100	2,500	20,500

Alkyl Ether Sulfates in Mixtures

In most products, alkyl ether sulfates are used in form of mixtures with other surfactants. Alkyl benzene sulfonate (LAS) is the most important anionic surfactant used in combination with alkyl ether sulfates. As a result, the properties of mixtures of alkyl ether sulfates and LAS are of special practical interest.

General Remarks. In the use of products containing alkyl ether sulfates, oily soil removal as well as dispersion plays an important role. The driving force responsible for the separation of oily soil from a substrate (Fig. 10) is the wetting tension j defined by equation (1):

$$ j = \gamma_{so} - \gamma_{sw} = - \gamma_{ow} \cos\theta_o \qquad (1) $$

The sketch in Fig. 10 shows the equilibrium of forces with an obtuse contact angle in the oil phase (θ_o). In this case the wetting tension, j, of the aqueous phase is positive, which means that the adhering oil droplet is pushed together by the aqueous phase. With the increase in j the tendency of an oil droplet to be cut off and removed from a solid substrate increases. Because of this, the impeding force for the removal of oil is the interfacial tension oil/water (γ_{ow}), which should be minimized. By minimization of the interfacial tension, moreover, the requirements for emulsification and stabilization of soil in the washing and cleaning liquid will be improved.

In general, at hydrophobic surfaces the final stage of the spontaneous complete oil removal will not be achieved, since a wetting equilibrium will be reached. The necessary additional work for complete removal of an oil droplet from the system must be added to the system in the form of mechanical energy.

In application-related problems the question may also be formulated in terms of minimizing the necessary additional work. From knowledge of the interfacial properties of surfactant mixtures the surface activity, tendency to form micelles, adsorption, etc., can be increased. The following effects may pertain:

a) The charged ionic groups of a surfactant in a mixed film may be shielded by the incorporation of a nonionic surfactant. The repulsion of the similarly charged groups is diminished and the impeding electrostatic forces of repulsion reduced.

b) If the hydrophilic head group of one surfactant of the mixture has a weak or strong charge opposite to

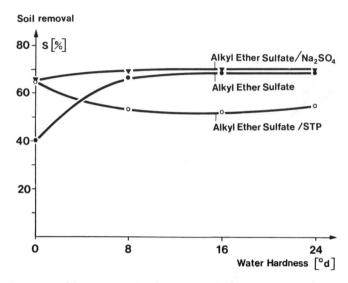

Fig. 9   Soil removal from wool by C_{12-14} 2EO sulfate

Test condition:
temperature: 30°C
electrolyte conc.:   1.5 g/l
surfactant conc.:   0.5 g/l
soil: sebum/pigment mixture

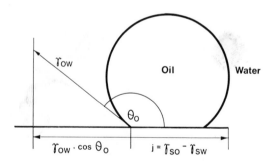

Fig. 10   Oil droplet on substrate in water

that of the other, not only shielding effects but also forces of attraction should be considered. (For example in mixtures of anionic and cationic surfactants, or anionic surfactants and amine oxides).

c) Inhibition of incorporation into an interfacial layer may be also reduced by an enlargement of the distances between the hydrophilic groups of the individual surfactants.

d) In addition to reduction of electrostatic repulsion, intensification of electrostatic attraction or van der Waal's forces of attraction may also be accomplished. This occurs by means of special interaction between the hydrophobic parts of the surfactant molecule. Such a special case was already discussed.

Selected Results. No mutual intensification of interfacial effects is observed between Na dodecyl sulfate as the first member of the homologous series of dodecyl ether sulfates and LAS. In Fig. 11, the oil/water interfacial tensions are shown. Neither the electrostatic nor the van der Waal's interactions of the mixtures are intensified.

However, this situation is changed significantly by the incorporation of a single oxyethylene group into the molecule of the alkyl sulfate. This is shown in Fig. 12 for various dodecyl ether sulfates. In contrast to the mixtures with dodecyl sulfate, a significant decrease of the interfacial tension with well defined minima at certain LAS/alkyl ether sulfate ratios are observed. With increasing number, m, of oxyethylene groups, this effect becomes more pronounced, and the minima of interfacial tension are shifted strongly in favour of LAS-rich solutions. Small additions of alkyl ether sulfates to LAS improve the interfacial properties to a significant extent. Intensification of other interfacial properties in LAS/alkyl ether sulfate mixtures is also to be expected. This is shown in Fig. 13 by means to two substrate specific and two substrate-nonspecific criteria of importance for the washing process. In the series of mixtures investigated, properties such as wetting tension on polyester, contact angle on polyester, olive oil/water, interfacial tension, and emulsification of olive oil all show a definite extreme point. This corresponds to the optimum surfactant mixture and should be also observed under application conditions.

In Fig. 14, the results of a dishwashing test close to practice are presented (41, 42). Plates soiled with fat are cleaned in a detergent solution at 45 °C and the decay of previously generated foam as well as

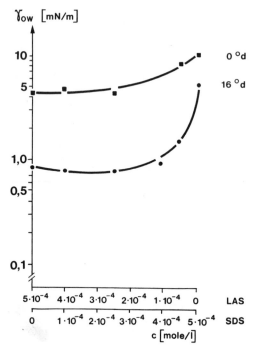

Fig. 11   Olive oil/water interfacial tension for
           LAS/SDS-mixtures

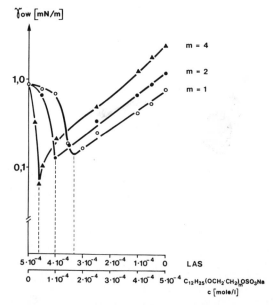

Fig. 12   Olive oil/water interfacial tension for
           LAS/Alkyl ether sulfate mixtures
           (purity: LAS- technical product, ether
           sulfates 98.0-99.5%)

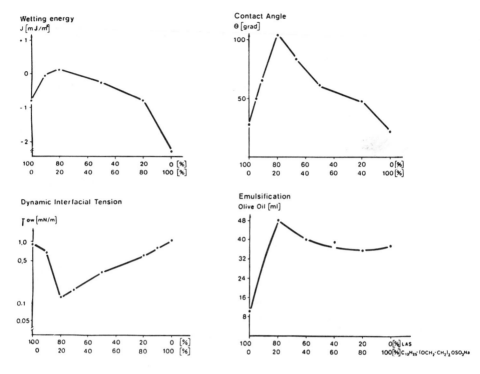

Fig. 13   Interfacial properties of
LAS/$C_{12}H_{25}(-O-CH_2-CH_2-)_2OSO_3Na$ mixtures

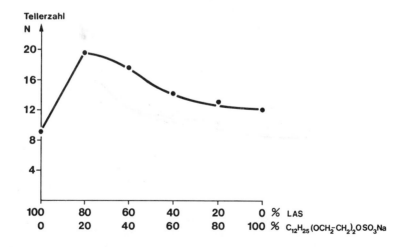

Fig. 14   Dishwashing by LAS/dodecyl 2EO sulfate
mixtures
N: number of plates washed at 45°C
(technical products)

possible deposition of fat are evaluated. As a measure
for the quality of a detergent the number, N, of plates
washed until the foam decays was used. The largest
number of perfectly clean dishes is obtained in the
case of an equal ratio of LAS and dodecyl ether sulfate,
as could be predicated from the measurement of interfacial properties. In addition, similar results where
obtained in washing experiments, where the maximum of
soil removal was found to be at the same ratio of surfactants (Fig. 15).

    Physicochemical and application results show that
mixtures of LAS and alkyl ether sulfates have especially positive properties. This makes it understand-
able why products containing both types of substances
are so widely used.

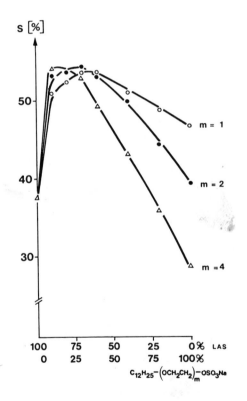

Fig. 15   Soil removal from wool by LAS/alkyl ether
           sulfate mixtures (technical products)
           Test condition:
           temperature: 30°C
           concentration: 5 x 10^{-3} mole/l
           soil: sebum/pigment mixture

Literature Cited

1. Glasl, J.; in: "Fettalkohole", Ed. by Henkel KGaA, Düsseldorf, 1981, p. 125
2. Grün, R. von der; Scholz-Weigel, S., Seifen, Öle, Fette, Wachse, 1982, 108, 121
3. Cox, M.F.; Matson, T.P., Soap Cosm. Chem. Spec., 1982, 58, 46
4. Felletschin, G.; J. Soc. Cosm. Chem., 1964, 15, 245
5. Shinoda, K.; Nakagawa, T.; Tamamushi, B.; Isemura, T., in: "Colloidal Surfactants" Academic Press, New York, 1963
6. Shinoda, K.; Becher, P., "Principles of Solution and Solubility", Marcel Dekker Inc. New York, 1978, p. 159
7. Lange, H.; Schwuger, M.J., Kolloid-Z. Z.-Polymere, 1968, 223, 145
8. Tartar, H.V.; Wright, K.A., J. Amer. Chem. Soc., 1938, 61, 539
9. Schwuger, M.J.; Kolloid-Z. Z.-Polymere, 1969, 233, 979
10. Weil, J.K.; Stirton, A.J.; Wrigley, A.N., in: Proc. 5th Int. Congr. on Surface Active Agents, Ediciones Unidas Sa, Barcelona 1969, Vol. I, p. 45
11. Shinoda, K.; Pure and Appl. Chem., 1980, 52, 1195
12. Hato, M.; Tahara, M.; Suda, Y., J. Coll. Interface Sci., 1979, 72, 458
13. Asinger, F.; Berger, W.; Fanhänel, E.; Müller, K.R. J. Prakt. Chem., 1965, 27, 82
14. Götte, E.; Schwuger, M.J., Tenside, 1969, 6, 131
15. Schwuger, M.J.; Chem. Ing. Techn., 1970, 42, 433
16. Lange, H.; in: Proc. 3rd Int. Congr. Surface Active Agents, Cologne 1960, Vol. I, p. 279
17. Rosen, M.J.; Cohen, A.W.; Dahanayake, M.; Hua, X.Y, J. Phys. Chem., 1982, 86, 541
18. Lange, H.; Kolloid-Z., 1965, 201, 131
19. Weil, J.K.; Bristline, R.G.; Stirton, A.J., J. Phys. Chem., 1958, 67, 1796
20. Götte, E.; in: Proc.3rd Int. Congr. Surface Active Agents, Cologne 1960, Vol. I. p. 45
21. Lange, H.; Tenside, 1975, 12, 27
22. Lange, H.; Schwuger, M.J., Colloid Polymer Sci., 1980, 258, 1264
23. Schick, M.J.; J. Phys. Chem., 1963, 67, 1796
24. Schick, M.J.; J. Amer. Oil Chemist's Soc., 1963, 40. 680
25. Tokiwa, F.; J. Phys. Chem., 1968, 72, 4331
26. Barry, B.W.; Wilson, R., Colloid Polymer Sci., 1978, 256, 251

27. Kratohvil, J.P.; J. Colloid Interface Sci., 1980, 75, 271
28. Tanford, C.; Nozaki, Y.; Rohde, M.F., J. Phys. Chem., 1977, 81, 1555
29. Herrmann, K.W.; J. Phys. Chem., 1962, 66, 295
30. Tokiwa, F.; J. Phys. Chem., 1968, 72, 1214
31. Tokiwa, F.; Ohki, K., J. Phys. Chem., 1967, 71, 1343
32. Schwuger, M.J.; Fette, Seifen, Anstrichmittel, 1970, 72, 25
33. Jakobi, G.; Schwuger, M.J., Chemiker-Z., 1975, 99, 182
34. Andree, H.; Krings, P., Chemiker-Z., 1975, 99, 168
35. Stüpel, H.; Scharer, D.H., in: Proc. 7th Int. Congr. Surface Active Agents, Moscow 1978, Vol. III, p. 202
36. Kravetz, L.; Scharer, D.H.; Stüpel, H., in: Proc. 7th Int. Congr. on Surface Active Agents, Moscow 1978, Vol. III, p. 192
37. Kästner, W.; Frosch, P.J., Fette, Seifen, Anstrichmittel, 1981, 83, 33
38. Lange, H.; unpublished data
39. Weiß, A.; unpublished data
40. Ekwall, P.; in: "Advances in Liquid Crystals" Ed. by G.H. Brown, Acad. Press, New York, 1975, Vol. I, p. 1
41. Lehmann, H.J.; Fette, Seifen, Anstrichmittel, 1972, 74, 163
42. Jakobi, G.; in: "Tensid-Taschenbuch", Ed. by H. Stache, Carl-Hanser-Verlag, München, 1981, p. 302

RECEIVED January 10, 1984

Effects of Structure on the Properties of Polyoxyethylenated Nonionic Surfactants

TSUNEHIKO KUWAMURA

Department of Synthetic Chemistry, Gunma University, Kiryu, 376 Japan

Data on the relationship of chemical structure to fundamental properties of polyoxyethylene(POE) non- ionics in aqueous solution are reviewed. These include : 1) the adsorption, micelle formation and thermody- namics of a series of highly purified POE n-alkyl monoethers, varying systematically in chain length of both alkyl and POE groups, 2) the effects on the aqueous properties of multi-chain and alicyclic structure in hydrophobe, 3) the evaluation of hydro- philicity and surface properties for a new class of nonionics, alkyl crown ethers, 4) the adsorption and dissolution behavior of long N-acyl α-amino acid POE monoesters having a short chain of homogeneous POE, $R^1CONR^2CHR^3COO(C_2H_4O)_mH$, with special reference to the structural effects of the α-amino acid residue.

The correct understanding of the relationships between chemical structure and properties in surfactants is most important to both their effective use in many applications and to molecular de- signing of new surfactants. Some reliable information is availa- ble on various structural effects in ionic surfactants. On the other hand, only a limited amount of reliable information is a- vailable for nonionics with much of the data in the literature being insufficient both in reliability and in the variety of structures dealt with, mainly because of the difficulty in ob- taining well-characterized compounds.

This paper will discuss data from the recent literatures and from our laboratory on structural effects for the usual and novel types of polyoxyethylene(POE) nonionics listed in Table I. These will be described for each type of nonionic in the order listed in Table I.

0097-6156/84/0253-0027$06.00/0
© 1984 American Chemical Society

Table I. Scope of Chemical Structure of the Nonionics mentioned
in this Paper

No of series	General formula[a]	Structural feature	POE[b] grade
1	$H(CH_2)_N \dot{-} O(EO)_m H$	Straight chain	Homogeneous
2	$[H(CH_2)_{N/2}]_2 CH \dot{-} O(EO)_m H$ $[H(CH_2)_{N/3}]_3 C \dot{-} O(EO)_m H$	Multi-chain in hydrophobe	Heterogeneous (Poisson distribution)
3	$(CH_2)_{N-1} CH \dot{-} O(EO)_m H$	Alicyclic hydrophobe	
4	$H(CH_2)_N \dot{-}$ (cyclic structure) $_{m-2}$	Cyclic POE[b]	Homogeneous
5	$H(CH_2)_{N-1} CO \dot{-} NCHR''CO \dot{} O(EO)_m H$ over R'	Amido ester inserted between alkyl and POE	

a) EO : $-CH_2CH_2O-$ b) POE : polyoxyethylene

n-Alkyl POE Monoethers (No 1 series)

Although a considerable number of reports have been published on
the surface and micellar properties of nonionics of this fundamental structure, there are only a few investigations using
highly purified compounds in which chain length of either alkyl
or POE group was varied systematically.

Meguro and coworkers recently reported the effect of variation in alkyl chain length(N) on the properties of homogeneous
materials of structure $H(CH_2)_N O(EO)_8 H$, where N was varied from 10
to 15, including both even and odd chain lengths (1, 2). They
found that most of the plots of surface tension at the critical
micelle concentration(cmc) vs. N at various temperature(15 - 40°C)
gave a zig-zag line, decreasing with increase in N and having a
convex break point at each odd number of N. They have observed a
similar tendency also for the plots of surface area per molecule
vs. N. From these results, they suggest that the packing of the
adsorbed film of the surfactant molecule with an odd number of
carbons is looser than that with even number of carbons. On the
other hand, plots of the log of the cmc vs. N gave a linear relationship, indicating no difference between even and odd carbon
number compounds in micellar properties.

They have studied also the N dependence of the thermodynamic
parameters of micellization, calculated from the cmc values and
their temperature dependence, as shown in Figure 1, where ΔGm is

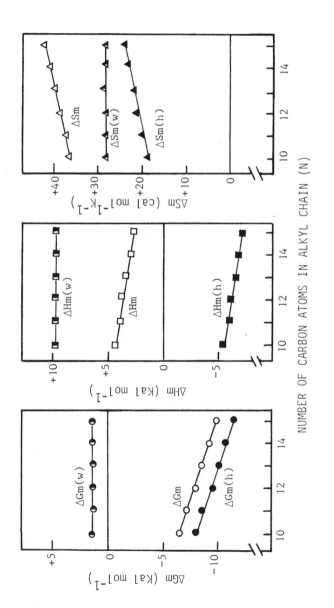

Figure 1. Thermodynamic parameters of micellization at 25°C as a function of alkyl chain length for $H(CH_2)_N O(EO)_8 H$ (from Meguro et al., Ref. 2. 1981)

the free energy, ΔHm the enthalpy and ΔSm the entropy of micellization. These parameters were separated into contributions from the hydrophobe and from the hydrophile, such as $\Delta G(h)$ and $\Delta G(w)$. Their conclusions are as follows : The hydrocarbon chain is an important factor in micellization. The hydrophobic part of the enthalpy, $\Delta H(h)$, is negative and $\Delta S(h)$ is positive, therefore both terms jointly contribute to the negative value of ΔGm. This contribution increases linearly with increasing N. On the other hand, the POE group opposes micellization, due to a large positive value of $\Delta H(w)$, suggesting dehydration of the hydrated monomeric POE chains on micellization.

Rosen and coworkers have also investigated similar properties for a homogeneous series of POE dodecyl monoethers, varying in the number of oxyethylene(OE) units from 2 to 8 ($\underline{3}$). Table II shows a part of their data on the thermodynamics of micellization. Their results indicate that a steady increase in ΔGm, (or cmc) with lengthening POE chain is predominantly due to the change in the ΔHm term, and suggest that desolvation of POE chain oxygens in micelle formation increases progressively with increasing OE unit number. They also determined the thermodynamic parameters of surface adsorption from cmc and surface tension data and discussed the value of ($\Delta Gm - \Delta Gad$), which is the work involved in transferring the nonionic molecule from a monolayer at zero surface pressure to the micelle. As shown in Table III, the positive value for the work of transfer increases with increasing OE unit number, mainly due to the enthalpy factor. Based on the results, they suggest that the motion of the alkyl chain in the surface film is not as restricted as in the interior of the micelle, and that dehydration of the POE chain required for micellization is greater than that for adsorption.

Rosen's group reported also the effect of POE chain length on wetting properties ($\underline{4}$).

Table II. Thermodynamic Parameters of Micellization for Nonionic of $n\text{-}C_{12}H_{25}O(EO)_mH$ at 25^0C, data from Rosen et at.($\underline{3}$)

m	$\Delta G^0mic/$ $(kJ\ mol^{-1})$	$\Delta H^0mic/$ $(kJ\ mol^{-1})$	$\Delta S^0mic/$ $(kJ\ mol^{-1}K^{-1})$
2	-25.6	$+4.2$	$+0.10_0$
3	-24.4	$+5.9$	$+0.10_2$
4	-23.9	$+8.8$	$+0.11_0$
5	-23.9	$+9.9$	$+0.11_3$
7	-23.3	$+12.5$	$+0.12_0$
8	-22.6	$+13.2$	$+0.12_0$

Table III.   Effects of POE Chain Length on Micellization and Adsorption for Nonionics of $n-C_{12}H_{25}O(EO)_mH$, data from Rosen et al.(3)

m	$\Delta G^0mic - \Delta G^0ad/$ (kJ mol^{-1})	$\Delta H^0mic - \Delta H^0ad/$ (kJ mol^{-1})	$T(\Delta S^0mic - \Delta S^0ad)/$ (kJ mol^{-1})
2	+9.6	+5.6	-3.9
3	+11.1	+6.2	-4.8
4	+12.0	+8.4	-3.6
5	+12.4	+9.3	-3.0
7	+13.6	+10.1	-3.6
8	+14.8	+9.9	-5.0

Multi-chain and Alicyclic Hydrophobe (No 2 and 3 series)

There have been few studies of the properties of pure compounds in these series of nonionics (5, 6). In our laboratory, the non-ionics shown in Table IV and V have been synthesized by the addition of ethylene oxide under the same conditions to pure ethylene glycol monoethers, rather than to secondary or tertiary alcohols. This method has been found to give the same Poisson distribution of OE units and to be suitable for evaluating quantitatively the structural effects of the hydrophobe (7).

Table IV shows a comparison of some aqueous solution properties of nonionics having the same total number (ten) of carbon atoms in the hydrophobe and a comparable POE chain length, but different hydrophobe structure. It can be seen that multi-chain hydrophobes bring about a striking decrease in the cloud point and in the surface tension at the cmc, and an increase in the cmc, while cyclic fixation of the alkyl chain causes a large increase in the cloud point, the cmc and in the surface tension at the cmc.

Table V shows the same structural effects for the higher homologues of these series. Here, molecular area and cmc of the highest members do not depend so much on the hydrophobe structure.

The effect characteristic of a multi-chain hydrophobe, that is, increase in the cmc and simultaneous decrease in the cloud point, appears to be inconsistent with the well-known HLB concept in surfactants. Tanford has pointed out that based on geometric considerations of micellar shape and size, amphiphilic molecules having a double-chain hydrophobe tend to form a bilayer micelle more highly packed rather than those of single-chain types (8). In fact, a higher homologue of α,α'-dialkylglyceryl polyoxyethylene monoether has been found to form a stable vesicle or lamellar micelle (9). Probably, the multi-chain type nonionics listed in

Table IV. Effects of Multi-chain and Alicyclic Structures of Hydrocarbon Group in R-O(EO)$_m$H

Pro-perties \ R-						
m	7.0	7.0	5.5	7.2	4.1	5.4
Cp[a] (°C)	75	22	17	58	80	>100
γcmc[b] (mNm^{-1})	36.0	28.3	---	33.9	47.3	48.0
cmc x10^3 (mol l^{-1})	0.85	6.0	4.8	6.0	20.5	27.0

a) Cp : cloud point   b) γcmc : surface tension at cmc

Table V. Effects of Multi-chain and Alicycli Structures of Hydrocarbon Group in the Higher Member of R-O(EO)$_m$H

R-	m	Cp (°C)	cmc x10^4 (mol l^{-1})	γcmc (mNm^{-1})	A x10^2 [a] (nm^2)
n-C$_{13}$H$_{27}$-	8.9	79	0.35	32.6	61
(n-C$_6$H$_{13}$)$_2$CH-	9.2	35	1.3	27.9	60
(n-C$_4$H$_9$)$_3$C-	9.2	34	5.8	27.5	71
n-C$_{12}$H$_{25}$-	9.4	84	0.94	32.5	62
(CH$_2$)$_{11}$CH-	9.2	75	15.0	36.9	86
n-C$_{16}$H$_{33}$-	12.2	97	0.15	34.3	58
(n-C$_8$H$_{17}$)$_2$CH-	11.6	36	0.11	28.8	63
(n-C$_5$H$_{11}$)$_3$C-	12.0	48	0.72	27.3	68
(CH$_2$)$_{15}$CH-	11.9	80	0.32	36.7	65

a) A : molecular area on the adsorption film

these Tables also form lamellar micelles above their cmc's, despite their high cmc values due to less hydrophobic character of

the hydrocarbon part. Such micellization should be accompanied
by closer packing of the POE moiety and greater dehydration of
POE chain than in micelle formation by the usual type of nonionic.
This seems to be related to the unexpectedly low cloud point of
the multi-chain type of nonionics.

Higher Alkyl Crown Ethers (No 4 series)

During the past ten years, there have been numerous reports on
the synthesis and the application of crown ethers of specific
character to various fields. There have been also a few studies
of the surface and micellar properties of crown ethers with hydrophobic groups (10 - 17). The author has called them surface
active crown ethers as a new class of surfactant possessing a
promising function (11).
   Okahara and coworkers have developed a one-step cyclization
of oligoethylene glycol (18, 19) and applied it to the convenient
synthesis of crown ethers with higher alkyl chains (20, 21).
They have prepared many series of n-alkyl crown ethers and N-alkyl
monoazacrown ethers varying in both alkyl chain length and OE
unit number, and compared their properties with those of the corresponding open chain compounds (14, 15).
   Figure 2 shows the plots of cloud point vs. alkyl chain
length for these compounds having the same number (six) of OE
units. It can be seen that cyclization of the POE chain lowers
the cloud point, indicating that the cyclic POE without the hy-
droxyl group has a lower hydrophilicity, and that replacing the
oxygen atom in the ring with a nitrogen atom considerably raises
the cloud point, implying stronger hydration of the amino group.
The same conclusions are also reached from the results of the relation of cloud point to OE unit number and from structural effects on the cmc.
   Table VI shows a comparison of the surface properties of the
crown and the corresponding open chain compound. The former has
a lower aqueous surface tension and a larger molecular area than
the latter, reflecting the lower hydrophilicity and the greater
rigidity of cyclic POE.
   Thus, an estimation can be made of the hydrophilicity of the
crown ring. The acetal-type crown ring obtained from hexaethyl-
ene glycol and a higher aliphatic aldehyde is estimated to be e-
quivalent to about four OE units in an alkyl POE monoether, from
our study of the cloud point (11). Moroi et al. concluded, from
a comparison of the cmc, that a diaza-18-crown-6 is equivalent to
20 OE units in the usual type of nonionic (12). Okahara's group
evaluated the effective HLB based on the cloud point, phenol
index and phase-inversion-temperature in emulsion of oil/water
system and they concluded that 18-crown-6 and monoaza-18-crown-6
rings with dodecyl group are approximately equivalent to 4.0 and
4.5 units, respectively, of OE chains with the same alkyl chain
(17).

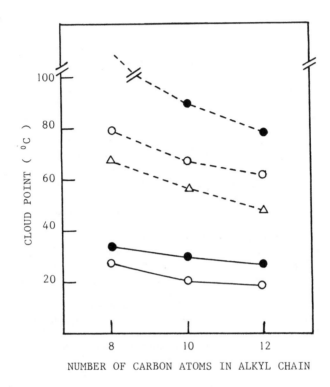

Figure 2. Effect of hydrophile-structure on cloud point (the data

points from Kuo, Ikeda and Okahara, Ref. 14. and 15. 1982).

—●—: N-alkyl monoaza 18-crown-6, ···●···: RN $(EO)_3H_2$, —O—:

alkyl 18-crown-6, ···◐···: $HOCHRCH_2O(EO)_5H$, ···△···: $RO(EO)_6H$

Table VI.   Effect of the Hydrophile-Structure on the Surface
Properties at $20\,^{0}$C, data from Okahara et al.($\underline{14},\underline{15}$)[a]

Structure	R-	Surface tension at cmc (mNm^{-1})	Area/molecule ($nm^2 \times 10^2$)
R~O~ (O)₃ crown structure	C_8	32.9 (-)	55 (-)
	C_{10}	33.0 (36.5)	56 (45)
	C_{12}	34.7 (-)[b]	56 (-)[b]
R-N~ (O)₃ crown structure	C_8	34.0 (36.5)	53 (47)
	C_{10}	34.0 (36.5)	53 (45)
	C_{12}	33.0 (35.5)	54 (47)

a) The data for the corresponding precursor of the crown ether
are shown in parentheses   b) Measured at $10\,^{0}$C

Surface-active crown ethers are distinctly differ from usual
type of nonionics in salt effect on the aqueous properties, due
to the selective complexing ability with cations depending on the
ring size of the crown. As shown in Figure 3 ($\underline{22}$), the cloud
point of the crowns is selectively raised by the added salts. This
indicates that the degree of cloud point increase is a measure of
the crown-complex stability in water ($\underline{23}$).

Long N-Acyl α-Amino Acid POE Monoesters (No 5 series)

The salts of long N-acyl amino acids have been gaining more interest and commercial use in the fields of detergents, cosmetics and
toiletries, due to good surface properties, mildness to skin and
high biodegradability. In order to develope a new type of nonionics possessing noteworthy character from these materials, we
have recently investigated the synthesis and the aqueous properties of homogeneous series of oligoethylene glycol esters of
various N-acyl α-amino acids ($\underline{25}$).
   The results show the effect of inserting typical α-amino acid
residues between a long alkyl group and a POE chain, based on a
comparison of properties with those of the usual type of nonionic.
   Table VII shows that glycine(G), and particularly glycyl-
glycine(G_2), residues raise the melting, Krafft and cloud points.
On the other hand, the melting and Krafft points of nonionics
having a G residue steadily decrease with increasing number of OE
units. L-Alanine(L-A) slightly raises the melting and Krafft
points, and sarcosine(S) has the least influence on the properties. This indicates that introducing a highly hydrogen bond-
forming secondary amido linkage into the nonionic molecule increases both the crystallinity and hydrophilicity of POE nonionic.

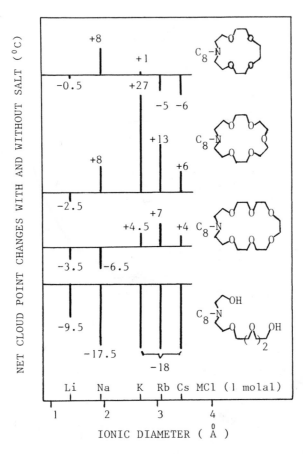

Figure 3. Salt effects on cloud points of N-alkyl monoaza crown
ethers and the corresponding open chain derivatives (from Kuo
and Okahara, Ref. 22. 1983)

Table VII. Melting Point, Krafft Point and Cloud Point of Nonionics

Compound[a]	Melting point(^{0}C)	Krafft point(^{0}C)	Cloud point(^{0}C)
$C_{12}OE_6$[b]	25	< 0	51
$C_{12}O_2E_6$[c]	21 - 23	< 0	54
$C_{12}GE_2$	59 - 61	40	78
$C_{12}GE_3$	56 - 58	24	46
$C_{12}GE_4$	44 - 46	< 0	74 - 76
$C_{12}G_2E_4$	102 - 103	52	>100
$C_{12}AE_3$	33 - 34	14	22 - 23
$C_{12}SE_3$	liq.	< 0	43 - 45

a) G : $-NHCH_2COO-$ E_m : $-(C_2H_4O)_mH$ G : $-(NHCH_2CO)_2O-$ A : $-NHCHMeCOO-$ (L-alanine)   S : $-NMeCH_2COO-$
b) $C_{12}H_{25}O(C_2H_4O)_6H$ c) $C_{11}H_{23}COO(C_2H_4O)_6H$

Table VIII. Cmc and Surface Properties of Nonionics (30^0C)

Compound	cmc (mol $l^{-1} \times 10^4$)	Surface tension at cmc (mNm^{-1})	Area per molecule ($nm^2 \times 10^2$)
$C_{12}OE_6$	0.80[a]	31.5[a]	50[a]
$C_{12}O_2E_6$	1.2[a]	31.5[a]	--
$C_{12}GE_3$	2.9	28.2	44
$C_{12}GE_4$	3.8	29.8	51
$C_{12}AE_3$	3.6	28.8	53
$C_{12}SE_3$	3.0	29.6	52

a) Measured at 20^0C

Table VIII shows that G, L-A and S residues cause the same degree of increase in the cmc and only slightly affect surface tension reduction and molecular area on the adsorption film. For each series of this type of nonionic, plots of log cmc vs. number of carbon atoms(8 - 16) in the acyl chain gave a straight line with a slope similar to that for the usual type of nonionic. However, the former types differ distinctly from the latter type in

Table IX.   Thermodynamic Data for Micellization of Nonionic
Surfactants at 25^0C

Compound	$-\Delta Gm$ (kJ mol^{-1})	ΔHm (kJ mol^{-1})	ΔSm (J $mol^{-1}K^{-1}$)
$C_{10}OE_{5.7}$	28.1	+5.12	112
$C_{10}SE_4$	24.3	+2.43	90
$C_{10}AE_4$	23.7	0.0	80
$C_{10}GE_4$	23.6	-2.10	72
$C_{12}OE_6$	33.3	+10.0	146
$C_{12}SE_4$	30.1	0.0	100
$C_{12}AE_4$	29.4	-3.35	88
$C_{12}GE_4$	29.7	-15.5	46

the temperature dependence of the cmc, as shown in Figure 4. The
nonionics having G, L-A or S residues all gave a line with negative slope and a break point. Such temperature dependence of cmc
has been reported not for the known POE nonionics, but for the
zwitterionic N-alkyl betaines (26), the molecules of which are intermediate in polarity between ionics and POE nonionics.
    Table IX shows the thermodynamic parameters of micellization,
calculated from the cmcs and their temperature dependence, for two
homologues of each of these types of nonionics. It can be seen
that ΔHm and ΔSm progressively decrease in the order : usual-type
> S > L-A > G type for both homologous series. This suggests that
the high polarity of the amido linkage causes either less dehydration or exothermic association of the hydrophilic part on
micelle formation.
    Figure 5 shows plots of cloud point vs. number of OE units for
each of these types of nonionics. These plots give four curves
roughly parallel to one another, except for a part of the G type.
From these curves, it is possible to estimate a contribution of
the inserted residue to hydrophilicity of the nonionics. Namely,
such contribution may be represented by the difference between
usual and other types in OE unit number required to give the same
cloud point in the temperature range 40 - 60^0C. Thus, the equivalent number of OE units is about 3 for both G and S, and 2 for
L-A, respectively. Only the G type gives an unusual curve with a
minimum, which has not previously been reported for POE nonionics.
A similar unusual behavior can also be observed regarding the acyl
chain length (N) dependence of the cloud point, as shown in Figure
6. Only the S type, with tertiary amido linkage, gave a curve de-
scending with increasing N, like the usual type. On the other

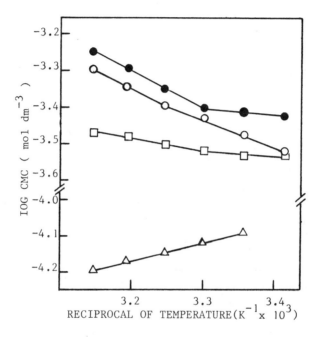

Figure 4. Plots of log cmc vs. the reciprocal of temperature for nonionics : $C_{12}GE_4$ (—o—), $C_{12}AE_4$ (—•—), $C_{12}SE_4$ (—□—) and $C_{12}OE_6$ (—△—)

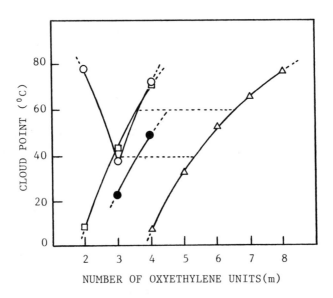

Figure 5. Plots of cloud point vs. number of oxyethylene units(m)
for nonionics : $C_{12}GE_m$(—o—), $C_{12}AE_m$(—•—), $C_{12}SE_m$(—□—) and
$C_{12}OE_m$(—△—)

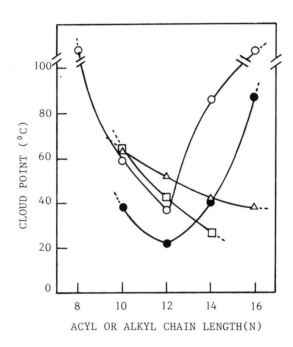

Figure 6. Plots of cloud point vs. number of carbon atoms in the acyl or alkyl group for nonionics : C_NGE_3 (—o—), C_NAE_3 (—•—), C_NSE_3 (—□—) and C_NOE_6 (—△—)

hand, the curves of G and L-A types having secondary amido
linkages have a sharp minimum at the same N(12). This problem
will be discussed in relation to the phase diagram of the G type
and water system.

Binary phase diagram of the G type-water system were determined by visual observations with polarizing microscopy and by
differential-scanning-calorimetry. Several photographic illustrations of the liquid crystal phase of a 50 wt% $C_{14}GE_3$ dispersion
in water over the temperature range 35 - 86°C are shown in Figure
7. We can see typical textures characteristic of a neat phase
above the Krafft point(38°C). The textures partially disappear at
63°C, but new liquid crystals appear again with ascending temperature up to 85°C. At 86°C, a portion of the liquid crystals begin
to melt and phase separation occures, but the other portion remains up to 97°C.

Figure 8 shows the binary phase diagram of the $C_{14}GE_3$-water
system. This indicates that this nonionic has no middle phase and
has a markedly wide area of neat phase over both composition and
temperature ranges. On the other hand, the S type with the same N
gives not so wide area of neat phase, and it has been reported
that the $C_{12}OE_6$-water system has middle and neat phases (27). The
results suggest that G type nonionics with a secondary amido
linkage and a long chain acyl group are particularly apt to form a
lamellar liquid crystals due to highly orienting molecules.

Figure 9 shows the effect of acyl chain length (N) on the binary phase diagram of G type nonionics. The area of neat phase
spreads with increasing N and the extension of the area takes
rapid strides between N of 12 and 14. This N is just in accord
with that of the sharp minimum in the cloud point - N curve of the
G type. The same thing has also been found for the relation between the effect of OE number on the area of the neat phase and
the minimum in the cloud point - OE number curve of $C_{12}GE_m$ (m = 2
- 4).

These facts may be explained as follows : The secondary amido
linkage located near the alkyl chain contributes, together with
hydrophobic bonding by the alkyl chain, to form a tight lamellar
aggregate of the nonionic molecule, due to its intermolecular
hydrogen-bonding ability. This is reflected in the wide area of
neat phase over a great range of both temperature and composition
for the higher members of the G type. Consequently, dehydration
of the hydrophilic part of the nonionic fixed in the lamellar
aggregate may be significantly restricted up to a high temperature. This results in raising the cloud point curve as a whole.

Concerning the effects of introducing amino acid residues into
POE nonionics, it can be concluded that 1) amido groups do not
lower surface-active properties, 2) amido groups generally increase the hydrophilicity of nonionics (giving a high cloud
point), 3) secondary amido groups increase the crystallinity of
nonionics (giving a high Krafft point and the ability to form lamellar liquid crystals). Thus, the insertion of amino acid

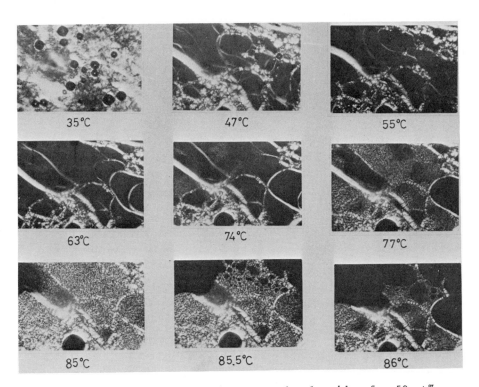

Figure 7. Photomicrographs with crossed polaroids of a 50 wt%
$C_{14}GE_3$ dispersion in water over the temperature range 35 – 86°C.
Magnification Ca x 100

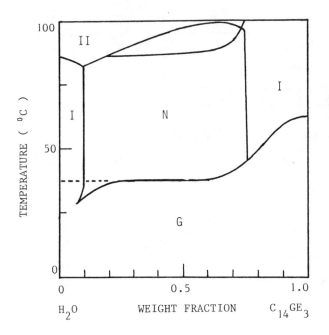

Figure 8.   Binary phase diagram of $C_{14}GE_3$-water system. G : gel
phase, N : neat phase, I : isotropic solution, II : two liquid
phase

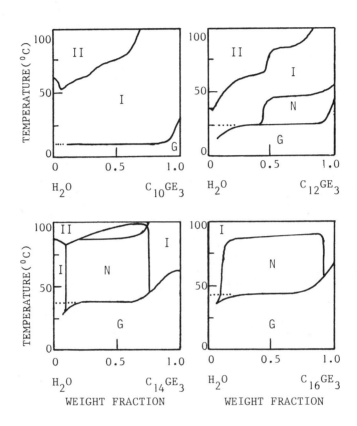

Figure 9. Effect of acyl chain length(N) on the binary phase diagram of $C_N GE_3$-water systems

residues becomes a new structural factor for control of two important properties of POE nonionics, hydrophilicity and crystallinity, which have so far been controlled only with hydrocarbon and polyoxyethylene chain lengths.

Acknowledgments

The author wish to thank Professor M. J. Rosen, Brooklyn College of the City University of New York, for his kind editing of the manuscript. The author is also grateful to the following co-workers who at some time were responsible for part of the work presented from our laboratory : T. Suzuki, H. Tanaka, S. Inokuma and Dr. H. Takahashi.

Literature Cited

1. Ueno, M.; Takasawa, Y.; Tabata, Y.; Sawamura, T.; Kawahashi, N.; Meguro, K. Yukagaku 1981, 30, 421.
2. Meguro, K.; Takasawa, Y.; Kawahashi, N.; Tabata, Y.; Ueno, M. J. Colloid Interface Sci. 1981, 83, 50.
3. Rosen, M. J.; Cohen, A. W.; Dahanayake, M.; Hua, X-Y. J. Phys. Chem. 1982, 86, 541.
4. Cohen, A. W.; Rosen, M. J. J. Am. Oil Chem. Soc. 1981, 58, 1062.
5. Elworthy, P. H.; Florence, A. T. Kolloid-Zeit. 1964, 195, 23.
6. Schüring, S.; Ziegenbein, W. Tenside Detergents 1967, 4, 161.
7. Kuwamura, T.; Akimaru, M.; Takahashi, H.; Arai, M. Reports of Asahi Glass Foundation for Industrial Technology 1979, 35, 45 ; Chem. Abstr. 1981, 95, 61394q.
8. Tanford, C. J. Phys. Chem. 1972, 76, 3020.
9. Okahata, Y.; Tanamachi, S.; Nagai, M.; Kunitake, T. J. Colloid Interface Sci. 1981, 82, 401.
10. Le Moigne, J.; Gramain, P. J. Colloid Interface Sci. 1977, 60, 566.
11. Kuwamura, T.; Kawachi, T. Yukagaku 1979, 28, 195.
12. Moroi, Y.; Pramauro, E.; Grätzel, M.; Pelizzetti, E.; Tundo, P. J. Colloid Interface Sci. 1979, 69, 341.
13. Kuwamura, T.; Yoshida, S. Nippon Kagaku Kaishi 1980, 427.
14. Kuo, P.-L.; Ikeda, I.; Okahara, M. Tenside Detergents 1982, 19, 4.
15. Kuo, P.-L.; Ikeda, I.; Okahara, M. Tenside Detergents 1982, 19, 204.
16. Inokuma, S.; Hagiwara, Y.; Shibazaki, K.; Kuwamura, T. Nippon Kagaku Kaishi 1982, 1218.
17. Kuo, P.-L.; Ikeda, I.; Okahara, M. Bull. Chem. Soc. Jpn. 1982, 55, 3356.
18. Kuo, P.-L.; Miki, M.; Okahara, M. J. Chem. Soc. Chem. Commun. 1978, 504.
19. Kuo, P.-L.; Kawamura, N.; Miki, M.; Okahara, M. Bull. Chem. Soc. Jpn. 1980, 53, 1689.

20. Ikeda, I.; Yamamura, S.; Nakatsuji, Y.; Okahara, M. J. Org. Chem. 1980, 45, 5355.
21. Kuo, P.-L.; Miki, M.; Ikeda, I.; Okahara, M. J. Am. Oil Chem. Soc. 1980, 57, 227.
22. Kuo, P.-L. Ph. D. Thesis, Osaka University, Osaka, 1983.
23. Okahara, M.; Kuo, P.-L.; Yamamura, S.; Ikeda, I. J. Chem. Soc. Chem. Commun. 1980, 586.
24. Bluestein, B. R.; Cowell, R. D. J. Am. Oil Chem. Soc. 1981, 58, 173A.
25. Kuwamura, T. et al. unpublished data which have been presented at the meeting of Division of Colloid and Surface Chemistry in Japan Chemical Society and at the Japan Oil Chemists' Society meeting, 1981 - 1983.
26. Swarbrick, J.; Daruwara, J. J. Phys. Chem. 1967, 73, 2627.
27. Balmbra, R. R. et al. Trans. Faraday Soc. 1962, 58, 1661.

RECEIVED January 10, 1984

Surface Properties of Zwitterionic Surfactants

1. Synthesis and Properties of Some Betaines and Sulfobetaines

M. DAHANAYAKE and MILTON J. ROSEN

Department of Chemistry, Brooklyn College, City University of New York, Brooklyn, NY 11210

Zwitterionic surfactants of structure $RN^+(CH_2C_6H_5)$-$(CH_3)CH_2COO^-$, where R is an alkyl chain of 10 or 12 carbon atoms, and $RN^+(CH_2C_6H_5)(CH_3)CH_2CH_2SO_3^-$, where R is 8, 10, or 12 carbon atoms, have been synthesized. From surface tension-concentration curves in aqueous solution at 10°, 25°, and 40°C, surface excess concentrations and areas/molecule at surface saturation, critical micelle concentrations, efficiency and effectiveness of surface tension reduction, and thermodynamic parameters of adsorption and micellization have been calculated. The areas/molecule indicate that the entire ionic head group in each series is lying flat in the aqueous solution/air interface. For the glycines, the standard free energies of micellization and of adsorption per methylene group at the aqueous solution/air interface are -2.80 kJ and -3.05 kJ, respectively; for the taurines, the standard free energy of adsorption per methylene is -3.15 kJ, all at 25°C.

The use of zwitterionic surfactants commercially has increased dramatically in recent years (1) because of their unique properties, such as compatibility and synergism when used in conjunction with most other types of surfactants. This type of surfactant is used in textile processing aids, cosmetic products, cleaning agents, and as antistatic agents. The sulfobetaines have been found to be very good lime soap disperants (2).

In spite of this wide applicability, a survey of the literature reveals that, compared to ionic and nonionic surfactants, there have been relatively few investigations of their surface and thermodynamic properties. Investigation has been hampered by the nonavailability of pure compounds and proper analytical techniques to determine their concentration in solution.

Tori and Nakagawa (3-7), in a series of papers, described the micellar properties of C_8, C_{10}, and C_{12} C-alkylbetaines of the type $(CH_3)_3{}^+N-CH(R)COO^-$, and N-octylbetaine, $C_8H_{17}{}^+N(CH_3)_2CH_2COO^-$. From studies of the temperature dependence of the cmc, they were able to calculate ΔH°_{mic}. Herrmann (8) studied C_{10}, C_{12}, and C_{16} N-alkyl-sulfobetaines of the type, $R-{}^+N(CH_3)_2(CH_2)_3SO_3^-$ with regard to the chain length and ionic strength variation on the cmc. He calculated the standard free energy contribution to micellization of a methylene group to be 0.61 kcal mol^{-1} and, thereby, concluded that the internal structure of the micelles of these zwitterionics is similar to all other ionic and nonionic surfactants studied. Thermodynamic parameters of micellization have also been investigated by Molyneux (9), and Swarbrick (10). They were able to estimate the standard free energy contribution to micellization of the head group of N-alkyl and C-alkylbetaines to be +3.3 and +2.7 Kcal mol^{-1}, respectively. Molyneux et al. found the plot of log cmc vs. 1/T for dodecyl N-methylbetaines to be linear, whereas Swarbrick et al. observed a minimum in the curve for the corresponding decyl and undecyl betaines. From these data, these workers were able to estimate the standard enthalpies and entropies of micellization and to compare these results with other nonelectrolyte amphophiles.

In contrast to this, there is little information available (11) on the thermodynamics of adsorption of alkylbetaines and no data on the thermodynamic parameters of adsorption or micellization for sulfobetaines.

In the present work, we have synthesized two betaines and three sulfobetaines in very pure form and have determined their surface and thermodynamic properties of micellization and adsorption. From these data on the two classes of zwitterionics, energetics of micellization and adsorption of the hydrophilic head groups have been estimated and compared to those of nonionic surfactants.

Experimental Section

N-alkyl N-benzyl N-methylglycines, $C_nH_{2n+1}N^+(CH_2C_6H_5)(CH_3)CH_2COO^-$.

Two homologues, in which n = 10 (C_{10}BMG) and 12 (C_{12}BMG), were synthesized by reacting N-methylbenzylamine (3 moles) and sodium chloroacetate (1 mole) in 95% ethanol overnight at 40°C. The resulting solution was treated with 0.5 moles of Na_2CO_3 and steam distilled to remove the excess N-methylbenzylamine. Water was removed by rotary evaporator and the crude residue N-methylbenzyl-glycine was recrystallized from isopropyl alcohol.

The tertiary amine thus obtained was dissolved in absolute ethanol and was refluxed for two days with five molar percent excess of the appropriate bromoalkane (97% Humphrey Chemical, North Haven, Conn.). Solvent was removed and the residue in aqueous Na_2CO_3 solution was extracted with hexane to remove any unreacted bromo-alkane. Next, the N-alkyl N-benzyl N-methylglycine was extracted into chloroform from the aqueous layer. Solvent was stripped off and the crude material was recrystallized thrice from carbon tetra-chloride and twice from THF/$CHCl_3$ (60:40 v/v) mixture. The yields of the purified betaines were about 75% of the theoretical.

Analytical data for the compounds were as follows:

	\multicolumn{3}{c}{calculated}	\multicolumn{3}{c}{found}
	C	H	N	C	H	N
$C_{10}BMG$	75.19	10.41	4.38	74.48	10.82	4.32
$C_{10}BMG$	76.03	10.73	4.03	75.70	10.76	3.92

N-alkyl N-benzyl N-methyltaurines, $C_nH_{2n+1}N^+(CH_2C_6H_5)(CH_3)CH_2CH_2SO_3^-$.
Three homologues in which n = 8 (C_8BMT), 10 ($C_{10}BMT$), and 12 ($C_{12}BMT$)
were synthesized by a procedure similar to that for the N-alkyl-
betaines. Here, the N-methylbenzylamine and sodium salt of 2-chloro-
ethanesulfonic acid were refluxed in 95% methanol for two days.
After treatment with 0.5 M Na_2CO_3, the resulting solution was steam
distilled to remove the excess N-methylbenzylamine. Water was
removed and the crude residue was recrystallized from ethanol.
    The tertiary amine thus obtained was dissolved in absolute
ethanol and refluxed for five days with five molar percent excess of
the appropriate bromoalkane. Thereafter, the procedure was similar
to that for the N-alkylglycines. Crude product was recrystallized
thrice from water and then from THF/CHCl$_3$ (50:50 v/v) mixture.
    Analytical data for the compounds were:

	\multicolumn{3}{c}{calculated}	\multicolumn{3}{c}{found}
	C	H	N	C	H	N
$C_{10}BMT$	64.99	9.55	3.79	65.20	9.92	3.74
$C_{12}BMT$	66.45	9.89	3.52	66.31	10.01	3.48

    The molar absorptivities for the two betaines and the three
sulfobetaines in aqueous solution are listed in Table I. Before
being used for surface tension measurements, aqueous solution of
surfactants were further purified by repeated passage (12) through
minicolumns (SEP-PAK C_{18} Cartridge, Waters Assoc., Milford Mass.) of
octadecylsilanized silica gel. The concentration of surfactant in
the effluent from these columns was determined by ultraviolet
absorbance, using the molar absorptivities listed in Table I.

Table I.   Molar Absorptivities for $R-N^+(CH_2C_6H_5)(CH_3)CH_2COO^-$ and
$R-N^+(CH_2C_6H_5)(CH_3)CH_2CH_2SO_3^-$

Compound	λ_{max}	$\varepsilon(dm^3mol^{-1}cm^{-1} \times 10^{-3})$
$C_{10}BMG$	263	3.80
$C_{12}BMG$	263	3.55
C_8BMT	263	3.88
$C_{10}BMT$	263	3.80
$C_{12}BMT$	210	12.12

Surface tension measurements. Solutions of the betaines were prepared with quartz-condensed, distilled water, specific conductance, 1.1×10^{-6} mho cm^{-1} at 25°C. All surface tension measurements were made by Wilhelmy vertical plate technique. Solutions to be tested were immersed in a constant-temperature bath at the desired temperature ±0.02°C and aged for at least 0.5 h before measurements were made. The pH of all solutions was > 5.0 (usually, in the range 5.5-5.9), where surface properties show no change with pH.

Results and Discussions

Plots of surface tension, γ, vs. the log of the molar concentration, C, of the surfactant in the bulk phase at 10°, 25°, and 40°C for the N-alkyl glycines and the N-alkyltaurines are shown in Figures 1 and 2 respectively.

Surface excess concentration, Γ, in mol cm^{-2}, and area/molecule, A, in nm^2, at the liquid/air interface were calculated from the relationships:

$$\Gamma = \frac{1}{2.303RT} \left(\frac{-\partial\gamma}{\partial \log C}\right)_T \quad \text{and} \quad A = \frac{10^{14}}{NT}$$

where $(\partial\gamma/\partial \log C)_T$ is the slope of the γ-log C curve at constant temperature, T, R = 8.31 J $mol^{-1}K^{-1}$, and N = Avogadro's number. Values of the critical micelle concentration (cmc), minimum area per molecules (A_{min}), π_{cmc}, the effectiveness of surface tension reduction (13), and pC_{20}, the efficiency of surface tension reduction (14), are listed in Table II.

The $C_{10}BMG$ and $C_{12}BMG$ were found to have high solubility in water, whereas the corresponding N-alkyltaurines were sparingly soluble in water. As a result of the poor solubility, the only cmc determined in this series was for $C_{10}BMT$ at 40°C. The cmc of C_8BMT was not determined due to its high cmc and insufficient material.

The areas per molecule for the glycines and for the taurines, when compared to the cross sectional areas of the compounds as obtained from molecular models, suggest that, at the aqueous solution/air interface, the ionic head groups, $-N^+(CH_2C_6H_5)(CH_3)CH_2CH_2SO_3^-$ (in the case of the taurines) and $-N^+(CH_2C_6H_5)(CH_3)CH_2COO^-$ (in the case of the glycines), are lying flat in the interface.

Although the efficiencies of surface tension reduction, pC_{20}, for the betaines and their corresponding sulfobetaines are almost the same, the former appear to show greater effectiveness in surface tenion reduction, as indicated by the π_{cmc} values. This may be due to the smaller areas per molecule of the betaines as compared to the corresponding sulfobetaines.

Standard Thermodynamic Parameters of Micellization. Standard free energies of micellization were calculated by the relationship:

$$\Delta G_{mic}^{\circ} = RT \ln CMC$$

Standard entropies and enthalpies of micellization, ΔS_{mic}° and ΔH_{mic}°, can be calculated from the relationships:

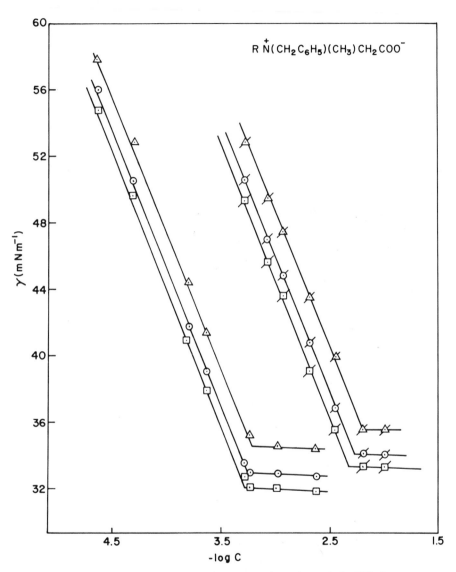

Figure 1. Surface tension vs. log concentration of C_{10}BMG in aqueous solution at 10° ⟁ , 25° ⟁ , and 40° ⟁ ; of C_{12}BMG at 10° △ , 25° ⊙ , and 40° ⊡ .

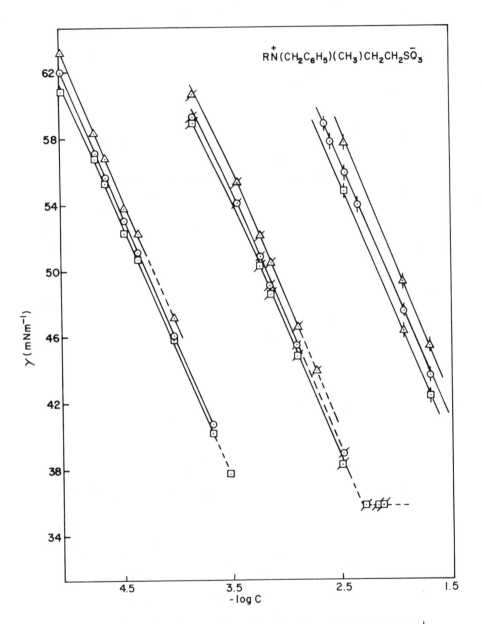

Figure 2. Surface tension vs. log concentration of C_8BMT at $10°$ ⬠ , $25°$ ⬠ , and $40°$ ⬠ ; of C_{10}BMT at $10°$ △ , $25°$ ⬠ , and $40°$ ⬠ ; of C_{12}BMT at $10°$ △ , $25°$ ⊙ , and $40°$ ⊡ .

Table II. Surface Properties of $R-N^+(CH_2C_6H_5)(CH_3)CH_2COO^-$ and
$R-N^+(CH_2C_6H_5)(CH_3)CH_2CH_2SO_3^-$

Compound	$T(°C)$	cmc mole dm^{-3}	A_{min} $nm^2 \times 100$	pC_{20}	π_{cmc} $mN\ m^{-1}$
$C_{10}BMG$	10°	6.31×10^{-3}	54.8	3.34	38.7
	25°	5.25×10^{-3}	56.9	3.36	38.0
	40°	4.36×10^{-3}	59.7	3.30	36.3
$C_{12}BMG$	10°	6.026×10^{-4}	56.2	4.42	39.7
	25°	5.49×10^{-4}	57.6	4.42	39.0
	40°	5.25×10^{-4}	59.7	4.32	37.6
C_8BMT	10°	---	54.0	2.26	--
	25°	---	60.9	2.23	--
	40°	---	63.4	2.17	--
$C_{10}BMT$	10°	---	55.8	3.4	--
	25°	---	60.9	3.34	--
	40°	4.57×10^{-3}	64.0	3.22	33.8
$C_{12}BMT$	10°	---	58.5	4.52	--
	25°	---	61.2	4.44	--
	40°	---	64.0	4.32	--

$$d(\Delta G°/dt) = -\Delta S°$$

and
$$\Delta H° = \Delta G° + T\Delta S°$$

The $\Delta S°_{mic}$ values in Table III are all positive, indicating increased randomness in the system upon transformation of the zwitterionic surfactant molecules into micelles. The $\Delta H°_{mic}$ values, too, are positive, due to the endothermic desolvation associated with micellization. Smaller $\Delta H°_{mic}$ and $\Delta S°_{mic}$ values at 25-40°C than at 10-25° are due to the smaller hydration of the monomers at the higher temperatures. In the temperature range studied, no minimum in the variation of $\Delta H°_{mic}$ with temperature was observed, in agreement with the work of Swarbrick et al. (10).

From the variation of the $\Delta H°_{mic}$ values for the two alkyl betaines, it is seen that, for the shorter alkyl chain compounds, the enthalpy change is a significant factor in the process of micellization while, for the longer chain compounds, the free energy change is due almost entirely to the entropy change.

From the standard free energy of micellization of the N-alkyl-glycines, the $\Delta G°_{mic}$ per methylene group at 25°C is -2.80 kJ. This

Table III.   Standard Thermodynamic Parameters of Micellization for
$R-N^+(CH_2C_6H_5)(CH_3)CH_2COO^-$

Compound	T(°C)	ΔG° (kJ mol^{-1})	ΔH° (kJ mol^{-1})	ΔS° (kJ mol^{-1}K^{-1})
C_{10}BMG	10	-11.5		
			+8.5	+0.070
	25	-13.0		
			+4.7	+0.044
	40	-13.9		
C_{12}BMG	10	-17.4		
			+4.6	+0.078
	25	-18.6		
			+0.9	+0.067
	40	-19.6		

is in close agreement with the corresponding value of -2.85 kJ at
20°C obtained by Molyneux et al. (9).

Standard Thermodynamic Parameters of Adsorption.   Table IV lists the
standard thermodynamic parameters of adsorption.   Values have been

Table IV.   Standard Thermodynamic Parameters of Adsorption for
$R-N^+(CH_2C_6H_5)(CH_3)CH_2COO^-$ and $R-N^+(CH_2C_6H_5)(CH_3)CH_2CH_2SO_3^-$

Compound	T(°C)	ΔG° (kJ mol^{-1})	ΔH° (kJ mol^{-1})	ΔS° (kJ mol^{-1}K^{-1})
C_{10}BMG	10	-24.7		
			+0.2	+0.087
	25	-26.0		
	40	-27.2	-2.6	+0.078
C_{12}BMG	10	-30.9		
			-7.1	+0.082
	25	-32.1		
	40	-33.2	-10.9	+0.071
C_8BMT	10	-18.7		
			+5.9	+0.083
	25	-20.0		
	40	-20.6	-9.2	+0.036
C_{10}BMT	10	-25.1		
			-2.7	+0.079
	25	-26.3		
	40	-27.0	-12.9	+0.045
C_{12}BMT	10	-31.5		
			-10.6	+0.074
	25	-32.6		
	40	-33.5	-14.8	+0.060

calculated by the relationship (15):

$$\Delta G^{\circ}_{ad} = RT \ln CMC - \pi_{cmc} \cdot A_{min}$$

From the data, ΔG°_{ad} per methylene group at 25°C is -3.05 kJ for the glycines and -3.15 kJ for the taurines. These are in agreement with values of -3.15 kJ for long-chain alchols and 1,3-diols (15). The ΔH°_{ad} values are less positive than the ΔH°_{mic} values for the same alkyl glycines. This shows less dehydration of the surfactant required for adsorption at the aqueous solution/air interface than for the process of micellization. This is consistent with previous observations on polyoxyethylenated nonionics and alkylpyridinium halides (16,17). The $\overline{\Delta S^{\circ}_{ad}}$ values are all slightly more positive than the ΔS°_{mic} values for the same compound, reflecting the greater restriction of space in the micelle than at the aqueous solution/air interface. The ΔS°_{ad} values and ΔH°_{ad} values are both more positive for the betaines than for the corresponding sulfobetaines. This shows that the sulfobetaines require less dehydration for adsorption at the aqueous solution/air interface.

Using $\Delta G^{\circ}(-CH_3) = \Delta G^{\circ}(-CH_2-) - 5.56$ kJ mol^{-1}, on the basis of solubility data (9,18) for liquid N-alkanes in water at 25°C, standard free energies of adsorption and micellization, $\Delta G^{\circ}_{ad}(-W)$ and $\Delta G^{\circ}_{mic}(-W)$ respectively, for the hydrophilic head groups, $-N^+(CH_3)(CH_2C_6H_5)CH_2CH_2SO_3^-$ and $-N^+(CH_2C_6H_5)(CH_3)CH_2COO^-$, were calculated. These values are listed in Table V together with the standard free energy values for the head groups $-CHOHCH_2-CH_2OH$ and $-OCH_2CH_2OH$ (19). The $\Delta G^{\circ}(-W)$ values for the two zwitterionics are comparable to each other. Both the $\Delta G^{\circ}_{ad}(-W)$ and $\Delta G^{\circ}_{mic}(-W)$ values for the two nonionics are less positive than for the two zwitterions, possibly due to the greater hydration of the zwitterions than of the ether oxygen and/or -OH groups.

From the solubility data of n-decane in water, the enthalpy for the process n-decane (H_2O) → n-decane (pure) at 25°C has been estimated by Boddard et al. (20) to be -5.85 kJ mol^{-1}. Substracting this value from the calculated $\Delta H^{\circ}(25^{\circ}c)$ values for $C_{10}BMG$ and $C_{12}BMT$, in Tables III and IV, the $\Delta H^{\circ}(-W)$ values for micellization and for adsorption at the aqueous solution/air interface at 25°C can be estimated. Values are shown in Table V.

From the $\Delta H^{\circ}_{ad}(-W)$ term for the two types of hydrophilic groups, it is evident that there is an exothermic effect in the transfer of the $-N^+(CH_2C_6H_5)(CH_3)CH_2CH_2SO_3^-$ from aqueous medium to the interface. This exothermic enthalpy term, together with a larger negative entropy term for the N-alkyltaurine head group, is possibly due to the partial neutralization of the oppositely charged groups in the hydrophilic heads due to their arrangement in checkerboard fashion aqeuous solution/air interface, as suggested by Beckett and Woodward (21). In the case of the N-alkylglycines, endothermic dehydration of the hydrophilic head may outweigh the neutralization effect, thus making $\Delta^{\circ}_{ad}(-W)$ positive.

Table V.   Standard Thermodynamic Parameters of Adsorption and Micellization of Various Head Groups (-W) at 25°C

(-W)	ΔG°_{ad} (kJ mol^{-1})	ΔH°_{ads} (kJ mol^{-1})	ΔS°_{ads} (kJ mol^{-1}K^{-1})	ΔG°_{mic} (kJ mol^{-1})	ΔH°_{mic} (kJ mol^{-1})	ΔS°_{mic} (kJ mol^{-1}K^{-1})
$-N^+(CH_2C_6H_5)(CH_3)CH_2COO^-$	+10.0	+4.6	-0.018	+20.5	+12.4	-0.027
$-N^+(CH_2C_6H_5)(CH_3)CH_2CH_2SO_3^-$	+10.6	-1.9	-0.042	--	--	--
$-CH(OH)CH_2CH_2OH$	+7.4	--	--	+10.0	--	--
$-OCH_2CH_2OH$	+8.8	--	--	+16.3	--	--

Literature Cited

1. Ernst, R.; Miller, E. J., Jr. "Amphoteric Surfactants";
   Bluestein, B. R.; Hilton, C. L., Ed.; Marcel Dekker: New York,
   1982; pp. 137-150.
2. Kaminiski, M.; Linfield, W. M. J. Am. Oil Chem. Soc. 1979, 56,
   771.
3. Tori, K.; Nakagawa, T. Kolloid - Z. Z. Polym. 1963, 50, 187.
4. Tori, K.; Nakagawa, T. Kolloid - Z. Z. Polym. 1963, 188, 47.
5. Tori, K.; Nakagawa, T. Kolloid - Z. Z. Polym. 1963, 189, 50.
6. Tori, K.; Nakagawa, T. Kolloid - Z. Z. Polym. 1963, 191, 42.
7. Tori, K.; Nakagawa, T. Kolloid - Z. Z. Polym. 1963, 191, 48.
8. Herrmann, K. W. Colloid Interface Sci. 1966, 22, 352.
9. Molyneux, P.; Rhodes, C. T.; Swarbrick, J. Trans. Faraday Soc.
   1965, 61, 1043.
10. Swarbrick, J.; Daruwala, J. J. Phys. Chem. 1969, 73, 2627.
11. Swarbrick, J., J. Pharm. Sci. 1969, 58, 147.
12. Rosen, M. J. J. Colloid Interface Sci. 1981, 79, 587.
13. Rosen, M. J. J. Colloid Interface Sci. 1976, 56, 32.
14. Rosen, M. J. J. Am. Oil Chem. Soc. 1974, 51, 461.
15. Rosen, M. J.; Aronson, M. Colloids Surfaces 1981, 3, 201.
16. Rosen, M. J.; Cohen, A. W.; Dahanayake, M.; Hua, X. J. Phys.
    Chem. 1982, 86, 541.
17. Rosen, M. J., Dahanayake, M.; Cohen, A. Colloids Surfaces 1983,
    5, 159.
18. McAuliffe, C. Nature (London) 1963, 200, 1092.
19. Kwan, C.; Rosen, M. J. J. Phys. Chem. 1980, 84, 547.
20. Goddard, E. D.; Hoeve, C. A.; Benson, G. C. J. Phys. Chem. 1957,
    61, 593.
21. Beckett, A. H.; Woodward, R. J. J. Pharm. Pharmcol. 1963, 15,
    422.

RECEIVED January 20, 1984

Surface Properties of Zwitterionic Surfactants

2. Effect of the Microenvironment on Properties of a Betaine

MILTON J. ROSEN and BU YAO ZHU[1]

Department of Chemistry, Brooklyn College, City University of New York, Brooklyn, NY 11210

The effect of pH and electrolyte on the surface properties of a betaine surfactant, $C_{12}H_{25}N^+(CH_2C_6H_5)(CH_3)CH_2COO^-$, were studied. The ΔG^o_{ad} of both the zwitterionic form of the betaine and of the cationic protonated betaine were calculated. Surface activity decreases slightly with decrease in pH. As expected, the cationic form has somewhat lower surface activity than the zwitterionic form. Although the pK_a of the protonated betaine is 2.8, the properties of both the zwitterionic betaine and the cationic protonated betaine determine the surface properties, even in distilled water (pH = 5.85). At pH = 5.85, the betaine interacts more strongly with Na_2SO_4 than with NaCl or $CaCl_2$; anionic surfactants also show stronger interaction with the betaine than do cationic surfactants. These effects cannot be attributed to the zwitterion, but to the presence of the protonated cationic form in equilibrium with the zwitterion.

There has been a recent revival of interest in zwitterionic surfactants (1-4) because of certain useful properties shown by these molecules, including: 1) mild behavior on the skin, 2) compatability with both anionics and cationics, 3) adsorption onto skin and hair, and 4) lime soap dispersing ability. Although this type of surfactant has been produced and used industrially for the last few decades, there have been few studies of the properties of well purified surfactants of this type (5-11) and almost all of these have been concerned with the micellar properties of these compounds rather than with their behavior at interfaces.

One of the problems associated with the study of the physico-chemical properties of these materials is the lack of analytical methods of determining accurately their concentrations in dilute solution. In order to permit such determinations, an N-betaine type surfactant was synthesized with a benzyl group attached to the quaternary nitrogen. This permitted analysis of dilute aqueous

[1]Current address: Colloid Chemistry Laboratory, Department of Chemistry, Peking University, Beijing, People's Republic of China

0097-6156/84/0253-0061$06.00/0
© 1984 American Chemical Society

solutions of the compounds by ultraviolet spectrometry. The present
work describes the effect of pH and various electrolytes on the
surface properties of the betaine, N-dodecyl-N-benzyl-N-methylgly-
cine, $C_{12}H_{25}N^+(CH_3)(CH_2C_6H_5)CH_2COO^-(C_{12}BMG)$.

Experimental

The synthesis and purification of $C_{12}BMG$ by the reaction of N-methylbenzylamine with sodium chloroacetate followed by the quaternization
of the resulting tertiary ammonioacetate with 1-bromododecane is
described elsewhere (12). Purification of aqueous solutions of the
surfactant for surface tension measurements and determination of the
surface tension of the solutions by the Wilhelmy method using a sand-
blasted platinum blade were by techniques previously described (13).
The concentration of $C_{12}BMG$ in aqueous solution was determined by
measuring its absorbance at 263 nm (ε = 350.5).

The ionization constant, K_a, of the protonated betaine,
$C_{12}H_{25}N^+(CH_3)(CH_2C_6H_5)CH_2COOH(BH^+)$, was obtained by adding V_{H^+} ml of
an aqueous hydrochloric acid solution of hydrogen ion concentration
$C_{H^+}^o$, in moles dm^{-3}, to V_B^o ml of the surfactant betaine solution of
concentration C_B^o at a pH of 5-6 and measuring the hydrogen ion
concentration, $[H^+]$, of the resulting mixture with a pH meter. The
ionization constant was calculated by use of the following relationship:

$$K_a = [H^+] \times \left[\frac{V_B^o \cdot C_B^o}{V_{H^+} \cdot C_{H^+}^o - [H^+](V_B^o + V_{H^+})} - 1 \right]$$

The average of 10 different measurements was 1.6 ±0.2 x 10^{-3} (pK_a =
2.8). The K_1 of N-propylglycine is 4.46 x 10^{-3} (14).

Results and Discussion

Effect of pH. Surface tension (γ) as a function of log of the molar
concentration of $C_{12}BMG$ (log C_B) in aqueous solution (25°C) at
various pHs is shown in Figure 1. Table I shows the effect of change
in the pH of the solution on the surface properties of the betaine.
With decrease in the pH of the solution, the material which, at a pH
of 5.85, is 99.9 mole percent in the zwitterionic form, $B^\pm$, is
converted more and more to the cationic protonated form, BH^+. From
the K_a value of BH^+, its solution phase concentration will exceed
that of the zwitterion, $B^\pm$, when the pH of the solution is below 2.8.
However, the cation, BH^+, is less surface-active than the zwitterion,
$B^\pm$, (see below) as is to be expected, and therefore there is little
change in some of the surface properties of the mixture until a pH
considerably below that value is reached. The smaller
activity of BH^+, compared to $B^\pm$, at the aqueous solution/air
interface is indicated by the steady decrease in the pC_{20}, the bulk
phase molar concentration of surfactant required to produce a surface

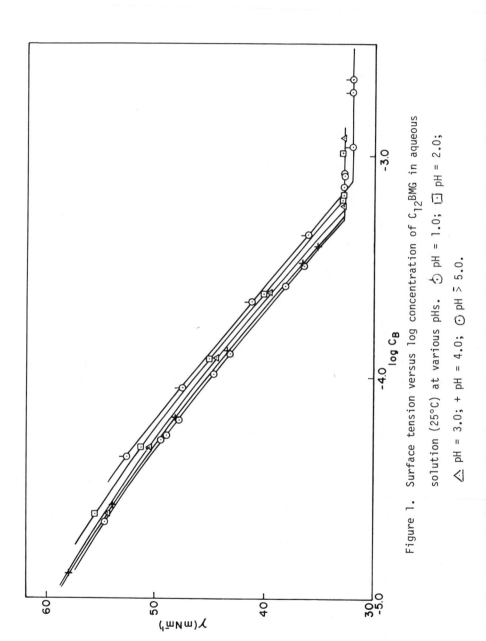

Figure 1.   Surface tension versus log concentration of $C_{12}BMG$ in aqueous solution (25°C) at various pHs. ⟠ pH = 1.0; ▢ pH = 2.0; △ pH = 3.0; + pH = 4.0; ⊙ pH > 5.0.

pressure, π, of 20 mN m^{-1}) with increase in the H$^+$ content of the solution. At constant surface area per molecule, the pC$_{20}$ value is a linear function of the $-\Delta G^{\circ}_{ad}$ of the mixture (15).

The critical micelle concentration (c.m.c.) of the material increases with decrease in pH of the solution below 5, as is to be expected as the ratio of BH$^+$ to B$^{\pm}$ increases. The increase in the c.m.c. is somewhat greater than the increase in the C$_{20}$ value with pH decrease as shown by the cmc./C$_{20}$ ratio, indicating somewhat greater inhibition of micellization than of adsorption at the aqueous solution/air interface as the BH$^+$/B$^{\pm}$ ratio increases. This may reflect some steric inhibition of micellization resulting from the increased size of the protonated hydrophilic head.

On the other hand, the value of Γ_{max} shows no significant change in the pH range investigated. Since an increase in the BH$^+$/B$^{\pm}$ ratio at the aqueous solution/air interface would be expected to cause an increase in the surface area per molecule, due to increased electrical repulsion between the cationic head groups of BH$^+$, the lack of change must be due to a compensatory compression of the electrical double layer surrounding these groups as a result of the increase in the ionic strength of the solution with decrease in the pH.

From molecular models, the minimum cross-sectional area of the C$_{12}$BMG molecule with an orientation normal to the interface is a trapezoid of 0.41 nm^2. The minimum rectangular area is 0.54 nm^2. The latter value agrees well with the experimental values shown in Tables I and II. Tori (7) obtained a value of 0.54 nm^2 for the C-dodecyl betaine, C$_{12}$H$_{25}$CH(COO$^-$)N$^+$(CH$_3$)$_3$ in water at 27°C.

The value of π_{cmc} shows almost no change until pH 1 is reached. Since the value of π_{cmc} is determined by the values of Γ_{max} and the cmc/C$_{20}$ ratio (16) and there is no change in Γ_{max} with change in pH, the increase in π_{cmc} reflects the sharp increase in the cmc/C$_{20}$ ratio at pH 1.

TABLE I. Effect of pH on the Surface Properties of C$_{12}$BMG

pH	cmc (mol dm^{-3} x 10^4)	Γ_{max} (mol dm^{-2} x 10^{10})	A_{min} (nm^2 x 10^2)	π_{cmc} (mN m^{-1})	pC$_{20}$	$\frac{cmc}{C_{20}}$
5.85*	5.11	2.96	56.1	39.0	4.45	14.2
5.0	5.11	2.96	56.1	39.0	4.45	14.2
4.0	5.33	3.00	55.3	39.2	4.43	14.4
3.0	5.78	2.96	56.1	39.1	4.41	14.8
2.0	6.37	2.94	56.4	39.2	4.37	14.8
1.0	7.71	2.94	56.4	40.1	4.32	16.1

* Data at pH = 9.0 indicate no change in surface properties above pH = 5.85.

<u>Effect of Electrolyte.</u> Surface tension-log C_B curves in solutions of different electrolytes at pH = 5.85 and pH = 3.0 (25°C) are shown in Figure 2 and 3, respectively. Table II shows the effect of various electrolytes on the surface properties of the betaine at the two different pHs. From the data, it is apparent that, at both pHs, the

TABLE II. Effect on Electrolyte on the Surface Properties of $C_{12}BMG$

electro-lyte conc.	pH	cmc (mol dm^{-3} x 10^4)	Γ_{max} (mol dm^{-2} x 10^{10})	A_{min} (nm^2 x 10^2)	Π_{cmc} (mN m^{-1})	pC_{20}	$\frac{cmc}{C_{20}}$
0	5.85	5.11	2.96	56.1	39.0	4.45	14.2
0.1 N NaCl	5.85	4.24	3.00	55.3	39.2	4.54	14.6
0.1 N CaCl$_2$	5.85	4.24	3.00	55.3	39.2	4.54	14.6
0.1 N Na$_2$SO$_4$	5.85	4.04	3.00	55.3	39.3	4.58	15.2
0	3.0	5.78	2.96	56.1	39.1	4.41	14.8
0.1 N NaCl	3.0	4.21	3.10	53.6	40.2	4.58	14.9
0.1 N CaCl$_2$	3.0	4.21	3.10	53.6	40.2	4.58	14.9
0.1 N Na$_2$SO$_4$	3.0	3.86	3.10	53.6	40.3	4.62	15.1

addition of electrolyte increases the surface activity of the material, i.e., it decreases the c.m.c. and increases both the pC_{20} value and π_{cmc}. However, the effect of electrolyte on the surface properties of the product is greater at the lower pH where the $BH^+/B^\pm$ ratio is larger. There is a larger decrease in the c.m.c., a larger increase in Γ_{max}, and a larger increase in π_{cmc}. The greater effect of electrolyte in depressing the cmc of C-octyl betaine $C_8H_{17}CH(COO^-)N^+(CH_3)_3$, at a lower pH was noted by Tori ($\underline{9}$).

In contrast to the effect of HCl in the absence of other added electrolyte, discussed above, the addition of electrolyte at constant pH results in a decrease in the surface area per molecule at the aqueous solution/air interface, presumably due to compression of the electrical double layer surrounding the ionic head groups. The effect is more pronounced at pH 3 the at pH 5.85. This slightly larger Γ_{max} at pH 3 in the presence of electrolyte accounts for the higher π_{cmc} value under those conditions, since the cmc/C_{20} ratio is virtually unchanged.

A noteworthy feature is that, in their effects on all these properties, equivalent amounts of NaCl and CaCl$_2$ are identical,

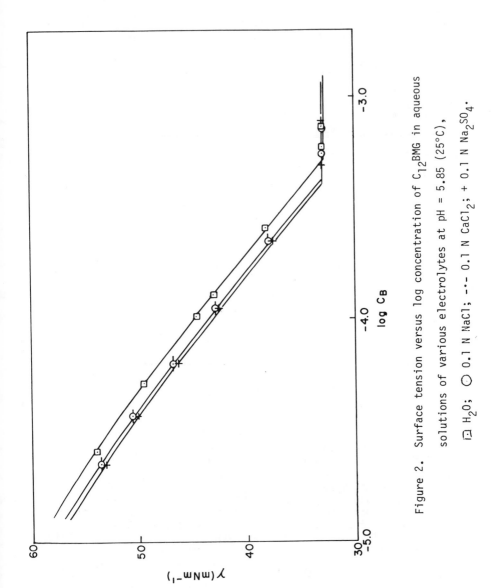

Figure 2.   Surface tension versus log concentration of $C_{12}BMG$ in aqueous solutions of various electrolytes at pH = 5.85 (25°C),

▣ H_2O; ◯ 0.1 N NaCl; –·– 0.1 N $CaCl_2$; + 0.1 N Na_2SO_4.

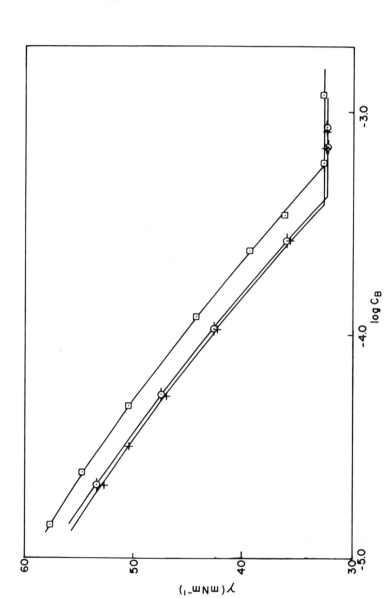

Figure 3. Surface tension versus log concentration of $C_{12}BMG$ in aqueous solutions of various electrolytes at pH = 3.0 (25°C).

□ H_2O; ⊙ 0.1 N NaCl; --- 0.1 N $CaCl_2$; + 0.1 N Na_2SO_4.

whereas an equivalent amount of Na_2SO_4 has a greater effort. This
indicates that the betaine, at both pHs, shows a stronger electrostatic interaction with anions than with cations. The greater
effect of Na_2SO_4 compared to NaCl or $CaCl_2$ on the c.m.c. of
C-octylbetaine, is seen also in the data of Tori (9).
   This greater interaction of the betaine with anions than with
cations is seen also in its interaction with other surfactants (17)
as measured by the molecular interaction parameter for mixed monolayer formation, β, using non-ideal solution theory (18).
Whereas the degree of interaction of $C_{12}BMG$ with the cationic
surfactant, $C_{12}H_{25}N(CH_3)_3Br$, yields a value of β = -1.3, not much
greater than that with the nonionic surfactant, $C_{12}H_{25}(OC_2H_4)_8OH$,
(β = -0.60), its interaction with the anionic surfactant, $C_{12}SO_3Na$,
is considerably greater (β =-5.7) and increases with decrease in the
pH of the solution.
   It is believed that this greater interaction of $C_{12}BMG$ with
anions than with cations is due, not to the zwitterionic betaine, $B^{\pm}$,
but to the cationic protonated betaine, BH^+, in equilibrium with it.
Although the concentration of BH^+ at pH = 5.8 is very small, it is
felt that strong electrostatic interaction between it and an anion
can displace the zwitterion-cation equilibrium sufficiently to cause
an appreciable effect.

<u>Bulk and Surface Phase Concentrations of $B^{\pm}$ and BH^+</u>. The mole
fractions of the zwitterion, $x_{B\pm}^b$, and of the protonated betaine, $x_{BH^+}^b$,
in the total surfactant in the bulk phase were calculated by
combining

$$K_a = \frac{C_{H^+} \times C_{B\pm}}{C_{BH^+}} \tag{1}$$

and

$$x_{B\pm}^b = \frac{C_{B\pm}}{C_{B\pm} + C_{BH^+}} \tag{2}$$

from which,

$$x_{B\pm}^b = \frac{K_a}{K_a + C_{H^+}} \tag{3}$$

and

$$x_{BH^+}^b = \frac{C_{H^+}}{K_a + C_{H^+}} \tag{4}$$

Values of $x_{BH^+}^b$ as a function of pH are tabulated below:

pH	5.85	5.0	4.0	3.0	2.0	1.0
$x_{BH^+}^b$	0.0009	0.006	0.059	0.385	0.86	0.984

The surface phase mole fractions of zwitterion, $x_{B\pm}^S$ $(= \frac{\Gamma_{B\pm}}{\Gamma_t})$,

and of protonated betaine, $x_{BH^+}^S$ $(= \dfrac{\Gamma_{BH^+}}{\Gamma_t})$, where $\Gamma_{B^\pm}$, Γ_{BH^+}, and Γ_t are
the surface (excess) concentrations of $B^\pm$, BH^+, and their mixture at
the aqueous solution/air interface were calculated from surface
tension data at constant $C_{B^\pm}$ at pH $\lesssim$ 3. The Gibbs absorption equation for a mixture of $B^\pm$ and BH^+ in dilute HCl solution at constant
$C_{B^\pm}$ is:

$$-d\gamma = RT(\Gamma_{BH^+}d \ln C_{BH^+} + \Gamma_{H^+}d \ln C_{H^+} + \Gamma_{Cl^-}d \ln C_{Cl^-}) \qquad (5)$$

Since H^+ is a non-surface-active ion of similar charge to BH^+, we can
assume that $\Gamma_{H^+}d \ln C_{H^+} \approx 0$. If we assume that $\Gamma_{BH^+} = \Gamma_{Cl^-}$ to
maintain electroneutrality, then

$$-d\gamma = RT\Gamma_{BH^+}(d \ln C_{BH^+} + d \ln C_{Cl^-}) \qquad (6)$$

When $C_{B^\pm} \ll C_{H^+}$ (i.e., at low pH), $d \ln C_{Cl^-} = d \ln C_{H^+}$. Moreover,
at constant $C_{B^\pm}$, from equation (1), $d \ln C_{BH^+} = d \ln C_{H^+}$. Therefore,

$$-d\gamma = 2RT\Gamma_{BH^+}d \ln C_{BH^+} \qquad (7)$$

or
$$\Gamma_{BH^+} = \frac{1}{2RT}\left(\frac{-\partial\gamma}{\partial \ln C_{BH^+}}\right)_{C_{B^\pm}} \qquad (8)$$

and
$$x_{BH^+}^S = \frac{\Gamma_{BH^+}}{\Gamma_t} \qquad (9)$$

$$\Gamma_t = \frac{1}{RT\left(\frac{-\partial\gamma}{\partial \ln C_B}\right)_{pH}} \qquad (10)$$

where C_B is the total concentration of C_{12}BMG in the aqueous phase.
$x_{BH^+}^S$ can, consequently, be determined from the data in Figure 1 by
selecting from the curves at pHs 1, 2, and 3, points at constant $C_{B^\pm}$,
calculating C_{BH^+} and plotting γ versus C_{BH^+}.
    Values of $x_{BH^+}^b$ and $x_{BH^+}^S$ at pHs 1 and 2 are tabulated below:

pH	1	2
$x_{BH^+}^b$	0.984	0.86
$x_{BH^+}^S$	0.345	0.257

It is apparent from these values that BH^+ is considerably less surface-active than $B^{\pm}$.

Standard Free Energies of Adsorption of $B^{\pm}$ and BH^+. The standard free energy of adsorption of $C_{12}BMG$ was calculated from the surface tension data in Figures 1 and 2 by use of the equation:

$$\Delta G^{\circ}_{ad} = RT \ln C_B - \pi A_{min} \tag{11}$$

where C_B is the bulk phase concentration of $C_{12}BMG$ to produce a given surface pressure, π, in the region just below the c.m.c. where the γ-log C curve is essentially linear. The standard state for the surface phase is a hypothetical monolayer at its closest packing ($A = A_{min}$) and at zero surface pressure (15).

The standard free energy of adsorption of the zwitterion, $\Delta G^{\circ}_{ad,B^{\pm}}$, was calculated from the surface tension data at pH $\geqslant$ 5, where the concentration of BH^+ is negligible. The value of $\Delta G^{\circ}_{ad,B^{\pm}}$ at 25°C is -32.04 kJ mol^{-1} from the data in water and -32.10 kJ mol^{-1} from the data in 0.1 N NaCl. The ΔG°_{ad} for a series of nonionics, $C_{12}H_{25}(OC_2H_4)_xOH$, containing the same $(C_{12}H_{25})$ hydrophobic group as $C_{12}BMG$, ranges from -35.2 (x = 2) to -37.4 (x = 8) kJ mol^{-1}, using the same standard state for the surface phase (13).

The standard free energy of adsorption of the protonated betaine, $\Delta G^{\circ}_{ad,BH^+}$, was calculated from the surface tension data at pHs 1 and 2 by use of the relationships:

$$\Delta G^{\circ}_{ad} = X^S_{BH^+} \cdot \Delta G^{\circ}_{ad,\,BH^+} + X^S_{B^{\pm}} \cdot \Delta G^{\circ}_{ad,B^{\pm}} \tag{12}$$

The average value of $\Delta G^{\circ}_{ad,BH^+}$ was -29.8 kJ mol^{-1} in aqueous hydrochloric acid solution of 0.1 N total ionic strength (pH = 1).

Acknowledgment

This material is based upon work supported by the National Science Foundation under Grant No. ENG-7825930.

Literature Cited

1. Hidaka, H. J. Amer. Oil Chem. Soc. 1979, 56, 914; 1980, 57, 382.
2. Takai, M.; Hidaka, H.; Ishikawa, S.; Takada, M.; Moriya, M.
    J. Amer. Oil Chem. Soc. 1980, 57, 183.
3. Takai, M.; Hidaka, H.; Moriya, M. J. Amer. Oil Chem. Soc. 1979,
    56, 537.
4. Weil, J. K.; Linfield, W. M. J. Amer. Oil Chem. Soc., 1976, 53,
    60; 1977, 54, 339; 1979, 56, 85.
5. Tori, K.; Nakagawa, T. Kolloid Z. Z. Polym., 1963, 187, 44.
6. Tori, K.; Wakagawa, T. Kolloid Z. Z. Polym., 1963, 188, 47.
7. Tori, K.; Nakagawa, T. Kolloid Z. Z. Polym., 1963, 191, 42.
8. Tori, K.; Kuriyama, K.; Nakagawa, T. Kolloid Z. Z. Polym., 1963,
    191, 48.
9. Tori, K.; Nakagawa, T. Kolloid Z. Z. Polym., 1963, 189, 50.

10. Molyneaux, P.; Rhodes, C. T.; Swarbrick, J. Trans. Faraday Soc. 1963, 61, 1043.
11. Evans, N. G.; Pilpel, N. J. Pharm. Sci. 1969, 1228.
12. Dahanayake, M.; Rosen, M. J. In Chapter 3, this book.
13. Rosen, M. J.; Cohen, A. W.; Dahanayake, M.; Hua, X. Y. J. Phys. Chem. 1982, 86, 541.
14. "Handbook of Chemistry and Physics"; 61st ed., CRC Press: Boca Raton, Florida, 1980; p. D-161.
15. Rosen, M. J.; Aronson, S. Colloids Surfaces 1981, 3, 201.
16. Rosen, M. J. Colloid Interface Sci. 1976, 56, 320.
17. Rosen, M. J.; Zhu, B. J. Colloid Interface Sci., in press.
18. Rosen, M. J.; Hua, X. Y. J. Colloid Interface Sci. 1982, 86, 164.
19. Rosen, M. J.; Dahanayake, M.; Cohen, A. W. Colloids Surfaces 1982, 5, 159.

RECEIVED January 20, 1984

Effect of Structure on Activity at the Critical Micelle Concentration and on the Free Energy of Micelle Formation

Ionic and Nonionic Surfactants

M. NAKAGAKI and T. HANDA

Faculty of Pharmaceutical Sciences, Kyoto University, Kyoto, 606 Japan

The slope of the linear relationship between log cmc and number of carbon atoms in the chain, n_c, is -0.5 for nonionic and zwitterionic surfactants, whereas the slopes are -0.3 and -0.25 for univalent and bivalent ionic surfactants, respectively. When the activity of the surfactants at cmc, cma, is used instead of cmc, the slopes are -0.5——-0.58 irrespective of head group. This results in a value of -680——-777 cal/mol for the free energy of micelle formation per CH_2 group. Furthermore, the linear relationship between log cmc and log[counter ion] shows a slope of -0.6 for potassium dodecanoate, -0.67 for sodium dodecyl sulfate, and -0.95 for disodium dodecyl phosphate. When the critical micelle activity, cma, is used instead of cmc, the slopes are -0.9 for the univalent surfactants and -1.8 for the bivalent surfactants. These results indicate that the cma is substantially constant regardless of the counter ion concentration.

Performance of surfactants is closely related to surface activity and to micelle formation. Both these are due to amphiphilic nature of the surfactant molecule. The molecule contains a nonpolar hydrophobic part, usually, a hydrocarbon chain, and a polar hydrophilic group, which may be nonionic, zwitterionic, or ionic. When the hydrophobic group is a long straight chain of hydrocarbon, the micelle has a small liquid like hydrocarbon core(1,2). The primary driving

force for micelle formation has been considered to be
the hydrophobic effect(1—4), that is, the tendency of
the hydrocarbon chains to associate together with themselves rather than to remain in the aqueous phase.   A
thermodynamic equilibrium is established between
micellar and monomeric states.
      To understand micelle formation quantitatively,
critical micelle concentrations (cmc) have to be
determined for a large number of surfactants(5).   When
the cmc values of the surfactants with the same
hydrophilic group (a homologous series) are examined,
a nearly 3-fold decrease in cmc is observed for
nonionic and zwitterionic surfactants (1,2) upon the
addition of a methylene group into the hydrocarbon
chain, whereas, a 2-fold or only 1.8-fold reduction in
cmc can be observed for univalent(1,2) and bivalent(6)
ionic surfactants, respectively.
      In this work, the critical micelle activity, cma,
which is the activity of the surfactant at the cmc, is
introduced and used instead of the cmc to investigate
the free energy of micelle formation.   It is found that
upon the addition of an extra methylene group into the
hydrocarbon chain, an approximately 3-fold reduction in
cma is observed, irrespective of the hydrophilic head
group.   The effect of added electrolyte on cmc is also
examined by the use of cma.

Experimental

Materials and Method.   Aqueous solutions of disodium
alkyl phosphates were prepared by dissolving the
corresponding acids in sodium hydroxide solutions.   The
alkyl phosphoric acids were synthesized by the reaction
of pyrophosphoric acid with the respective alcohol in
benzene at room temperature for 4 days.   Details of the
purification procedures are given elsewhere(7).
      The capillary-rise method was employed to measure
the surface tension of aqueous solutions of disodium
alkyl phosphate at 25 °C.   The cmc values of the
solutions were obtained from the discontinuity in the
surface tension - concentration curves(7).
      Other cmc data than those of disodium alkyl
phosphates investigated in this study are taken from
the extensive tables by Mukerjee and Mysels (5).

Theoretical Background

Nonionic Micelle Formation.   Micelle formation in
aqueous solution was first considered to be an
equilibrium between monomer and micelle(1).   The law
of mass action controls the equilibrium

$$n \, L \rightleftharpoons M \tag{1}$$

where L is the surfactant monomer, M is the micelle, and n expresses the number of monomers associated in the micelle (the association number). The equilibrium constant, K, for the process can be written as

$$K = f_M[M]/(f_L[L])^n \tag{2}$$

Here f_M and f_L are the activity coefficients of micelle and monomer. The free energy of micelle formation ΔG is, therefore,

$$\Delta G = -(RT/n) \ln K \tag{3}$$

and from equation 2, ΔG is rewritten as

$$\Delta G = -(RT/n) \ln(f_M[M]) + RT \ln(f_L[L]) \tag{4}$$

At the critical micelle concentration, [L] = cmc. The first term of the right hand side of equation 4 is usually negligible when n is large (n = 50——100). If f_M and f_L are regarded to be unity, Equation 4 reduces to

$$\Delta G = RT \ln[L] = RT \ln cmc \tag{5}$$

Equations 4 and 5 are derived by completely ignoring electrostatic contributions to micelle formation, and can be applied only to nonionic and zwitterionic micelle formation.

On the other hand, micelle formation has sometimes been considered to be a phase separation of the surfactant-rich phase from the dilute aqueous solution of surfactant. The micellar phase and the monomer in solution are regarded to be in phase equilibrium and cmc can be considered to be the solubility of the surfactant. When the activity coefficient of the monomer is assumed to be unity, the free energy of micelle formation, ΔG, is calculated by an equation similar to equation 5 (8——11). Detailed examinations of micelle formation have indicated the phase separation model to be only an approximation (4,12).

In Figure 1, the log of the cmc is shown as a function of the number of carbon atoms in the hydrocarbon chain, n_c. It is clear that the nonionic surfactants (hexaethyleneglycol alkyl ethers) and the zwitterionic surfactants (N-alkyl betaines) exhibit linear relations with similar slopes of about -0.5

(hexaethyleneglycol alkyl ethers: -0.517; N-alkyl
betaines: -0.49), while the ionic surfactants without
added electrolyte show different slope values.
    The free energy of micelle formation of a
methylene group, $\Delta\Delta G(CH_2)$, is

$$\Delta\Delta G(CH_2) = \Delta G(C_nH_{2n+1}-Y) - \Delta G(C_{n-1}H_{2n-1}-Y)$$
$$= 2.303 \ RT X \ (\text{slope of log cmc vs. } n_c) \qquad (6)$$

$\Delta\Delta G(CH_2)$ thus calculated is -705 cal/mol and -657
cal/mol for hexaethyleneglycol alkyl ethers and N-alkyl
betaines, respectively, as shown in Table 1.
    For the transfer of one methylene group in a
hydrocarbon chain from aqueous to hydrocarbon liquid
environment, the free energy have been calculated to
be -825 cal/mol (at 25° C) (4). For the transfer
(adsorption) of the same group from the aqueous phase to
air-water and hydrocarbon-water interfaces (sparsely
covered interfaces), the free energies are -620 and
-820 cal/mol, respectively(13). The free energy gained
by removing a methylene group in the hydrocarbon chain
of the surfactant from the aqueous phase and placing it
in the micelle may be less than that of complete
transfer to bulk liquid hydrocarbon,because one end of
the hydrocarbon chain is anchored to the hydrophilic
head group in the micellar surface and there is
restricted freedom of motion inside the micelle(3).
Values of $\Delta\Delta G(CH_2)$ obtained here for nonionic and
zwitterionic micelle formation are , therefore,
reasonable. Values between -650 and -720 cal/mol are
indicated by Mukerjee(4) and also by Tanford(3). Values
of $\Delta\Delta G(CH_2)$ for various transfers are shown in Figure 2.

Ionic Micelle Formation. Even for ionic micelle
formation, the free energy of transfer of one methylene
group to a micelle should have a value similar to those
for nonionic and zwitterionic micelle formation. The
negative slope of log cmc vs. n_c is, however, smaller
for nonionic surfactants, as shown in Figure 1. The
univalent ionic surfactants, alkyk trimethylammonium
bromide, sodium alkyl sulfate, and sodium alkyl
carboxylate give slopesof about-0.3. The bivalent
anionic surfactant, disodium alkyl phosphate has the
slope of -0.25. These values correspond to the 3-, 2-,
and 1.8-fold lowering of the cmc for nonionic and
zwitterionic , univalent ionic, and bivalent ionic
surfactants,respectively, upon addition of one
methylene group to the hydrocarbon chain.
    The direct application of Equation 6 therefore
results in $\Delta\Delta G(CH_2)$ values of -400 cal/mol for uni-

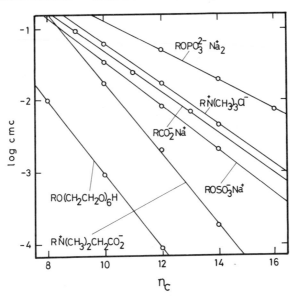

Figure 1. Log cmc vs. n_C

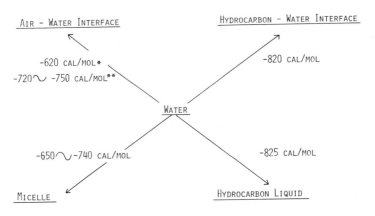

Figure 2. The free energy of transfer of a methylene group
in a hydrocarbon chain
* The value for the adsorption to the sparsely
covered surface.
** The value for the adsorption to the monolayer
at its minimum surface area per molecule (17).

Table 1. Slopes of log cmc and log cma vs. n_c and the Free Energy of Micelle Formation of a Methylene Group, $\Delta\Delta G(CH_2)$

Surfactant	Slope (log cmc vs. n_c)	Slope (log cma vs. n_c)	$\Delta\Delta G(CH_2)$ (cal/mol)
Hexaethyleneglycol alkyl ether (non-ionic)	-0.517		-705**
N-Alkyl betaine (zwitterionic)	-0.49		-657**
Alkyl trimethyl-ammonium bromide (cationic)	-0.29	-0.54	-736***
Sodium alkyl sulfate (univalent anionic)	-0.30	-0.57	-777***
Sodium alkyl carboxylate (univalent anionic)	-0.28	-0.52	-708***
Disodium alkyl phosphate (divalent anionic)*	-0.25	-0.50	-681***

Data used here are taken from reference 5.
* Data taken from references 6 and 7.
** Calculated by Equation 6 (25 °C)
*** Calculated by Equation 21 (25 °C)

valent and -350 cal/mol for bivalent ionic surfactants.
These differences could be due to electrical interactions in ionic micelle formation. The exact
evaluation of the electrostatic part of the free energy
of micelle formation is considered to be a very
complicated problem and our present position seems to
be far removed from that necessary to offer a well
founded calculation of it (3,4).
     We may begin the examination of ionic micelle
formation by reviewing the main theories already
presented. First of all, the mass action law is
extended to ionic micelle formation(14——16) as

$$n \ L^- \ + \ m \ X^+ \ \underset{\longleftarrow}{\overset{\longrightarrow}{\rule{1cm}{0pt}}} \ M^{(n-m)-} \tag{7}$$

Considerations analogous to those used in the derivation
of Equation 5 lead to Equation 8, if f_L and f_X are
assumed to be unity.

$$\Delta G = RT \ ln[L] + \frac{m}{n} RT \ ln[X] \tag{8}$$

Here, X indicates the counter ions of the ionic surfactant. When there is no added electrolyte,$[L] = [X] =$
cmc and
$$\Delta G = (\ 1 \ + \ \frac{m}{n} \ ) \ RT \ ln \ cmc \tag{9}$$

     Althogh, ΔG evaluated by Equation 9 takes into
account the loss in translational entropy of counter
ions upon micellar association(3,4), it is doutfull that
the term (m/n) RT ln[X], can include all the effects of
interionic interaction in micelle formation.
     On the basis of Equation 9, $\Delta \Delta G(CH_2)$ is derived as

$$\Delta \Delta G(CH_2) = 2.303(\ 1 + \frac{m}{n} \ ) \ RTX(\text{slope of log cmc vs. } n_c) \tag{10}$$

To obtain -650——-720 cal/mol for $\Delta \Delta G(CH_2)$ from the
calculation with the slope of -0.3, the value of (m/n)
should be in the range between 0.62 and 0.80. The
values of (m/n) (the degree of association of counter
ion) have been experimentally determined from the
measurements of electromotive force(18——22), electric
conductivity(23), and light scattering(24,25) of
micellar solution. Though, these measurements gave
0.75——0.85 for the degree of association of counter
ion, (m/n), the theoretical vagueness inherent in
Equations 8 and 9 have not been setteled yet.
     In contrast to Equations 7 and 8, based on the
binding of counter ion to micelle, an equation
including a form for the electrostatic part of free
energy of micelle formation, F_{el}, evaluated from double

layer theory, has been used to calculate ΔG. In this
method, the equilibrium between micellar and monmer
states is represented as

$$n\ L^- \ \underset{\longleftarrow}{\longrightarrow}\ M^{n-} \tag{11}$$

The free energy of micelle formation , ΔG, is written
as
$$\Delta G = RT \ln cmc - F_{el} \tag{12}$$

The phase separation model for nonionic micelle
formation has been modified for ionic micelle formation
to give an equation close to Equation 12 for ΔG (26).
In this modification, the ionic micelle has been
considered as the charged phase, which has difficulties
from the thermodynamic viewpoint. The precise
measurement of the surface tension of aqueous sodium
dodecyl sulfate solutions revealed the cotinuous decrease of surface tension above the cmc and indicated
that the charged phase separation model is not correct
(27).

Hobbs(28), and later, Shinoda(26) investigated F_{el}
by assuming the charged micellar surface to be flat.
The assumption of very high electrical potential gave
rise to the relationship between cmc and counter ion
concentration, [X]:

$$\log cmc = -\log[X] + k \tag{13}$$

The direct application of Equation 13 did not explain
experimental results. Hobbs, and also Shinoda,
introduced a factor, K_g, whose meaning has not been
theoretically elucidated(26,28), yielding

$$\log cmc = - K_g \log[X] + k \tag{14}$$

Tanford examined the application of Debye-Hückel
theory and found the theory not to be valid because the
high charge density generated by the closely spaced head
groups leads to substantial charge neutralization by
counter ions(3). Alternatively, he equated the work of
compression of the charged surfactant monolayer at a
hydrocarbon-water interface to F_{el}(29).

Later, Stigter evaluated F_{el} on the basis of the
Stern-Göuy model of ionic double layer with a
substantial part of the counter ions in a regular
distribution between ionic head groups and the remaining
counter ions in a diffuse layer. In this calculation,
he also introduced the correction for the discreteness
of charge at the micellar surface. Finally, the free
energy of micelle formation contributed by hydrophobic
interaction was related to the change of hydrocarbon -

water contact area in micelle formation and the value of $\Delta \Delta G(CH_2)$ was estimated as -740 cal/mol(30,31).

As mentioned above, a substantial part of the electrical charge of the micelle surface has been shown to be neutralized by the association of the counter ions with the micelle. In the calculation based on Equation 12, however, the loss in entropy arising from this counter ion association is not taken into account. This is by no means insignificant in comparison to F_{el} of Equation 12 (4). A major part of the counter ions are condensed on the ionic micelle surface and counteract the electrical energy assigned to the amphiphilic ions on the micellar surface. The minor part of the counter ions, in the diffuse double layer, are also restricted to the vicinity of the micellar surface.

More complete neutralization by counter ions has sometimes been observed in micellar solutions with added electrolyte(32). In explaining the specific binding of counter ions, Stigter has used the "image force" resulting when the counter ion approachs the micellar core which has a low dielectric constant(30). On the other hand, Lindman proposed hydrogen bonding between (polarized) water molecules in the primary hydration sheath of the bound counter ion and the head group of the surfactant to account for the sequence of counter ion binding to micelle(33,34).

Application of Activity at cmc. The above consideration suggested us to propose a new treatment for ionic micelle formation. According to thermodynamics, the micelle-monomer equilibrium is achieved when the chemical potential of surfactant in the micelle is equal to that in the bulk solution. The free energy of micelle formation can be represented by the use of the critical micelle activity, cma, which is the activity of surfactant at the cmc, as

$$\Delta G = RT \ln cma \tag{15}$$

The cma is calculated as

$$cma = a_{\pm}^{\nu} = f_{\pm}^{\nu} [L]^{\nu_-} [X]^{\nu_+} \tag{16}$$

where ν_- and ν_+ are the numbers of anions and cations in the surfactant molecule, $\nu (= \nu_+ + \nu_-)$ is the total number of ions in the surfactant molecule, and $a_{\pm}$ is the mean ionic activity of the surfactant at the cmc. According to Equation 15,

$$\Delta G = \nu RT \ln f_{\pm} + \nu_- RT \ln[L] + \nu_+ RT \ln[X] \tag{17}$$

If c' is the concentration of added electrolyte that has the same cation as the surfactant, the cma is expressed as follows

$$cma = f_{\pm}^{\nu} cmc^{\nu-} (cmc + c')^{\nu+} \qquad (18)$$

Eor univalent ionic surfactants, e.g., alkyl trimethyl-ammonium bromide and chloride, sodium alkyl sulfate, and sodium alkyl carboxylate,

$$cma = f_{\pm}^{2} cmc (cmc + c') \qquad (18a)$$

and for bivalent ionic surfactant, e.g., disodium alkyl phosphate

$$cma = f_{\pm}^{3} cmc (cmc + c)^{2} \qquad (18b)$$

In Equations 16——18, $f_{\pm}$ represents the mean ionic activity coefficient of the surfactant and may be calculated according to the Güntelberg approximation of the Debye-Hückel equation(6,7),

$$\log f_{\pm} = - \frac{A |z+ z-| \sqrt{I}}{1 + \sqrt{I}} \qquad (19)$$

where
$$A = 1.825 \times 10^{6} \times (DT)^{3/2} \qquad (20)$$

Here, I is the ionic strength based on the free ions in the micellar solution and D is the dielectric constant of solvent.

On the basis of Equation 15, the increment in ΔG by the addition of a methylene group to the hydrocarbon chain of surfactant, $\Delta \Delta G(CH_2)$, is written as

$$\Delta \Delta G(CH_2) = 2.303RT \times (slope\ of\ \log\ cma\ vs.\ n_c) \qquad (21)$$

in which cmc in the Equation 6 has been replaced with cma.

Results and Discussion

In Figure 3, the logarithm of cma is represented as a function of the number of carbon atoms in the hydro-carbon chain, n_c. It is found that the log cma values for the various surfactants at the same n_c value, except N-alkyl betaine, become closer compared to log cmc, and ΔG calculated by Equation 15 is -4.91—— -6.54 cal/mol for n_c = 12. For N-alkyl betaines, higher values of ΔG compared to other surfactants have been recognized in the literature and were ascribed to the ordering of the zwitterionic head group on the micellar surface(35,36). Also the contact of the hydrocarbon chain with water molecules at the micellar surface has been considered to be an important factor(3,30,31).

On the other hand, the slopes obtained were -0.5 for disodium alkyl phosphate, -0.54 for alkyl trimethyl-ammoniumchloride,-0.52 for sodium alkyl carboxylate, and -0.58for sodium alkyl sulfate. These values are close to those for nonionic and zwitterionic surfac-tants, obtained from the linear relation of log cmc and n_c. The $\Delta\Delta G(CH_2)$ values calculated from Equation 21 are in the range of -708— -777 cal/mol and have good agreement with the values for nonionic and zwitterionic surfactants, as shown in Table 1.

In the calculation of ΔG by Equation15, the cma values forahomologous series of surfactants are evaluated at different ionic strengths and the results obtained for $\Delta\Delta G(CH_2)$,therefor,indicate that ΔG thus calculated is independent of the ionic strenght of the micellar solution. Equation 17,derived from Equation 15 , is rewritten as

$$\nu_- \log cmc + \left(\frac{\nu}{\nu_-}\right) \log f_\pm = -\left(\frac{\nu_+}{\nu_-}\right) \log[X]$$
$$+ G/(2.303\nu_- RT) \qquad (22)$$

In Figure 4, the values of log cmc are presented as a function of the logarithm of total concentration of counter ion, log[X]. The slopes of these plots are -0.67 for sodium dodecyl sulfate, -0.6 for potassium dodecanoate, and -0.9 for disodium dodecyl phosphate. In accordance with Equation 22, the values of log cmc + (ν/ν_-) log $f_\pm$ are also plotted against log[X] in the same figure. Thelinear relations obtained now give slopes of -0.9 for sodium dodecyl sulfate and potassium dodecanoate, and -1.8 for disodium dodecyl phosphate, which are in fairly good agreement with the theoretical values of -1 and -2, respectively. The results obtained here suggest that the condensation of counter ions on the ionic micellar surface andthe reduction of electrostatic energy play important roles in the ionic micelle formation.

The application of the activity of the surfactant has been examined also for the surface tension and adsorption of disodium alkyl phosphate(6,7), sodium dodecyl sulfate(37), alkyl trimethylammonium bromide(35), and sodium perfluorooctanoate(13) solutions. These studies show that the surface tension and theadsorption amount are controlled by the activity of surfactant, irrespective of the added electrolyte concentration.

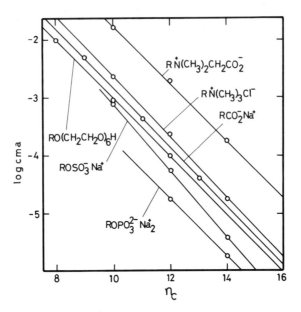

Figure 3. Log cma vs. n_c

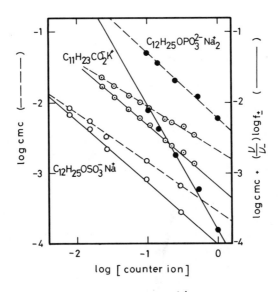

Figure 4. Log cmc and log cmc + $(\frac{\nu}{\nu_-})$ log f± as a function of log[counter ion]

It is also noteworthy that the surface tension of mixed solutions of anionic and cationic surfactants is controlled by the activity of the salt(38). These results indicate that, in adsorption of an ionic surfactant at the solution surface, the accompanying adsorption of the counter ion also plays an important role.

In conclusion, micelle formation of an ionic surfactant is found to take place at a constant activity,i.e., critical micelle activity, cma, irrespective of counter ion concentration.

Literature Cited

1. Hartley, G.S. "Aqueous Solutions of Paraffin Chain Salts" Hermann et Cie : Paris, 1936.
2. Debye, P. J.Colloid Sci. 1948, 3, 407.
3. Tanford, C. "The Hydrophobic Effect" 2nd Ed., John Wiley & Sons: New York, 1980.
4. Mukerjee, P. Advan. Colloid Interface Sci. 1967, 1, 241.
5. Mukerjee, P; Mysels, K.J. "Critical Micelle Concentration of Aqueous Surfactant Systems" NSRDS-NBS 36, US Nat. Bur. Stand., 1971.
6. Nakagaki, M; Handa, T; Shimabayashi,S. J.Colloid Interface Sci. 1973, 43, 521.
7. Nakagaki, M; Handa, T. Bull. Chem. Soc. Jpn. 1975, 48, 630.
8. Stainsby, G; Alexander, A.E. Trans.Faraday Soc. 1950, 46, 587.
9. Hutchinson, E; Inaba. A; Bailey, L.G. Z.Physik. Chem. 1955, 5, 344.
10. Matijevic, E; Pethica, B.A. Trans.Faraday Soc. 1958, 54. 587.
11. Herrman, K.W. J.Phys.Chem. 1962,66, 295.
12. White, P; Benson,G.C. Trans.Faraday Soc. 1959. 55, 1025.
13. Mukerjee, P; Handa, T. J. Phys.Chem. 1981, 85, 2298.
14. Murray,R.C; Hartley, G.S. Trans.Faraday Soc. 1935, 31, 183.
15. Mysels, K. J. Colloid Sci. 1955, 10, 507.
16. Mukerjee,P. J.Phys. Chem. 1962, 66, 33.
17. Rosen, M.J; Aronson,S. Colloid Surf. 1981, 3, 201.
18. Shedlovsky, L; Jakob, C.W; Epstein, M.B, J.Phys. Chem. 1963, 67, 2075.
19. Botre, C; Cresenzi,V.L; Mele, A. J.Phys.Chem. 1969, 63, 650.
20. Ingram, T; Jones, M.N. Trans. Faraday Soc. 1969, 65, 297.
21. Mathews, W.K; Larsen, J.W; Pikai, M.J. Tetrahedron Letter, 1972, 6, 513.

22. Malik, W.U; Srivastava, S.K; Gupa, D. J.Electro
    -annal. Chem. 1072, 34, 540.
23. Anderson, J.E; Taylor, H.   J.Pharm. Pharmacol.1971,
    23, 311.
24. Philips, J.N; Mysels,K.J. J.Phys.Chem. 1955, 59,325.
25. Barry, B.W; Morrison, J.C. Pikai, M.J. Tetrahedron
    Letter, 1972, 6, 513.
26. Shinoda, K; Nakagawa, T; Tamamushi, B; Isemura, T.
    "Colloid Surfactant" Academic Press; New York,
    1963; Chap.1.
27. Elworthy, P.H; Mysels, K. J.Colloid Sci. 1966, 21,
    331.
28. Hobbs, M.E. J.Phys.Chem. 1951, 55, 675.
29. Tanford, C. Proc.Nat. Acad. Sci. U.S.A. 1974, 71,
    1811.
30. Stigter,D. J.Phys.Chem. 1975, 79, 1008, 1015 and
    1020.
31. Stigter,D. J.Colloid Sci.1974, 47, 473.
32. Kresheck, G.C. in "water"; Franks,F.Ed.; plenum
    Press: New York, 1975, Vol.4, Chap.2.
33. Gustavsson, H; Lindman,B. J.Am.Chem.Soc. 1975, 97,
    3923.
34. Lindman, G; Lindman,B ; Tiddy, G.T.J. J.Am.Chem.Soc.
    1978, 100, 2299.
35. Molyneux,P; Rhodes,C.T; Swarbrick, J. Trans.Faraday
    Soc. 1965, 61, 1043.
36. Beckett,A.H; Woodward,R.J. J.Pharm. Pharmacol. 1963,
    15, 422.
37. Lucassen-Reynder, E.H. J.Phys.Chem. 1966, 70, 1977.
38. Rosen,M.J. Friedman,D; Gross,M. J.Phys.Chem. 1964,
    68, 3219.

RECEIVED January 10, 1984

EFFECT OF STRUCTURE
ON PERFORMANCE
IN MICELLES
AND MICROEMULSIONS

Relationship of Solubilization Rate to Micellar Properties

Anionic and Nonionic Surfactants

Y. C. CHIU, Y. C. HAN, and H. M. CHENG

Department of Chemistry, Chung Yuan Christian University, Chung-Li, Taiwan 320, Republic of China

This paper presents a new finding that the oil solubilization rate is a function of surfactant aggregate
size. Light scattering and conductance measurements
were used in the experiments. Alcohol ethoxylates
and SDS were used as surfactants. The aggregate size
was changed by changing the surfactant structure or
by adding chemicals. The solubilization rate shows
a maximum at a certain aggregate size for a given surfactant and a given oil. Thus, we have found a measurable and controllable factor (size) in the process
of oil solubilization. A theory was proposed to relate solubilization rate with micellar properties and
surfactant structure. By using this theory, we can
explain the performance of petroleum sulfonate in enhanced oil recovery and improve current formulation
in achieving ultra-low interfacial tension. We can
also explain the nonionic detergent performance as
a function of surfactant structure.

This paper presents a new finding that the oil solubilization rate
is a function of the surfactant aggregate size. This idea origi-
nated from Chiu's observation on solubilization phenomena in tertiary oil recovery.

During experiments with petroleum sulfonates in surfactant
flooding, it was found that the surfactant solutions in the optimum
electrolyte region, contaning large surfactant aggregates, are
effective in oil recovery (1). These solutions give fast solubilization of oil and exhibit ultralow interfacial tension when they
are in contact with oil (1). A theory was proposed that the large
surfactant aggregates are important in obtaining rapid solubilization and ultralow interfacial tension (1). In order to test this
theory, medium molecular weight alcohols were used to replace
electrolyte in increasing surfactant aggregate size. The resulting
solutions also gave good oil recovery (2). This theory has been

0097-6156/84/0253-0089$08.00/0
© 1984 American Chemical Society

used to explain the detergent performance of alcohol ethoxylates with respect to surfactant structure (3).

Although the proposed theory has been used effectively in several practical applications, no experimental proof has been given that the oil solubilization rate is a function of surfactant aggregate size. In view of the importance of solubilization and the existence of practical methods of measuring and controlling surfactant aggregate size, we decided to correlate the solubilization rate with micellar properties for some anionic and nonionic surfactants.

Although solubilization (4) has been a subject of many investigations, most of the studies were made on the final equilibrium solubilization. Only a few studies concern about the kinetics and mechanism. In recent publications, Carrol (5) measured solubilization rate of nonpolar oils by nonionic surfactant solutions using a microscopic observation of the change of oil droplet adhering on a fiber. Chan et al (6,7) studied the kinetics and mechanism of solubilization in detergent solutions by radio tracer technique. These methods are either tedious or requiring safety precaution. For large scale laboratory operation using simple equipments, we have developed a light scattering technique to measure the solubilization rate and particle size.

Experimental

Light scattering technique was used in determining the oil solubilization rate. Debye's equation (8) was used in the interpretation. The basic principle involves the measurement of the surfactant aggregate size during the solubilization. As the oil goes into the surfactant micelle, the increased size will be reflected by the turbidity of the solution.

The instrument used in the turbidity measurement was Hatch Model 2100 A Turbidimeter. A Hotech Shaker Bath, Model 901 (Hotech Instruments Corp.) was used in mixing the oil and surfactant solution. The nonionic surfactants, Newcol 1102, 1103 and 1105 were produced by Sino-Japan Chemical Co., Ltd. The active ingredient is dodecanol ethoxylate. Sodium dodecyl sulfate (SDS, No. L. 5750, Sigma Chemical Co. 95% active, containing 65% C_{12}, 27% C_{14} and 6% C_{16}) was used as the anionic surfactant. Oleic Acid (Extra pure reagent, Kanto Chemical Co., Tokyo, Japan), Triolein (glycerol trioleate ($C_{17}H_{33}COO)_3C_3H_5$, Technical, BDH Chemicals, England) and n-decane (E. Merck, G.C., 95%) were used as oil. Sodium chloride (E. Merck, purity 100 ± 0.05%) was used as electrolyte.

The experiment was done by adding a given amount of oil to 14.0 g of 0.05% nonionic surfactant solution. The oil was first added to the surfactant solution and dispersed into tiny droplets by hand and was then mixed with the surfactant solution by the Hotech Shaker at 120 cycles/min for 20 minutes. The turbidity of the solution was measured. The solution was shaken and turbidity

was measured repeatedly until the turbidity reached a constant
value. For anionic surfactant experiments, 15.0 g of 0.5% surfactant solution was used and the shaking time was 5 minutes.

Some experiments concerning the solubilization of triolein
in SDS solutions were done by adding triolein dropwise to the surfactant solution. 15.0 g of 0.5% SDS solution was used. One drop
of triolein (0.006 ± 0.001 g) was added to the surfactant solution.
The oil was dispersed, shaken and turbidity measured as it was
mentioned above. When turbidity reached a constant value, another
drop of triolein was added and the process repeated until the
turbidity value did not change with the addition of triolein.

It should be mentioned here that the nonionic surfactant solutions were used within 1-10 days after the preparation. The
anionic surfactant solutions were used after aging for two days.
All experiments were done at room temperature, $25 \pm 1\ ^{\circ}C$.

Result and Discussion

A Proposed Theory. In earlier publications (1-3), a theory was
proposed to correlate solubilization rate, interfacial tension
and size of the surfactant aggregate: (1) the interfacial tension
lowering between the oil-surfactant solution interface is a
function of the rate of solubilization of oil, and (2) the rate of
solubilization ($\Delta S/\Delta t$) is a function of the effective volume for
solubilization:

$$\Delta S/\Delta t = k\ n\ V_{eff} \qquad (1)$$

Where   k = constant.
             n = number of aggregates in unit volume of surfactant
                solution.
         V_{eff} = effective volume for solubilization by an aggregate.
                = f (accessible volume of the hydrocarbon core,
                chemical nature of the surfactant molecule, chemical
                nature of the oil).

The effective volume for solubilization may or may not be
proportional to the geometrical size of the aggregate. It depends
on the packing of the molecules in the aggregate and the mutual
compatibility of the surfactant and oil molecules. In most cases,
V_{eff} is proportional to the size of the micelle (or aggregate).
When the aggregate size is too large and the packing of monomer
becomes too tight, V_{eff} may decrease with the aggregate size.

Interpretation of Light Scattering. We used Debye's equation (8)
for micellar solution as a basis for the light scattering
measurement:

$$\tau = \frac{32\pi^3}{3}\ \frac{\mu_0^2}{N\lambda^4}\ \left(\frac{d\mu}{dC}\right)^2 M\ (C-C_0) \qquad (2)$$

Where τ = Turbidity.
       μ = refractive index of the solution.
       μ_o= refractive index of the solvent.
       C = concentration (g/ml).
       C_o= critical micellar concentration (CMC).
       M = aggregate weight.
       N = Avogadro's number.
       λ = wave length of the light.

When $(d\mu/dC)^2(C-C_o)$ becomes constant, τ would be proportional to
M. For nonionic surfactant, we used Newcol 1102, 1103 and 1105.
These surfactants contain dodecanol ethoxylate. The last digit in
the Newcol number represents the ethylene oxide (EO) number. The
CMC values for pure dodecanol ethoxylate (3) with EO number from
3 to 5 are in the concentration range of 0.001-0.003%. C value in
our experiment is 0.05%. Therefore, $C-C_o$ can be considered as
constant. Values (3) of μ as a function of C also show that $d\mu/dC$
is almost constant. Thus in the nonionic surfactant measurement
in this paper, τ is considered to be proportional to M.
     For anionic surfactant, we used sodium dodecyl sulfate (SDS).
The CMC values were measured by conductance method. The CMC
values were taken from the breaks of curves from plots of K/C
versus $N^{\frac{1}{2}}$. Where K is the specific conductance, C is molar concentration and N is the equivalence. Figure 1 shows the CMC
values of SDS at 25 ^{o}C. The curve showing in the lower left side
represents data taken from literature for pure SDS. The curve
showing in the upper right side represents measurements for our
impure sample. Table I shows some values of $(d\mu/dC)^2(C-C_o)$ for
pure SDS at 25 ^{o}C. The values for NaCl concentrations of 0.03 M
to 0.50 M are not far from constant. Therefore, in this concentration region, τ is also considered to be proportional to M.
The C value used in our experiments is 0.5% (0.0171 M, average
molecular weight was taken as 293).

Table I.   Some SDS Properties at 25 ^{o}C

System	Data from Fig.1 Lower Left Curve $C - C_o$	Data from Ref.12 λ = 4358 $(d\mu/dC)^2$	$(d\mu/dC)^2(C-C_o)$
H_2O	0.0091	0.01426	0.000130
0.03M NaCl	0.0141	0.01421	0.000200
0.20M NaCl	0.0163	0.01339	0.000218
0.50M NaCl	0.0162	0.01199	0.000194

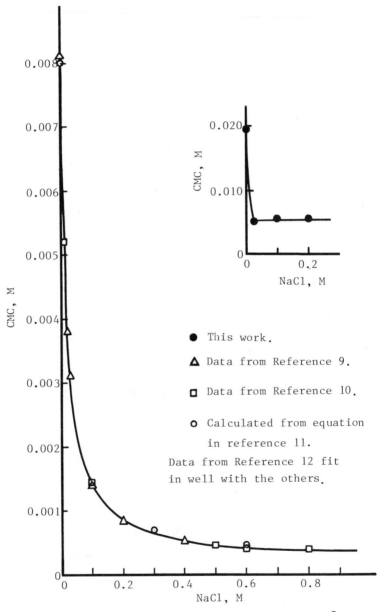

Figure 1.   CMC Values of SDS at 25 °C.

During the solubilization experiment, oil continues to solubilize in the surfactant micelle and the M value continues to
increase. The change of M is reflected by the change of τ. And
$d\tau/dt$ (t = time) is taken as the rate of solubilization in our
experiments.

The contribution to τ by emulsified oil in our experiments is
considered negligible in the nonionic surfactant solutions due to
the very low CMC values. In SDS solutions, the emulsification
occurs at the very beginning when no NaCl is added to the solution
and the turbidity introduced by emulsification does not change
with time. When NaCl is added to SDS solutions, the CMC becomes
low and emulsification becomes unimportant as it will be shown in
the following sections.

Solubilization in Nonionic Surfactant. Figure 2 shows the solubilization of oleic acid in Newcol nonionic surfactants. Turbidity
was plotted against shaking time. The first number on the curve
represents the surfactant. 1102 means dodecanol ethoxylate
containing 2 EO. The second number on the curve represents the
amount of oleic acid added to the surfactant solution.

For dodecanol ethoxylate, when EO number is larger than 8, the
aggregate weight decreases with the increase in EO number (13).
For the low EO members, water solubility becomes low. For example,
when EO = 4, the cloud point (3) is about 8 $^{\circ}$C. When EO = 5, the
cloud point (3) is about 30 $^{\circ}$C. In Figure 2, the turbidity in
zero shaking time reflects the aggregate weight in the surfactant
solution before the addition of oil. The turbidity is very high
in 1102 indicating large aggregates in the solution. The turbidity
for the original solution is slightly higher for 1103 than 1105
indicating larger aggregates existing in 1103 solution.

$d\tau/dt$ (slope of the curve) in Figure 2 represents solubilization rate and the steady turbidity showing at the end of each
curve signifies the solubilization of oil at that particular
condition. Among these 6 curves, the 1102 curve should be discussed separately. Since our experiments were carried out at
25 $^{\circ}$C, the temperature is far above the 1102 cloud point. Although the aggregate size is large, the aggregate is packed tight
and should have low solubilization volume. The addition of a
small amount of oleic acid (0.002 g) increases the aggregate size
tremendously. Further addition of oleic acid results in coagulation and decreases turbidity. For curves representing Newcol 1103
and 1105, several trends are shown in Figure 2: (1) for surfactant
solutions containing the same amount of oleic acid, $d\tau/dt$ is
higher for 1103 than 1105, showing good agreement with Equation 1
and (2) when the same surfactant is used, $d\tau/dt$ is higher for more
oil addition. This is expected from kinetics rules.

Figure 3 shows the solubilization of triolein in Newcol surfactants. Figure 4 shows the solubilization of n-decane in the
same surfactant solutions. The general characteristics of the
curves in Figures 3 and 4 are the same as those shown in Figure 2.

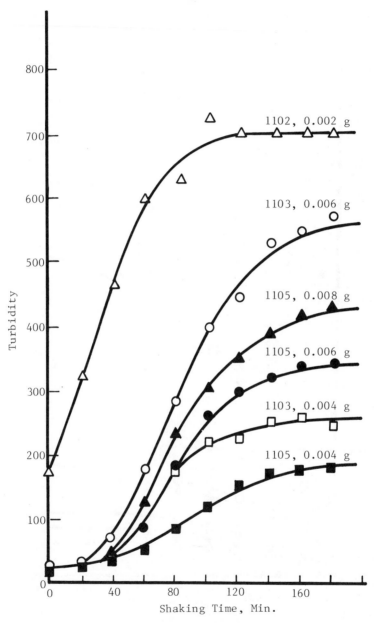

Figure 2. Solubilization of Oleic Acid in Nonionic Surfactant.

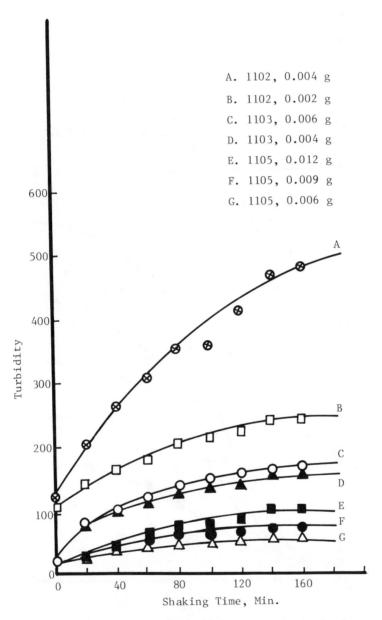

Figure 3.   Solubilization of Triolein in Nonionic Surfactant.

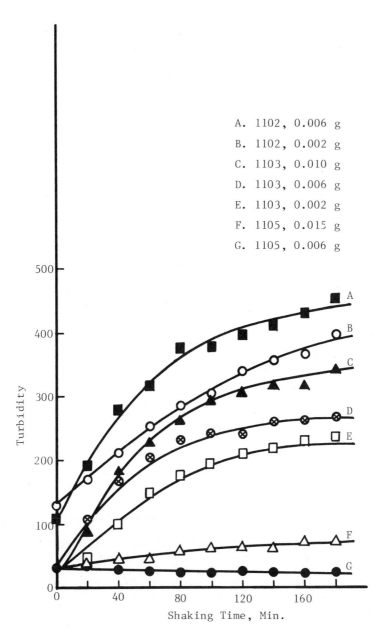

Figure 4. Solubilization of n-Decane in Nonionic Surfactant.

From the lower turbidity values shown in Figures 3 and 4, one may estimate that the solubilization of oleic acid is higher than triolein or n-decane in Newcol surfactant solutions. It is diffi- cult to make further distinction between triolein and n-decane from Figures 3 and 4.

Solubilization in Anionic Surfactant. Figures 5 and 6 show the solubilization of oleic acid in SDS solutions. The surfactant aggregate size is varied by changing NaCl concentration in the surfactant solution. Table II shows the aggregation number of SDS micelles at 25°C as reported by various authors. The aggregation number is in general increases with the NaCl concentration. When the NaCl concentration is above 0.4M, the aggregation number in- creases more rapidly. The micellar shape changes from spherical to rod (14,15).

Table II. Aggregation Number of SDS Micelles at 25°C

Solution	Ref.9	Aggregation Number Ref. 10	Ref. 12	Ref. 14
Water	80	–	62	–
0.01M NaCl	–	70-77	–	–
0.02M NaCl	94	–	–	–
0.03M NaCl	100	–	72	–
0.10M NaCl	112	97-101	–	–
0.15M NaCl	–	–	–	90
0.20M NaCl	118	–	101	–
0.30M NaCl	–	–	–	110
0.40M NaCl	126	–	–	–
0.45M NaCl	–	–	–	200
0.50M NaCl	–	148	142	–
0.55M NaCl	–	–	–	600
0.60M NaCl	–	174-528	–	940
0.80M NaCl	–	1630	–	–

The turbidity values at zero shaking time reflects the turbidity for the SDS solutions before the addition of oil. Usually, hand dispersion of oil does not increase turbidity of the surfactant solution. Turbidity increases only when mechanical shaking is applied. But in SDS solutions when no NaCl is added, the dispersion of oleic acid by hand increases turbidity and subsequent mechanical shaking causes no further change. The initial turbidity for SDS solution containing 0-0.4 M NaCl before the addition of oil ranges from 1-3 and is hardly distinguishable in the drawing. Figures 5 and 6 show similar characteristics. Each curve in Figure 5 represents the solubilization of 0.050 g oleic acid in SDS solution containing the specified NaCl. In Figure 6, 0.062 g oleic acid is used. In general, higher turbi- dity is observed in Figure 6 than in Figure 5.

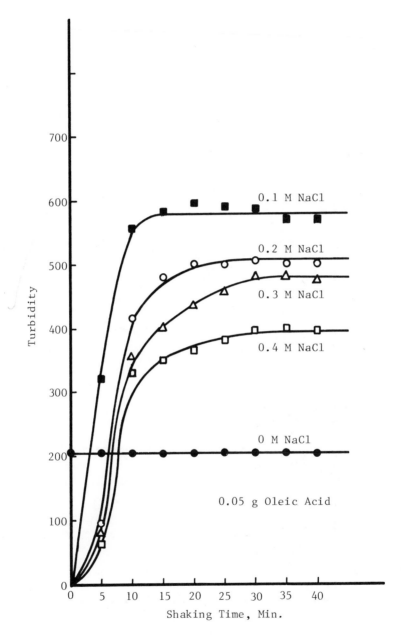

Figure 5.   Solubilization of Oleic Acid in Anionic Surfactant.

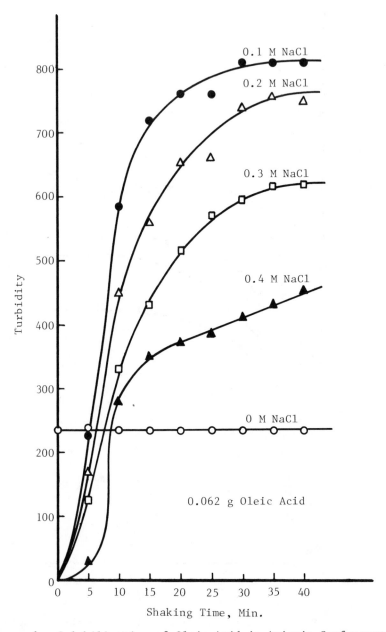

Figure 6.   Solubilization of Oleic Acid in Anionic Surfactant.

Among the SDS solutions with different NaCl concentrations, the micellar properties in these solutions require some discussion. From Figure 1, one can see the CMC values of our SDS solutions in the upper right side. Since the conductance method for CMC determination can be used only in electrolyte concentration of not more than 0.1 M, we may use the values obtained only in this concentration region. In SDS solution without NaCl, the CMC value is 0.56-0.60%. For NaCl concentration of 0.025-0.1 M, CMC values are around 0.15%. From literature data showing in the same figure, one may predict only small decrease of CMC would occur in NaCl concentration range of 0.1-0.4 M. In the SDS solutions we used in the experiments, the surfactant concentration is 0.5%. This value lies below the CMC when no NaCl is added. In the presence of 0.1-0.4 M NaCl, 70% of the surfactant will be in the micellar form.

When no NaCl is added to the SDS solution, the turbidity increase with slight dispersion of oleic acid is probably due to the monomer emulsification of the oleic acid. The process does not seem to require as much time and energy as solubilization. In the SDS solutions containing 0.1-0.4 M NaCl, the turbidity increase is mainly due to micellar solubilization. The solubilization rate ($d\tau/dt$) of the solutions seems to be in this order: 0.1 M > 0.2 M > 0.3 M > 0.4 M NaCl. The final solubilization of the solutions is also in the same order. The size of the pure SDS micelles in solutions containing NaCl has been shown to increase with the NaCl concentration (Table II). Our instrument is not sensitive enough to distinguish the size between these small micelles at different NaCl concentrations. In the solubilization of oleic acid in 0.5% SDS solutions, the maximum V_{eff} seems to occur when the NaCl concentration is around 0.1 M. The maximum may occur in NaCl concentration below 0.1 M or between 0.1 M and 0.2 M which we have not studied.

Figure 7 shows some turbidity data as a function of triolein addition in SDS solutions containing varying amount of NaCl. Each point was taken from experiments such as described in the experimental section by dropwise addition of triolein. The turbidity increases with the addition of triolein to a steady value for all SDS solutions containing different amount of NaCl. The highest turbidity (or the highest solubilization) and the highest solubilization rate ($d\tau/dt$) occur at 0.5 M NaCl. At higher NaCl concentrations, coagulation occurs with shaking and turbidity decreases. Figure 7 also shows some data taken from experiments performed in the same way as described above for the solubilization of oleic acid in SDS solutions. In these experiments, 0.050 g triolein was added to the SDS solution containing a specified amount of NaCl. The final turbidity values were plotted along the line corresponding to 0.050 g triolein addition with the specified NaCl concentration. These values are not too far from the values obtained by dropwise addition of triolein as shown in the same diagram.

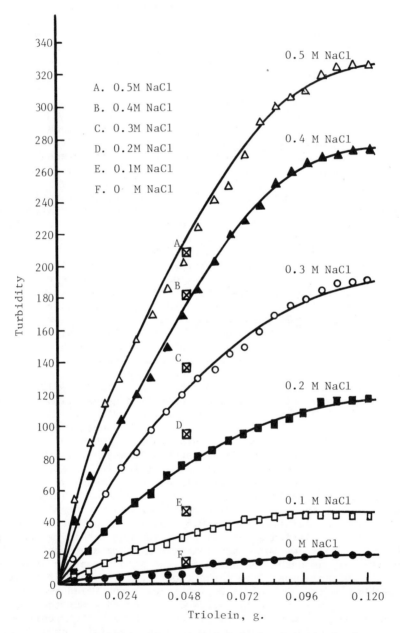

Figure 7.   Solubilization of Triolein in Anionic Surfactant.

Up to this moment we have studied the solubilization rate of several oily materials in some nonionic and anionic surfactants. In general we found that the oil solubilization rate is a function of the surfactant aggregate size. The maximum V_{eff} for a series of surfactant solutions seems to occur at the condition that the surfactant associates to the maximum aggregate volume without increasing the density of the aggregate. The V_{eff} value seems to parallel the final solubilization value of the surfactant.

All the results we obtained are qualitative in nature. The materials used in the experiments are mostly not 100% pure but should be usable in comparing performance result of industrial applications. And the conclusions we obtained do not have the limitation of a pure oil (or surfactant) or a particular oil (or surfactant). One important factor we have not discussed is the electrical effect. This may contribute significantly in anionic surfactant solubilization with polar oil. We have neither discussed the mechanism of solubilization nor the specific effect of a certain oil to a certain surfactant. To understand clearly such specific effects and to derive more quantitative relations governing the result of solubilization, we are planning to conduct future experiments with pure sample.

Relation Between Structure and Performance of Surfactants. In a detergent paper (3), we have repeated the detergency work of The Proctor and Gamble Company of some narrow range dodecanol ethoxylates with EO number from 2 to 8. A detergency maximum at EO number 4 for 24 °C (at EO number 5 for 38 °C) was found for triolein removal. For oleic acid removal, a steady increase of detergency was found for dodecanol ethoxylates from EO number 2 to EO number 5 at 38 °C and no significant change from EO number 5 to EO number 8. From cloud point measurement, it was found that the temperature at which the indicated surfactant showing the maximum detergency is 7-15 °C above the surfactant cloud point. Since triolein is relatively non-polar, its removal depends mainly on solubilization. The large aggregates formed at temperature somewhat above the cloud point are obviously very effective in solubilization. We have attributed this to the increase of solubilization rate with the surfactant aggregate size. The evidence has been shown by this paper.

Carroll (5) studied the kinetics of solubilization of nonpolar oil by nonionic surfactant solutions and found that the solubilization rate is strongly temperature dependent in the region of the nonionic cloud point: 15 °K below the cloud point, the rate is extremely small relative to that at the cloud point. Nakagawa and Tori (16) have found a tremendous increase in aggregate weight for nonionic surfactants near the cloud point and the increase in aggregate weight correlated with increased solubilization of long chain alkyl compounds at constant total surfactant concentration. These independent studies show more evidences to what we have just described in this paper.

Since oleic acid is relatively polar, it may become emulsified by the surfactant monomer. The removal of oleic acid comes mainly from two contributions: monomer emulsification and micellar solubilization. Although the V_{eff} has been decreased with increasing EO number in dodecanol ethoxylates, in higher EO numbers than 5, this factor has been compensated by the increase of monomer with increasing EO number (CMC decreases with EO number). The levelling of detergency of dodecanol ethoxylates from EO number 5 to EO number 8 has been interpreted by these reasons. The monomer emulsification of oleic acid has been clearly shown in this paper in SDS solution. The nonionic surfactants we used here have low EO numbers and show mainly the effect of solubilization.

It also appears that larger solubilizate prefers larger micelles for solubilization. This is shown in the SDS solubilization. Triolein solubilizes more effectively in SDS solutions containing higher NaCl concentration while oleic acid solubilizes better in lower NaCl concentration. The distinction of the size preference between different solubilizates is not so obvious in the nonionic surfactant solutions reported here. With EO number between 2-5, the sizes of the nonionic surfactant aggregates are so much larger than the SDS micelles (or the oil molecules) that even the smallest nonionic aggregates are more effective solubilizers than the SDS micelles. Thus they show a levelling effect for the different solubilizates. The electrical charge on anionic surfactants may also have some contribution to the difference in solubilization of oleic acid and triolein.

Another example showing the preference for large surfactant aggregates is demonstrated in tertiary oil recovery (1). When a 5% Bryton 430 petroleum sulfonate solution was used, the ultralow interfacial tension between oil and water and fast solubilization of crude oil appeared around 0.3 M NaCl. By Coulter counter measurement, surfactant aggregates around 1 μ in size was found. Below 0.3 M NaCl, the aggregate sizes were smaller and the solubilization rates were slower. Above 0.3 M NaCl, the aggregates became unstable and tended toward separation from water. We have not had an opportunity to measure the solubilization rate of petroleum sulfonate solutions. The statement made here is mainly from observation. The oil solubilization in 5% Bryton 430 containing 0.3 M NaCl is too fast to be measured by ordinary methods. In order to bring the solubilization rate in a measurable range, we used much smaller surfactant aggregates (SDS) in the experiments. The phenomenon exhibited by triolein solubilization resembled the phenomenon observed in Bryton 430 except the rate was much slower.

It is well known that SDS and many commercial surfactants can not be used to recover oil. In our opinion these surfactants can not associate to form large enough aggregates is one important reason. In most cases, when electrolyte is added to the surfactant solution to increase the size of the aggregate, surfactant separation occurs before large enough aggregates can be built up.

This paper presents a very basic principle in surfactant solubilization. More quantitative measurement in correlating solubilization rate with micellar properties and more applications of this principle to improve performance of various solubilization processes remain the subject of our investigation.

Acknowledgments

Shu-Mei Ann, Venice Hu, Dao-Shinn Hwang, Peter Hwang, Jing-Ming Hsu, Tian-Tsain Wu, Huann-Jang Hwang and Ruey-Jing Cheng have made contribution to this paper.

Literature Cited

1. Chiu, Y. C. in "Solution Behavior of Surfactants: Theoretical and Applied Aspects"; Mittal, K. L.; Fendler, E. J., Ed.; Plenum: New York, 1982; Vol. II, pp. 1415-1440.
2. Chiu, Y. C. Oilfield and Geothermal Chemistry Symposium, Denver, Co., U.S.A., June 1-3, 1983; SPE paper 11783.
3. Benson, H.L.; Chiu, Y. C. "Relationship of Detergency to Micellar Properties for Narrow Range Alcohol Ethoxylates", Technical Bulletin, SC: 443-80, Shell Chemical Co. Houston, Texas, U.S.A., 1980.
4. McBain, M.E.; Hutchinson, E. "Solubilization and Related Phenomena"; Academic Press: New York, 1955.
5. Carroll, B. J. J. Colloid and Interface Sci. 1981, 79, 126.
6. Chan, A. F.; Evans, D. F.; Cussler, E. L. AICHE 1976, 22, 1006.
7. Shaeiwitz, J. A.; Chan, A. F-C.; Cussler, E. L.; Evans, D. F. J. Colloid and Interface Sci. 1981, 84, 47.
8. Debye, P. Ann. N. Y. Acad. Sci. 1949, 51, 575.
9. Phillips, J. N. Trans. Faraday Soc. 1955, 51, 561.
10. Hayashi, S.; Ikeda, S. J. Phys. Chem. 1980, 84, 744.
11. Emerson, M. F.; Holtzer, A. J. Phys. Chem. 1967, 71, 1898.
12. Mysels, K. J.; Princen, L. H. J. Phys. Chem. 1959, 63, 1696.
13. Becher, P. J. Colloid Sci. 1961, 16, 49.
14. Mazer, N. A.; Benedek, G. B.; Carey, M. C. J. Phys. Chem. 1976, 80, 1075.
15. Missel, P. J.; Mazer, N. A.; Benedek, G. B.; Young, C. Y.; Carey, M. C. J. Phys. Chem. 1980, 84, 1944.
16. Nakagawa, T.; Tori, K. Koll. Z. 1960, 168, 132.

RECEIVED January 20, 1984

Hydrotropic Function of a Dicarboxylic Acid

STIG E. FRIBERG and TONY D. FLAIM

Chemistry Department, University of Missouri–Rolla, Rolla, MO 65401

The hydrotropic action of a dicarboxylic acid is
discussed against the general features of hydrotropic action; the liquid crystal/isotropic solution
equilibrium. It is shown that the hydrotropic action
of the dicarboxylic acid in question, 8-[5(6)-carboxy-
4-hexyl-cyclohex-2-enyl] octanoic acid, depends on its
conformation at an interface.

The word hydrotrope was introduced 67 years ago by Neuberg (1).
In his treatment and in the subsequent ones the hydrotropes were
investigated for their solubilizing power in aqueous solutions
(2-6).

A change in the perception of their mechanism of action came
in the sixties when Lawrence (7) pointed out that short chain surfactants would delay the gelling to a liquid crystalline phase
which takes place at high surfactant concentrations. Friberg and
Rydhag (8) showed that hydrotropes, in addition, prevent the formation of lamellar liquid crystals in combinations of surfactants
with hydrophobic amphiphiles, such as long chain carboxylic acids
and alcohols. The importance of this finding for laundry action
was evident.

The hydrotropes in this era were short chain aromatic sulfonates, with the p-xylene sodium sulfonate as a typical example.
Their action is preventing the formation of liquid crystals is
easily understood from a direct comparison of their molecular
geometry (Fig. 1).

The short bulky aromatic compound does not pack well in a
lamellar liquid crystalline structure, the mutual stabilizing
action of the straight hydrocarbon chains is lost, and instability
results.

0097-6156/84/0253-0107$06.00/0
© 1984 American Chemical Society

During the 1970's, a new kind of hydrotrope was introduced, WESTVACO DIACID® (9), (Fig. 2). It is a dicarboxylic acid with a total of twenty-one carbon atoms. Its basic properties have been investigated by Matijevic and collaborators (10), who determined its cmc and the association constants for the diacid in water.

This new hydrotrope posed an intriguing problem; the explanation of the hydrotropic action of such a long chain dicarboxylic acid. The problem was solved in a series of articles (11-14) which contain the essential experimental information for this article. These articles each provided a part of the total framework of the problem and we considered a unified treatment essential in order to present a systematic pattern for the different aspects involved.

The analysis of hydrotropic action is conveniently divided into two parts: (1) The isotropic liquid/liquid crystalline equilibrium between essential phases and (2) The molecular mechanism for the influence of the hydrotrope on the equilibrium.

The Isotropic Liquid/Liquid Crystalline Equilibrium

Hydrotropes with their gel-prevention action are an essential part of liquid cleaners for which they provide two essential functions: (a) they allow high surfactant concentrations in the formulation by preventing its gelling at the low water concentrations employed, and (b) they prevent gel formation in extremely water-rich systems under laundering conditions.

Low Water Content

The gel-prevention action in the water-poor part of the system is illustrated by a comparison between the dicarboxylic acid and a monocarboxylic acid of approximately the same carboxylic/methylene group ratio.

The carboxylic acids were combined with hexylamine and water in order to study the association structures formed. The hexylamine was chosen because it did not by itself form a liquid crystalline phase with water, Fig. 3A. Water dissolves in the amine to a maximum of 60% to form an isotropic solution. The liquid crystal is formed first at a certain octanoic acid: amine ratio, approximately 0.1. The liquid crystalline phase forms a large region reaching to a weight fraction of 0.61 of the acid, corresponding to a 1:1 molar ratio of the two species.

The combination of the hexylamine with the dicarboxylic acid on the other hand, Fig. 3B, does not give a liquid crystalline phase for any combination between the acid and the amine or the water. The entire area, Fig. 3B, is an isotropic liquid.

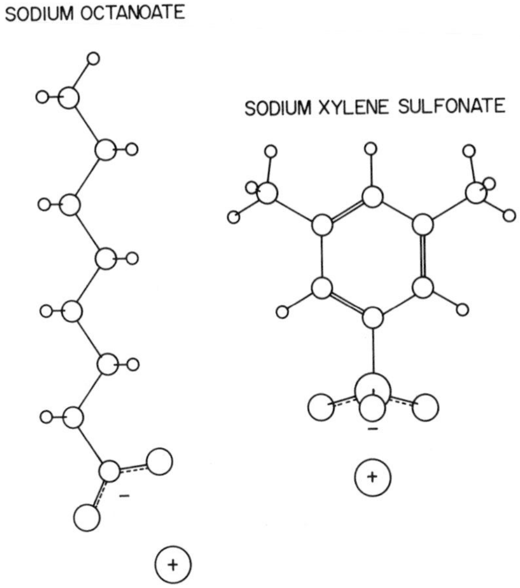

SODIUM OCTANOATE

SODIUM XYLENE SULFONATE

Figure 1.   The structure of a conventional hydrotrope (right) inhibits the formation of a lamellar structure with the common straight-chain surfactants.

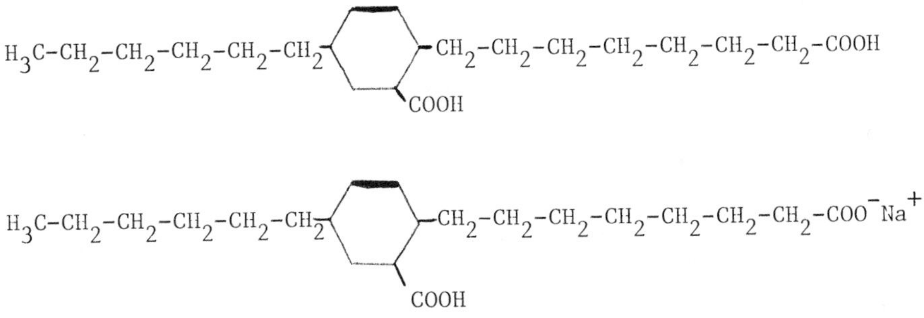

Figure 2.   The structure of 8-[5-carboxy-4-hexyl-cyclohex-2-enyl] octanoic acid (top) and its monosoap (bottom).

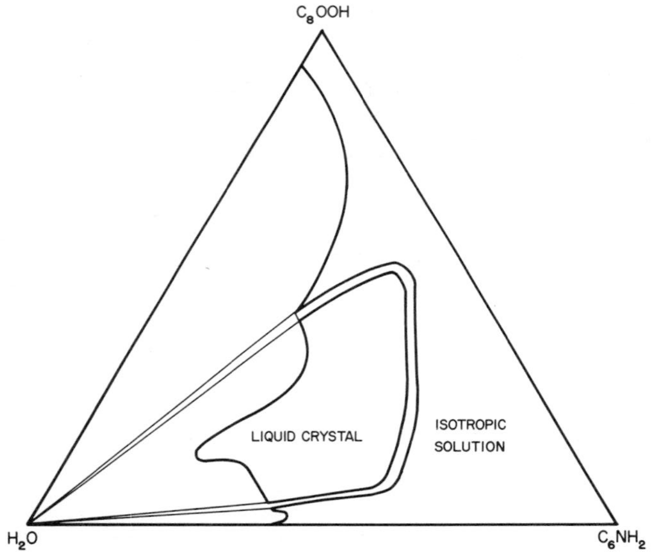

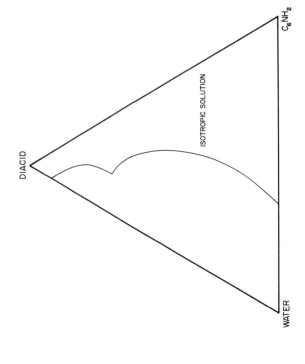

Figure 3A.

The combination of water and hexylamine forms only an isotropic liquid phase. In combination with octanoic acid, a large liquid crystalline area is found.

Figure 3B.

A combination of water and hexylamine with the dicarboxylic acid, 8-[5(6)-carboxy-4-hexyl-cyclohex-2-enyl] octanoic acid, gives no liquid crystalline phase.

These results showed an important feature of the dicarboxylic acid in question. Contrary to the corresponding monocarboxylic acid, it prevents the formation of a liquid crystalline phase in water-poor systems. The implications of this fact for the formulations of liquid cleaners is obvious.

High Water Content

The question of gelling in the water-rich systems, which are typical of laundering conditions, was probed using a simplified model system. The surfactant used was of shorter chain length than normally applied under laundry conditions. The eight carbon chain of sodium octanoate gives a cmc considerably higher than that of the normally used twelve to eighteen carbon chain varieties. The choice was made with the view of facilitating the experimental situation. The phase equilibria at the cmc and concentrations below become simpler to evaluate with a wider concentration range to explore. Octanol was used as a model for oily dirt, following the use of decanol in a recent contribution from Unilever (15).

The phase condition for concentrations in the range close to the cmc are found in Fig. 4A. For the lowest soap concentrations, a liquid isotropic alcohol solution separated, when the solubility limit of the alcohol was exceeded. This was changed at concentrations approximately one half the cmc, when a lamellar liquid crystalline phase appeared instead. After the relatively narrow three-phase region had been transversed, this liquid crystalline phase was the only phase in equilibrium with the aqueous solution. Solubilization of the long chain alcohol increased at the cmc, as expected.

Replacing part of the soap with the dicarboxylic acid at a sufficiently high pH value to ensure its complete ionization gave the results in Fig. 4B. At the site for the onset of the 3-phase area in Fig. 4A with the liquid crystal, the alcohol and the aqueous solution, solubilization of the alcohol now showed a sudden maximum.

The results are straight forward and the interpretation immediately evident. The liquid crystalline phase formed in these extremely water rich systems was destabilized by the dicarboxylic acid and transformed to an isotropic solution. The conclusion that the hydrotropic action of the dicarboxylic acid is intimately related to its capacity to destabilize a liquid crystalline phase also under the water-rich conditions during actual laundering conditions appears well justified.

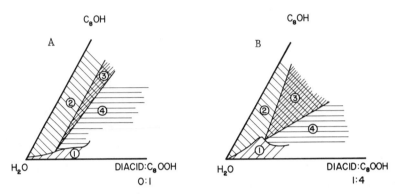

Figure 4.   Addition of the dicarboxylic acid to a water/

detergent combination prevents (B) the gelling

caused by a model dirt (octanol, C_8OH) below the

cmc (A). A The model dirt octanol (C_8OH) forms a

liquid crystalline phase with water and sodium

octanoate (C_8OOH at pH 10) in area 3 and 4

Partial substitution of the sodium octanoate with

the diacid soap (pH 10) leads to an increase of

solubilization of the octanol (B). 1: aqueous

solution, 2: two-phase region of 1 + octanol

solution, 3: three-phase region of 1 + alcohol

solution + lamellar liquid crystal and 4: two-

phase region of 1 + lamellar liquid crystal.

The Molecular Mechanism For The Hydrotropic Action

The remaining problem of the molecular mechanisms of this action
was judged to be related to the conformation of the dicarboxylic
acid at the interface. This conformation is usually determined
directly with the use of a Langmuir trough (16-18). The disadvantage of such a method for the present problem lies with the
restrictions of the environment of the molecule to be investigated.
The basic requirement is that the molecule must be virtually insoluble in the liquid substrate on which the monolayer is supported. For the dicarboxylic acid in question, this meant a pH
value as low as 2 and also a high electrolyte content in the
aqueous substrate.
    Although it may be argued that the conformation of the di-
carboxylic acid under such conditions may be of pronounced scientific interest, several factors suggested a different approach.
The interest in the conformation of the hydrotrope is primarily
related to its behavior at an oil/water interface in conjunction
with surfactant molecules. In addition, molecules from an "oily
dirt" may be present. Finally, it is essential to realize that
the action of the hydrotrope is to destabilize a liquid crystalline phase and transform it to an isotropic liquid.
    With these factors in mind, a new method to evaluate the
conformation of an amphiphilic molecule at the site of interest
was introduced. The method is built on the fact that the determination of interlayer spacings of a lamellar liquid crystal using
low angle X-ray diffraction methods in combination with density
measurements will provide sufficient information to calculate the
cross-sectional areas occupied by each amphiphile (19).
    This means that the partial molar area may directly be deter-
mined from the change in molecular area, when an amphiphilic molecule is introduced into a host liquid crystalline pattern. Of
course, this area is the change of area per molecule at the introduction of one molecule of the substance in question and may be
influenced by the interaction between the host molecules and the
guest molecules. Since this interaction is an essential part of
the present problem, it appears obvious that the method exactly
meets the requirements.
    The host liquid crystalline matrix was composed of water,
sodium octanoate and octanol. This combination was chosen in
order to create an environment as closely matching the specific
requirements of the problem as possible. In the first instance,
the surfactant was identical to the one used for the solubilization determinations (12) and the alcohol was present in order
to resemble actual laundering conditions with "oily dirt" molecules present (12).

The conformational distinction which was sought was the
location of the middle carboxylic group relative to the interface.
If the middle carboxylic group were found at the interface, the
interlayer spacing as influenced by the dicarboxylic acid would
be comparable to the one from sodium octanoate. A straight conformation of the diacid, on the other hand, would give an increment to the interlayer spacing similar to the one from sodium
oleate.

The influence of sodium oleate on the interlayer spacing is
found in Fig. 5A. The distance between layers is increased by an
average of 2.5A and 0.5 moles of the sodium oleate are added to
the liquid crystal. The corresponding addition of the monosoap
of the dicarboxylic acid gives only an insignificant increase of
the distance (Fig. 5B). This result shows the length of the
monosoap of the diacid to be comparable to the length of octanoic
acid.

The evaluation of the conformation of the diacid monosoap
from these results is unambiguous. Both the polar groups of the
monosoap are located at the water surface, giving a conformation
such as the one in Fig. 6. This conformation readily provides an
explanation for the hydrotropic action of the dicarboxylic acid.
When acting at an interface, the molecule does not posess the
extended conformation of Fig. 2; its conformation is as shown in
Fig. 6. Hence, the destabilizing action may be intuitively understood in the same manner as for the traditional aromatic hydrotrope, Fig. 1.

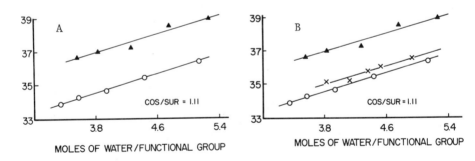

Figure 5.   Sodium oleate (▲) added to a lamellar liquid

crystalline phase of water, sodium octanoate, and

octanol (0) increases the interlayer spacing (A).

Addition of the dicarboxylic acid monosoap (X)

does not give the same change (B).

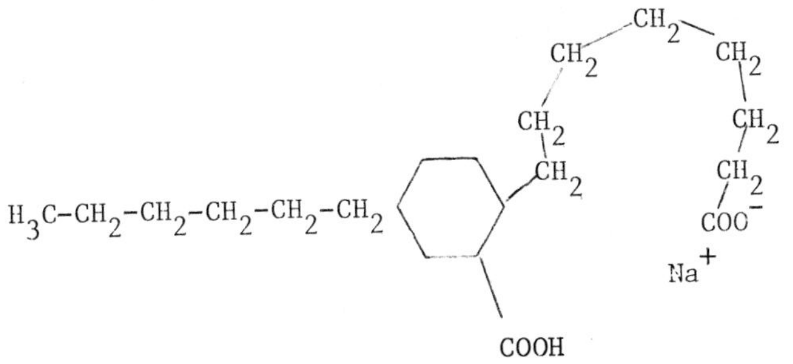

Figure 6.   The conformation of the dicarboxylic acid monosoap at the interface.

Acknowledgments

The authors express their gratitude to Westvaco Co. for financial support and stimulating interactions.

Literature Cited

1. Neuberg, C., Biochem. Z. 76, 107 (1916).
2. Durand, R., C. R. hebd. Scances Acad. Sci. 225, 409 (1948).
3. Krasnic, I. Wiss. Z. Martin-Luther Univ. Halle-Wittenberg math-naturwiss k. VIII, 205 (1959).
4. Booth, H. S. and H. E. Everson, Ind. Enggn. Chem. 41, 2627 (1959).
5. Rath, H., Tenside 2, 1 (1965).
6. Winsor, P. A., Trans. Faraday Soc. 44, 376-398, 451-471 (1948).
7. Lawrence, A.S.C., B. Baffey and K. Talbot, Proc. Internat. Congr. Surface Active Substances, Brussel, II, 673 (1964).
8. Friberg, S and L. Rydhag, Tenside 7, 80 (1970).
9. Ward, B. F. Jr., C. G. Force, A. M. Bills and F. E. Woodward, J. Am. Oil Chem. Soc. 52, 219 (1975).
10. Mino, J., E. Matijevic and L. Meites, J. Colloid Interface Sci. 60, 148 (1977).
11. Friberg, S. and L. Rydhag, J. Am. Oil Chem. Soc. 48, 113 (1971)
12. Cox, J. M. and S. E. Friberg, J. Am. Oil Chem. Soc. 58, 743 (1981).
13. Flaim, T., Friberg, S., Force, C. G. and Bell, A., Tenside Detergents 20, 177 (1983).
14. Flaim, T. and Friberg, S., J. Colloid Interface Sci. (In press).
15. Kielman, H. S. and P. J. F. van Steen "Surface Active Agents" Soc. Chem. Ind. London, 1979; p. 191.
16. Gershfeld, N. L. and Pak, Y. G., J. Colloid Interface Sci. 23, 215 (1967).
17. Bruun, H. H., Acta Chem. Scan. 9, 1721 (1955).
18. Parreira, H. C., J. Colloid Interface Sci. 20, 742 (1965).
19. Fontell, K. "Liquid Crystals and Plastic Crystals", Ellis Horwood Ltd., 1974; p. 80.

RECEIVED March 6, 1984

Aqueous Solution Properties of a Fatty Dicarboxylic Acid Hydrotrope

A. BELL

Westvaco Corporation, Charleston Research Center, North Charleston, SC 29406

K. S. BIRDI

Fysisk–Kemisk Institut, Technical University of Denmark, DK–2800 Lyngby, Denmark

Aqueous solution properties of the twenty-one carbon dicarboxylic acid 5(6)-carboxyl-4-hexyl-2-cyclohexene-1-yl octanoic acid (C_{21}-DA) in salt form - alone and in the presence of a nonionic, anionic or cationic detergent - are reported. Membrane osmometry results indicate that C_{21}-DA alkali salt forms low molecular weight aggregates or micelles, its aggregation behavior appearing to resemble that of certain polyhydroxy bile salts. In the presence of detergent, small aggregates are also formed provided the weight fraction of C_{21}-DA salt in the micelles exceeds ca. 0.5. Phase equilibrium studies show that C_{21}-DA (as the dipotassium or full triethanolamine salt) acts as a hydrotrope above certain concentration levels in concentrated detergent solutions, retarding build-up of anisotropic aggregates responsible for mesophase formation, in accordance with previous investigations by Friberg and co-workers.

In recent studies, Friberg and co-workers ([1,2]) showed that the 21 carbon dicarboxylic acid 5(6)-carboxyl-4-hexyl-2-cyclohexene-1-yl octanoic acid (C_{21}-DA, see Figure 1) exhibited hydrotropic or solubilizing properties in the multicomponent system(s) sodium octanoate (decanoate)/n-octanol/C_{21}-DA aqueous disodium salt solutions. Hydrotropic action was observed in dilute solutions even at concentrations below the critical micelle concentration (CMC) of the alkanoate. Such action was also observed in concentrates containing pure nonionic and anionic surfactants and C_{21}-DA salt. The function of the hydrotrope was to retard formation of a more ordered structure or mesophase (liquid crystalline phase).

0097–6156/84/0253–0117$06.00/0
© 1984 American Chemical Society

As is well known, hydrotropes are incorporated in liquid detergent formulations in order to produce transparent (isotropic) solutions at high solids concentrations which will be stable under varying conditions of temperature and composition. Depending upon its structure, the hydrotrope may also assist in solubilizing components added in small amounts (e.g., perfumes, colorants, bactericides) and, depending upon the nature of the soil, enhance detergency.

In order to better understand the molecular basis of hydrotropic action, it is useful to investigate phase properties of multicomponent regions of interest, specifically the micellar or isotropic regions. Recently, a powerful technique has again emerged for characterizing micellar formation in detergent solutions - membrane osmometry (3-6). With this technique, the effect of additives (electrolytes, organic solutes, etc.) can also be explored (7,8).

Here, utilizing membrane osmometry, we report on micelle formation in solutions of C_{21}-DA alone (in dilute electrolyte) and in the presence of surface active ingredients incorporated in commercial liquid detergent formulations. Phase diagrams of 3-component blends (detergent/C_{21}-DA salt/H_2O) are also presented.

Experimental

Materials. The dicarboxylic acid was Westvaco DIACID ^{R} 1550, or H-240, its partially neutralized 40% solids aqueous potassium salt solution (9,10). Linear fatty alcohol ethoxylate (LE-9), prepared from alcohols containing 12-15 carbon atoms and containing an average of 9 ethylene oxide groups per molecule, was obtained commercially. Sodium dodecylbenzenesulfonate (SDBS, technical grade) and cetyltrimethylammonium bromide (CTAB) were obtained from Fluka, triethanolamine (TEA) from Fisher. Inorganic chemicals were reagent grade. Water was Milli-Q ^{R} deionized.

Membrane Osmometry. The apparatus has been described in detail elsewhere (5). The concentration of detergent in the solution compartment was many times the CMC (10 to 100) whereas the concentration in the solvent compartment was somewhat above (i.e., 5 to 10 times) the CMC. Under these conditions, the following limiting equation applies (3-6):

$$\frac{\pi}{(C-C')}\bigg|_{C \rightarrow C'} = \frac{RT}{M_n} + 2\ RTB(C'-CMC)$$

where π is the osmotic pressure, C the detergent concentration in the solution compartment, C' the detergent concentration in the solvent compartment, R the universal gas constant, T the absolute temperature, M_n the number average molecular weight, and B the second virial coefficient. As shown previously (3-5), the term involving B is small compared to RT/M_n, and thus a plot of $\pi/RT(C-C')$ versus $(C-C')$ gives the value $1/M_n$ at the intercept. Recent fluorescent probe studies (11) indicate that in pure detergent solutions, M_n does not change appreciably in dilute electrolyte medium as concentration is increased well above the CMC. This suggests that in aqueous micellar regions containing mixtures of C_{21}-DA salt, detergent and dilute electrolyte, the micellar aggregation number should remain essentially constant (except perhaps near phase boundaries) at a given weight ratio of hydrotrope to detergent. A recent investigation by Kratohvil and co-workers (12) on the concentration-dependent aggregation of conjugated bile salts (especially taurodeoxycholate) in concentrated electrolyte media ($> 0.1M$ NaCl), however, appears to indicate otherwise.

CMC values were obtained from dye (azobenzene) solubilization and surface tension measurements. Values of π used in the above equation were obtained via extrapolation (4-5). For very low values of M_n, there is concern about leakage through the membrane (13). Assuming that leakage did occur within the timeframe of the measurements, M_n (and hence N_n) values would be lower than those calculated. For this reason, all values of M_n and N_n are reported as being apparent.

Phase Regions. Phase regions were determined by visual inspection of blended components stored in tightly capped vials. The blends were prepared by mixing the components together, with stirring, at 60-100°C, followed by cooling in air to 25°C. The C_{21}-DA dipotassium salt was fully neutralized (with 45% KOH) Westvaco H-240 (acid value equivalent), the final solids content being ca. 45% by weight. Higher concentrations of this salt were obtained via evaporation at 90-100°C. Anhydrous C_{21}-DA full triethanolamine salt is an isotropic liquid at 25°C. Anisotropic regions were determined by viewing the samples between crossed polarizers (1,2) or under a polarizing microscope. In the vicinity of phase boundaries, many weeks were often required before equilibrium was attained.

Results

CMC values in 0.05M electrolyte at pH 10 are given in Table I. Apparent number average molecular weights M_n of aqueous detergent mixtures in the same medium are listed in Table II. The maximum concentration of surface active agents in any given system was 50 g/l. Figure 2 is a plot of $\pi/RT(C-C')$ versus $(C-C')$ for the system CTAB:C_{21}-DA salt. Figure 3 shows the variation of N_n, the apparent micellar aggregation number (calculated by dividing M_n by the average molecular weight of the detergent mixture), as a function of weight fraction of detergent in the micelle, assuming that the ratio of detergent to C_{21}-DA salt in the micelle is equal to that of the overall surfactant inventory. This assumption is essentially valid well above the CMC of the mixture (14).

Results of phase equilibrium studies are shown in Figures 4-6.

Table I. Results of CMC Determinations (0.05M NaCl or NaBr, pH 10, 25°C)

Weight Ratio Detergent in Solution	CMC Value (g/l; obtained via extrapolation)		
	LE-9	SDBS	CTAB
0.00	0.4	0.4	0.4
0.25	0.4	0.25	0.3
0.50	0.3	0.15	0.2
0.75	0.25	0.25	0.15
1.00	0.2	0.2	0.1

Table II. Results of Membrane Osmometry (0.05M NaCl or NaBr, pH 10, 25°C)

Weight Ratio Detergent in Micelle	Apparent M_n Values (g/mol)		
	LE-9	SDBS	CTAB
0.00	2400	2400	2400
0.25	3000	3100	2700
0.50	2600	3000	6800
0.75	7000	10800	14400
1.00	48000	14000	90000

Discussion

The aggregation behavior of C_{21}-DA salt in dilute electrolyte medium appears to resemble that of certain polyhydroxy bile salts (15,16). That C_{21}-DA, with a structure quite different from bile acids, should possess solution properties similar to, e.g., cholic acid is not entirely surprising in light of recent conductivity and surface tension measurements on purified (i.e., essentially monocarboxylate free) disodium salt aqueous solutions, and of film balance studies on acidic substrates (17). The data in Figure 3 suggest that C_{21}-DA salt micelles incorporate detergents - up to an approximate weight fraction of 0.5 -much like cholate incorporates lecithin or soluble

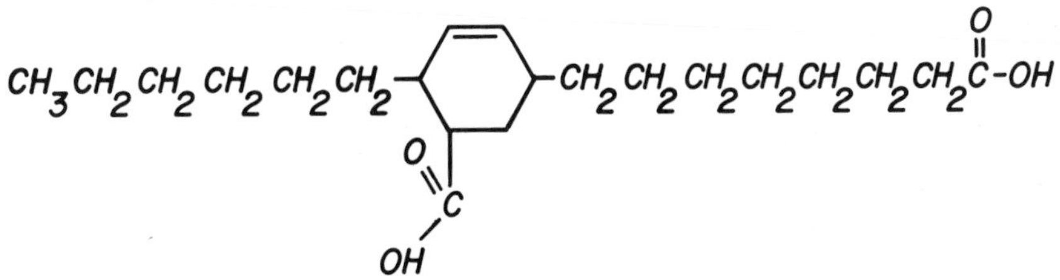

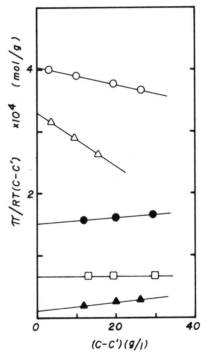

Figure 1.   Structure of the dicarboxylic acid 5-carboxy-4-hexyl-2-cyclohexene-1-octanoic acid (C_{21}-DA).

Figure 2.   Plots of   /RT(C-C') vs. (C-C') for mixtures of cetyltrimethylammonium bromide (CTAB) and C_{21}-DA alkali salt in 0.05M NaBr, pH 10, 25°C.   Weight ratio CTAB:C_{21}-DA salt 0:1 (0); 1:3 (▲); 1:1 (●); 3:1 (□); 1:0 (▲).

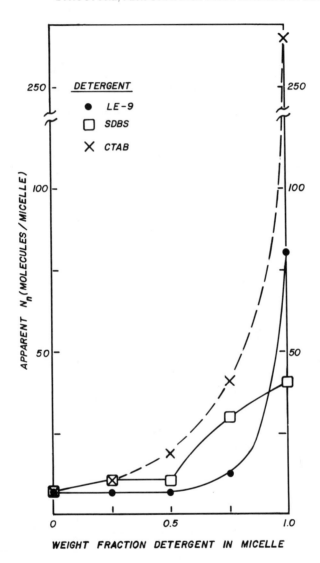

Figure 3.   Apparent aggregation number (N_n) of mixed micelles containing detergent and/or C_{21}-DA alkali salt.   Aqueous medium: 0.05 M NaCl or NaBr, pH 10, 25°C.   LE-9, linear fatty alcohol ethoxylate;   SDBS,   sodium   dodecylbenzenesulfonate;   CTAB, cetyltrimethylammonium bromide.

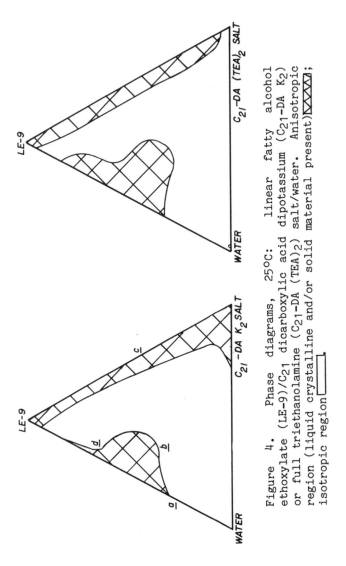

Figure 4. Phase diagrams, 25°C: linear fatty alcohol ethoxylate (LE-9)/C$_{21}$ dicarboxylic acid dipotassium (C$_{21}$-DA K$_2$) or full triethanolamine (C$_{21}$-DA (TEA)$_2$) salt/water. Anisotropic region (liquid crystalline and/or solid material present)$\boxtimes$; isotropic region $\boxed{}$.

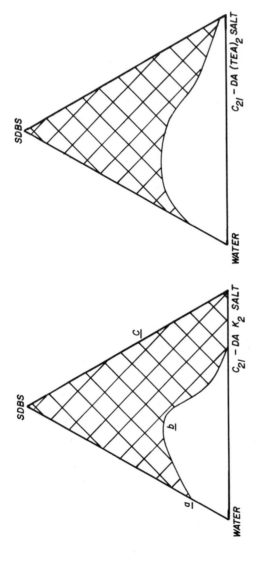

Figure 5. Phase diagrams, 25°C: sodium dodecylbenzenesulfonate
(SDBS)/C_{21} dicarboxylic acid dipotassium (C_{21}-DA K_2) or full
triethanolamine (C_{21}-DA (TEA)$_2$) salt/water.

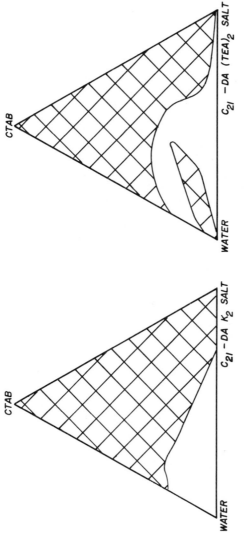

Figure 6. Phase diagrams, 25°C: cetyltrimethylammonium bromide (CTAB)/C_{21} dicarboxylic acid dipotassium (C_{21}-DA K_2) or full triethanolamine (C_{21}-DA (TEA)$_2$) salt/water.

amphiphiles (e.g., oleate, monolein) in micellar solutions (15,18). As the weight fraction of detergent approaches unity, the size (and presumably the shape) of the aggregates is altered, reflecting the influence of the main constituent.

It is probable that at higher weight fractions of C_{21}-DA salt, the shape of the mixed micelles is roughly spherical at concentrations not too far removed from the CMC (16). However, as the weight fraction of detergent and/or total solids concentration is increased, it is likely that the shape changes. Whether Small's model (15) (wherein all the C_{21}-DA salt molecules would be at the micelle exterior) or the "mixed-disc" model (18) (wherein some C_{21}-DA salt molecules would also enter the interior of the micelle) is not known. In the case of bile salt(cholate)/detergent micelles, recent calorimetric studies (19) suggest that the "mixed-disc" model is favored, with hydrogen bonding taking place in the interior. If this model applies to detergent/C_{21}-DA salt micelles (with accompanying hydrogen bonding), the less acidic C_{21}-DA secondary carboxyl group (20) should be only partially ionized within the micelle.

Concentrated systems are now considered. Here, we define a hydrotrope as a substance capable of increasing the absolute amount of detergent in an isotropic formulation. For both the nonionic (Figure 4) and anionic (Figure 5) systems, addition of C_{21}-DA K_2 salt to the onset of the water-rich liquid crystalline region (composition a) results in an increase of solubilized detergent up to a point where the weight ratio detergent: C_{21}-DA salt is ca. 1:1 (composition b). (Note that at composition c, obtained via extrapolation of the line joining a and b, the ratio is also ca. 1:1). Comparison with the results of membrane osmometry (Table II) suggests that there is a correlation between dilute solution properties and the tendency to form anisotropic aggregates or mesophases in water-rich concentrated systems.

In the nonionic system (Figure 4), C_{21}-DA K_2 salt also acts as hydrotrope in the water-poor (reversed) micellar region (i.e., in the vicinity of composition d). Indeed, in this region, its efficiency appears to be greater than in the water-rich region.

As might be expected, there is a greater interaction in concentrated systems between cationic detergent (CTAB) and the negatively charged C_{21}-DA K_2 salt, resulting in a diminution of the isotropic region or in the case of the full TEA salt, formation of an anisotropic phase within the region (Figure 6). The maximum hydrotropic action is observed when the ratio detergent:C_{21}-DA salt is ca. 0.8:1, again in line with osmometry results (Table II).

Employing x-ray methods, Flaim and Friberg (21) studied the conformation of C_{21}-DA as the acid or monosoap in a lamellar liquid crystalline matrix. At low water concentrations, the conformations are the same (extended). At somewhat higher water concentrations, the monosoap becomes coiled, occupying a larger area within the matrix, thus tending to disrupt close packing; under similar conditions, the acid remains extended. These findings point to the influence of structure on hydrotropic action.

Summarizing, past and present investigations indicate that C_{21}-DA salt acts as a hydrotrope in both concentrated and dilute systems by retarding build-up of anisotropic aggregates or micelles responsible for mesophase formation (1,2,21). Hydrotropic action appears to be pronounced in the system containing nonionic detergent (water-rich and water-poor regions). In liquid systems containing both anionic and cationic detergents, C_{21}-DA salt could serve to enhance compatibility, especially near phase boundaries, as a result of its unusual hydrotropic properties.

Acknowledgments

The authors are grateful to Professor S. E. Friberg for helpful comments, to Mr. R. A. Stanton for experimental assistance, and to Mrs. L. M. Baldwin for typing the manuscript.

Literature Cited

1. Cox, J. M.; Friberg, S. E. J. Amer. Oil Chemists' Soc. 1981, 58, 743.
2. Flaim, T.; Friberg, S. E.; Force, C. G.; Bell, A. Tenside Detergents 1983, 20, 177.
3. Coll, H. J. Phys. Chem. 1970, 74, 520.
4. Attwood, D.; Elworthy, P. H.; Keyne, S. B. J. Phys. Chem. 1970, 74, 3529.
5. Birdi, K. S. Kolloid Z. u. Z. Polymere 1972, 250, 731.
6. Birdi, K. S. In "Proceedings of the International Conference on Colloid and Surface Science", Wolfram, E., Ed.; Academic Kado: Budapest, 1976; p.473.
7. Birdi, K. S.; Backlund, S.; Sorensen, K.; Krag, T.; Dalsager; S. J. Colloid Interface Sci. 1980, 76, 2035.
8. Birdi, K. S.; Dalsager, S.; Backlund, S. J. Chem. Soc. Faraday I 1980, 76, 2035.
9. Ward, B. F, Jr.; Force, C. G.; Bills, A. M.; Woodward, F. E. J. Amer. Oil Chemists' Soc. 1975, 52, 219.
10. "DIACID^R Surfactants"; Westvaco Chemical Division: Charleston Heights, S. C., 1981.
11. Lianos, P.; Zana, R. J. Colloid Interface Sci. 1981, 84, 100.
12. Kratohvil, J. P.; Hsu, W. P.; Jacobs, M. A.; Aminabhavi, T. M.; Mukunoki, Y. Colloid & Polymer Sci. 1983, 261, 781.
13. Billmeyer, F. W., Jr. "Textbook of Polymer Science", 2nd ed.; Wiley-Interscience: New York, 1971; pp.72-73.

14. Holland, P. R., paper presented at this Symposium.
Funasaki, N.; Hada, S. J. Phys. Chem. 1979, 83, 2471.
15. Small, D. M. In "The Bile Acids"; Nair, P. P.; Kritchevsky,
     D., Eds.; Plenum Press: New York, 1971; Vol. I, Chap. 8.
16. Carey, M. C.; Small, D. M. Am. J. Med. 1970, 45, 590.
17. Pethica, B. A.; Pallas, N. R.; Bell, A., unpublished data.
18. Mazer, N. A.; Benedek, G. B.; Carey, M. C. Biochemistry
     1980, 19, 601.
19. Zimmerer, R. O., Jr.;Lindenbaum, S. J. Pharm. Sci. 1979,
     68, 581.
20. Mino, J.; Matijevic, E.; Meites, L. J. Colloid Interface
     Sci. 1977, 60, 148.
21. Flaim, T.; Friberg, S. E., paper presented at this
     Symposium.

RECEIVED February 3, 1984

Interaction of Long Chain Dimethylamine Oxide with Sodium Dodecyl Sulfate in Water

DAVID L. CHANG and HENRI L. ROSANO

Department of Chemistry, City College, City University of New York, New York, NY 10031

Lauryl, myristy, cetyl and steary-dimethylamine
oxide (LDAO) aqueous solutions alone and in combination with sodium dodecyl sulfate (SDS) have
been investigated by surface tension, viscosity
and pH measurements. 1) LDAO solutions: At any
particular pH, the amine oxide is in equilibrium
with its protonated form. The viscosity of the
solution depends strongly on the degree of ionization (β) and maximum viscosity was observed when
$\beta = 0.5$, due to the formation of elongated structure. 2) LDAO/SDS solutions: Progressive addition of SDS to dodecyl or myristyl DAO results in
pH and viscosity increases, both values reach
their maximum values at a 3:1 LDAO/SDS molar ratio.
However, the change in pH remains constant over a
wide range of molar ratios, while the viscosity
value depends strongly on that ratio. The viscous
mixture is non-Newtonian in nature, and exhibits
liquid crystalline behaviour. In the case of
cetyl- and stearyl-DAO/SDS mixtures, solutions are
turbid with low viscosity. This difference in
behaviour is attributed, in part, to the compatibility in chain length of the LDAO and SDS.

Binary mixtures of surfactant solutions have been studied extensively (1-5). Such mixtures can exhibit appreciable surface and
bulk synergistic effects. In general, cationic/anionic surfactant mixtures produce the largest effects. Surface activity
related synergistic phenomena are of technological and marketing
interests (6). In the present study, the interaction between a
long chain dimethylamine oxide (LDAO) and sodium dodecyl sulfate
(SDS) in aqueous solution has been investigated.
   The aggregation characteristics of LDAO in aqueous solutions (7-12) and their interaction with anionic surfactants

0097-6156/84/0253-0129$06.00/0

(1,2,13) have been studied previously. Goddard and Kung (14) have investigated the monolayer properties of docosyldimethyl amine oxide. LDAO, a weak base having about the same basicity as the acetate anion (15), can exist in both nonionic and cationic form depending upon the pH of the solution. At any particular pH, an equilibrium is established between the cationic and nonionic form, viz.,

$$LDAOH^+ \rightleftharpoons LDAO + H^+ \qquad [1]$$

According to Kolp et al. (1), K_a, the monomeric equilibrium constant has been determined by potentiometric titration to be

$$K_a = \frac{[H^+][LDAO]}{[LDAOH^+]} = 10^{-4.9} \qquad [2]$$

where the quantities in brackets are the activities of the indicated species relative to a standard state such that, at infinite dilution, the activity equals the molar concentration.
Tokiwa and Ohki (10) have shown that 1) the K_a value of the micellized LDAO is different from the molecular form, viz., $10^{-5.9}$ 2) while the values of K_a are independent of the degree of protonation (β) below the critical micelle concentration (CMC), they are dependent on β at concentrations above the CMC. These authors did not explain why the micellar pKa (5.9) at $\beta = 0$ is different from the molecular value (pKa = 4.9).

Difference between micellar and molecular pKa's has been reported for other systems (16,17), however, no explanations were provided. Recently, Mille (18) has concluded from a theoretical approach that for the case of LDAO, assuming only a charge-charge interaction in the nearest-neighbor model can not explain the difference in pKa's satisfactorily and that additional attractive interaction must be included, suggesting possible hydrogen bonding formation.

EXPERIMENTAL

Long chain dimethylamine oxides with alkyl chain length = 12, 14, 16, 18 carbons (Onyx Chem. Co., Jersey City, N.J.) and laboratory grade sodium dodecyl sulfate (Fisher Chem. Co., Fairlawn, N.J.) were used without further purification.
For the titration of LDAO, solutions were made with their pH adjusted to 12 with NaOH and titrated with HCl. Binary mixtures of LDAO and SDS solutions were prepared by mixing different volumes of the surfactant solutions at equal molar concentration.

pH Measurement. For amine oxide titration, pH was monitored with an Orion Research Ionalyzer (Model 801 A, Orion Res. Inc., Cambridge, Mass.). All other pH measurements were made with a Photovolt pH meter (Photovolt Corp., New York, N.Y.).

Surface Tension Measurement. Surface tension was determined
with a Rosano Tensiometer (Arenberg-Sage Inc., Jamaica Plain,
Mass.) utilizing a sand-blasted platinum blade.

Viscosity Measurement. A Brookfield Viscometer (Model LVT,
Brookfield Engineering Lab., Inc., Stoughton, Mass.) was
employed for the measurement of relative viscosity.

Electrical Resistance and Percent Light Transmittance. Low
frequency electrical resistance measurements were made on a
conductivity bridge (Model RC-18, Industrial Instrument, Cedar
Grove, N.J.) at a line frequency of 1 KC. Beckman conductivity
cell with cell constant 1.0 cm^{-1} was used. The percent transmission was also monitored for each of the mixtures at 490 nm
(Spectronic 20, Bausch & Lomb Co., Rochester, N.Y.).

Wettability Test of Glass Surface. Glass surface in water is
negatively charged. When the corner of a cleaned and water
covered microscope cover slide touches the surface of a solution
containing cationic surfactant just below its CMC, instantaneous
dewetting occurs; the positive charge of the surfactant interacts electrostatically with the negative charge of the glass
surface, the solid surface is then covered by a thin layer of
the surfactant molecules with their hydrophobic tails pointing
outward and the surface is no longer wettable. This effect is
not observed with anionic and nonionic surfactant solutions.
     In addition, viscoelasticity of solution was determined
qualitatively by the simple method of swirling a vial containing
the solution and observe the recoiling of air bubbles entrapped
in the solution (19).

RESULTS

Titration of LDAO. Fig. 1 represents the titration curves of
$C_{14}DAO$ at 0.2 M, 8 x 10^{-4}M, and 4 x 10^{-5}M concentration and
$3C_{14}DAO/1SDS$ mixture at 0.2 M total concentration with HCl at
25°C. For comparison, the four curves are plotted on a same
HCl equivalent scale.
     The CMC of $C_{14}DAO$ is about 1 x 10^{-3}M at 25°C. Below the
CMC a typical buffering action is observed (4 x 10^{-5}M), above
the CMC the titration curves are slanted toward lower pH's with
increasing HCl concentration; 0.2 M having a steeper slope than
8 x 10^{-3}M. Addition of SDS to a solution of $C_{14}DAO$ affects the
HCl titration curve markedly and will be discussed later.
     Also plotted is the viscosity variation with HCl added for
0.2 M $C_{14}DAO$. Maximum viscosity is observed when half of the
amine oxide molecules are in the cationic form. However, in
the absence of NaOH the change in viscosity with HCl added is

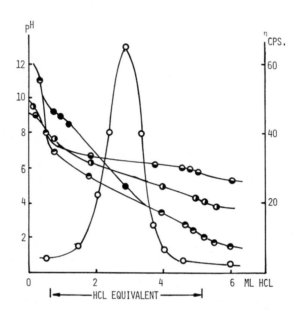

Figure 1.

Titration curves of $C_{14}DAO$ @ 0.2M, ◓ ; 8 x 10^{-4}M, ◐ ;
4 x 10^{-5}M, ◒ and $3C_{14}DAO/1SDS$ mixture @ 0.2M, ● .
Relative viscosity (η:○) vs. ml HCl added is also plotted
for $C_{14}DAO$ @ 0.2M.

much less pronounced (by about an order of magnitude). $C_{12}DAO$ gives similar results. The fact that an increase in viscosity of the solution is observed in the vicinity of a 1:1 molar ratio of the nonionic form to the cationic form of the LDAO is similar to that observed in an acid-soap (20).

LDAO-SDS Interactions. Mixtures of C_{12}- or C_{14}-DAO with SDS show a surface tension minimum at an 1:1 molar ratio, as shown on Fig. 2 for $C_{14}DAO$. Also shown is the variation in pH for different mixing ratios. The increase in pH of the mixed solution seems to indicate that the addition of SDS to a LDAO solution favor the protonation of the amine oxide, water being the proton donor. This point will be discussed more fully below. The change in viscosity of the mixture at different compositions is plotted in Fig. 2 as well, the maximum of which corresponds to a $3C_{14}DAO/ISDS$ association. Similar behavior is observed for $C_{12}DAO/SDS$ mixtures.

Minimum surface tension at a 1:1 cationic/anionic ratio has been observed in the system containing SDS and dodecyltrimethyl ammonium bromide (3). In addition, 1:1 association has been observed for $C_{12}DAO$ with long-chain sulfonates (1,2). The latter two cases, hydrogen bonding between the protonated amine oxide and the anionic sulfonate has been shown to exist in the solid salt. The salt may separate out in crystalline form, as reported in (1,2,20), but no precipitation is observed in this work when the pH of the solution is kept above 9.

Figs. 3 and 4 show the interaction between $C_{16}DAO$ and $C_{18}DAO$ with SDS respectively. For LDAO/SDS ratio greater than 1, these mixed solutions are turbid and birefringent; addition of SDS results in production of filament-like structures. When the molar amount of SDS is equal to or greater than that of the LDAO, the solutions become isotropic and clear. AT 3:1 LDAO/SDS molar composition, the change in pH reaches its maximum value. Minimum surface tension is reached when the LDAO content is still in excess, unlike the cases of C_{12}- and C_{14}-DAO. This observation indicates that the surface of the solution has already been saturated with a mixed molecule formed between C_{16}- or C_{18}-DAO and SDS, although the bulk concentrations of the individual species are different, and that the composition of the bulk must be the same throughout the low surface tension region. Enhancement in viscosity is still observed with $C_{16}DAO$. However, $C_{18}DAO$ shows a rapid decrease in viscosity upon mixing with SDS, the value of which reaches that of the SDS solution at equal molar composition.

The 3:1 C_{12}- or C_{14}-DAO/SDS mixture exhibits non-Newtonian fluid characteristics; its viscosity decreases with increased shear rate. At high pH ($\sim$12), the mixture is still non-Newtonian and its viscosity is enhanced by about an order of magnitude (at high shear rate) to two orders of magnitude (at low shear rate) compared to the values at its natural pH ($\sim$10). The

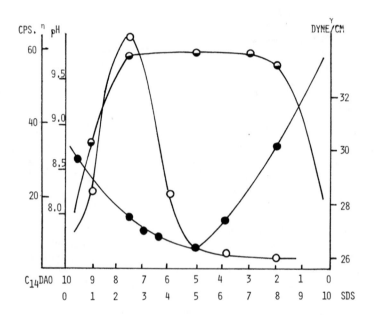

Figure 2.

Relative viscosity (η:◯), surface tension (γ:●) and pH (◓) variations at different $C_{14}DAO/SDS$, molar ratios. The total concentration equals 0.5M.

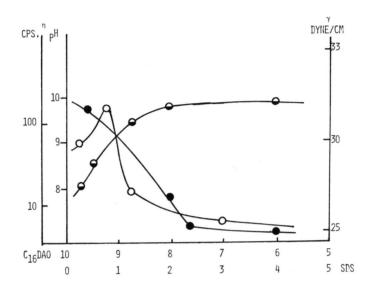

Figure 3.

Relative viscosity (η:◯), surface tension (γ:●) and
pH (◖) variations at different $C_{16}DAO/SDS$ molar ratios.
The total concentration equals 0.05M.

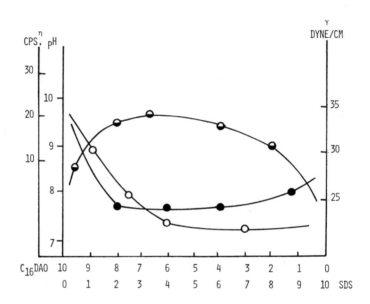

Figure 4.

Relative viscosity (η:$\bigcirc$), surface tension (γ:$\bullet$) and

pH ($\ominus$) variations at different $C_{18}DAO/SDS$ molar ratios.

The total concentration equals 0.005M.

solution exhibits birefringency when the solution is shaken.
Upon cooling, the solution becomes milky, indicating possible
crystal formation, and this behavior is reversible when the
temperature is raised. These observations suggest that the
mixed solution can probably be categorized as liquid crystal.
Below pH 9, precipitation occurs. Analysis of the ir, nmr and
mass spectra of the precipitate reveals that its composition is
$CH_3(CH_2)_{13}N(CH_3)_2^+OH^-O_3SO(CH_2)_{11}CH_3$, with the proton of the
cationic amine oxide involved in hydrogen bonding with the
sulfonate anion.

The 3:1 LDAO/SDS mixture becomes viscoelastic and rheopectic when a small amount of NaCl is added. Its viscosity
shows a reversible increase with time of shearing at constant
shear rate. The rheopectic behavior is probably due to long
thread-like micelles that are aligned parallel to the flow in
weakly bound clusters, as in the case of cetyltrimethyl ammonium
bromide and monosubstituted phenol mixed solutions (21).

DISCUSSION

Titration of LDAO. When hydrogen ions are added to a LDAO
solution at high pH, the excess hydroxide ions are neutralized
first. Then two reactions occur: protonation of the monomer,
and protonation of the micelle. The former is probably more
favored; the buffering-action region for concentrations below
the CMC occurs at a higher pH value than for concentrations
above the CMC. Tokiwa and Ohki (10) have shown that for $C_{12}DAO$,
the $pK_{micelle}$ decreases from 5.95 to about 4 with increasing
degree of protonation (β), implying that the micelle is even
less favored as a proton acceptor compared to its monomer
(which has a pK = 4.95) with diminishing nonionic character.
They calculated the surface potential of $C_{12}DAO$ micelles as a
function of β and found an increase from 0 to +120 mV as β is
varied from 0 to 1. These authors concluded that the micelles
were becoming more and more positively charged. The light-
scattering data obtained by Ikeda et al. (12) suggest the
formation of large rod-like micelles of $C_{12}DAO$ at β = 0.5 in
the presence of NaCl, while at high and low pH's, the micelles
are small and probably spherical. There is a spherical →
elongated → spherical transition of the amine oxide micelles as
β is varied from 0 to 0.5 and then to 1. As has been pointed
out by Nagarajan and Ruckenstein (22), transition from spherical
to cylindrical micelle is favored by factors which contribute to
an increase in the attractive component or a decrease in the
repulsive component of the free energy of micellization. In the
case of ionized amine oxide, the ionizable proton is capable of
hydrogen bonding to the neighboring oxygen thereby decreasing
the average distance between adjacent head groups effectively.
As a result, the tail groups have to readjust spatially thus
leading to a breakage of the spheroid. In addition, protona-

tion of the monomer can also lead to a breakage of the spheroid
if the assumption is made that the concentration of the monomer
remains constant above the CMC (23). Both mechanisms contribute
to the production of elongated structure.

From the study of monolayer properties of a long chain
(C_{22}) amine oxide, Goddard and Kung (14) have found that above
pH 7.5 of the aqueous substrate, the monolayer is quite
expanded, (55 $Å^2$/molecule at 10 dyne/cm surface pressure), while
the area per molecule equals 33 $Å^2$ for the same surface pressure
at pH = 4.8; LDAO yields a less expanded film when some of the
molecules are ionized than when they are all nonionic. They
remarked that the quaternary ammonium group in protonated amine
oxide is unusual in possessing a hydroxy group; the strong
electron-attracting, positively charged nitrogen substantially
augments the polarization of this OH group.

With increased proton character, the H^+ has a tendency to
hydrogen bond. Below pH 7.5, the appearance of the cationic
form has a marked condensing effect in association with un-
ionized amine oxide and results in the production of elongated
structures replacing progressively the small and spherical
micelles of LDAO in the nonionic form. Below pH 4.5, the elongated structures are progressively converted into low viscosity,
small spherical micelles made of LDAOH.

The present results also show that there is a discrepancy
between the monomeric pKa values found in this work and the
previous studies; ours being higher (6.6 vs 4.9, respectively).
The cause of this is believed to be due to the presence of
bicarbonate ions in the solution (27).

LDAO/SDS Interaction. Mixing of cationic and anionic surfactant
solutions results in the formation of a mixed species that is
more surface active than the individual species. The enhanced
synergistic effect has been explained (2,3) by showing that a
close-packed adsorption of electroneutral R^+R^- takes place (R^+
and R^- represent the long chain cation and anion respectively).
In the case of C_{12}- and C_{14}-DAO, a 1:1 LDAO/SDS molar ratio
produces a minimum in surface tension and is accompanied by an
increase in pH in the bulk solution; the association seems to be
of the type R^+R^-, and the absence of visible precipitate may be
attributed to the solubilization of the R^+R^- complex in the
solution. In the region where LDAO is in excess, the structure
is probably [cationic ($LDAOH^+$) anionic (SDS)] nonionic (LDAO),
while [cationic ($LDAOH^+$ anionic (SDS)] anionic (SDS) is formed
when SDS is in excess. Equal molar concentration results in
cationic ($LDAOH^+$) anionic (SDS) complex which should favor
precipitation. However, at pH >9, there is no indication of
precipitation (even when the total solute concentration is
0.35 M). When the pH is below 9, then precipitation will take
place.

Another possibility is the interaction between neutral LDAO

and SDS with adsorption of H_3O^+ ions on the aggregate surface. This type of combination not only explains an increase in the pH but also explains the absence of precipitation for any given mixing ratio in the concentration range studied. This interpretation is supported by the titration behavior of, for example, a 3:1 LDAO/SDS mixture (see Fig. 1) above pH 12 addition of HCl neutralizes the OH^- ions in the bulk, when more H^+ ions are added, the concentration of H^+ ions increases until surface neutralization occurs below pH 10, and below pH 9 precipitation of the 1:1 $LDAOH^+$/SDS complex takes place. Since the LDAO is in excess, below pH 7 when all the SDS molecules have been reacted, protonation of LDAO occurs, reminiscent of the titration of pure LDAO solution.

Another feature of surfactant-water systems is that they can also aggregate into lyotropic liquid crystalline phases when intermicellar interactions are significant. Typically, non-Newtonian behavior is usually found for these liquid crystalline phases. For the 3LDAO/1SDS mixed system, all evidence suggests that they do form liquid crystalline phase.

The difference in properties when the aliphatic chain of amine oxide contains more than 14 carbons is attributed to the mismatch of the hydrophobic chain with that of the SDS. The extra terminal segment results in a disruptive effect on the packing of the surface active molecules. The observed association behavior in the case of C_{12}- or C_{14}-DAO with SDS is then also due to the maximum cohesive interaction between hydrocarbon chains in addition to the reduced electrostatic repulsion in the head groups. Solubilization of the 1:1 association is also determined by this chain length compatibility effect which may contribute to the absence of visible precipitation in C_{12}/C_{12} and C_{12}/C_{14} mixtures. Chain length compatibility effects in different systems have been discussed by other investigators (24,25,26).

Acknowledgments

The authors are grateful to Dr. M. Aronson and Dr. E. D. Goddard for helpful discussions. D. L. C. also wishes to thank Lever Brothers Company and the Gillette Corporation for their financial support.

Literature Cited

1. Kolp, D. G.; Laughlin, R. G.; Krause, R. P.; Zimmere, R. E. J. Phys. Chem. 1963, 67, 51.
2. Rosen, M. J.; Friedman , D.; Gross, M. J. Phys. Chem. 1964, 68, 3219.
3. Lucassen-Reynders, E. H.; Lucassen, J.; Giles, D. J. Colloid Interface Sci. 1982, 81, 150.

4. Al-Saden, A. A.; Florence, A. T.; Whateley, T. L.;
    Puisieux, F.; Vaution, C. J. Colloid Interface Sci.
    1982, 86, 51.
5. Hua, X. Y.; Rosen, M. J. J. Colloid Interface Sci. 1982,
    90, 212.
6. Rosano, H. L.; U.S. Patent Application 473078 07, 1983.
7. Herrmann, K. W. J. Phys. Chem. 1962, 66, 295.
8. Ibid., 1964, 68, 1540.
9. Benjamin, L. J. Phys. Chem. 1964, 68, 3575.
10. Tokiwa, F.; Ohki, K. J. Phys. Chem. 1966, 70, 3437.
11. Ikeda, S.; Tsunoda, M. A.; Maeda, H. J. Colloid Interface
    Sci. 1978, 67, 336.
12. Ibid., 1979, 70, 448.
13. Tokiwa, F.; Ohki, K. J. Colloid Interface Sci. 1968, 27,
    247.
14. Goddard, E. D.; Kung, H. C. J. Colloid Interface Sci.
    1973, 43, 511.
15. Nylen, P. Z. Anor. Allgem. Chem. 1941, 246, 227.
16. Tokiwa, F.; Ohki, K. J. Phys. Chem. 1967, 71, 1824.
17. Corkill, J. M.; Gemmel, K. W.; Goodman, J. F.; Walker, T.
    Trans. Faraday Soc. I. 1970, 66, 1817.
18. Mille, M. J. Colloid Interface Sci. 1981, 81, 169.
19. Gravsholt, S. J. Colloid Interface Sci. 1976, 57, 575.
20. Rosano, H. L; Brendel, K.; Schulman, J. H.; Eydt, A. J.
    J. Colloid Interface Sci. 1966, 22, 58.
21. Hyde, A. J.: Maguire, D. J.; Stevenson, D. M. Proc. VIth
    Intern. Congr. Surface Active Substances, Zurich, 1972,
    Vol. II(2).
22. Nagarajan, R.; Ruckenstein, E. J. Colloid Interface Sci.
    1979, 71, 580.
23. Hutchinson, E.; Sheaffer, V. E.; Tokiwa, F. J. Phys. Chem.
    1964, 68, 2818.
24. Schick, M. J; Fowkes, F. M. J. Phys. Chem. 1957, 61, 1062.
25. Fort, T., Jr. J. Phys. Chem. 1962, 66, 1136.
26. Cameron, A.; Crouch, R. F. Nature (London) 1936, 198, 475.
27. Chang, D. C.; Rosano, H. L., to be published.

RECEIVED February 3, 1984

Effects of Surfactant Structure
on the Thermodynamics of Mixed Micellization

PAUL M. HOLLAND

Miami Valley Laboratories, The Procter & Gamble Company, Cincinnati, OH 45247

Calorimetric measurements are used to examine
interactions between different surfactant
components in nonideal mixed micelles and assess
the effects of surfactant structure on the thermodynamics of mixed micellization. Results for
some anionic/nonionic surfactant mixtures show
that variations in surfactant structure can have
important effects on heats of mixing in the micelles
and significantly influence the critical micelle
concentration (cmc) of the mixed surfactant systems.
Here, both the heats of mixing and deviations of
the cmc from ideality are smaller for alkyl ethoxylate sulfates than alkyl sulfates when mixed with
alkyl ethoxylate nonionics. The calorimetric
results for these systems are also used to examine
the appropriateness of the regular solution approxi-.
mation used in pseudo-phase separation models for
treating mixed micellization. The failure of the
regular solution approximation to account for the
observed heats of mixing in these systems suggests
that the net interaction parameter of the nonideal
mixed micelle models be interpreted as an excess
free energy parameter in such cases.

The formation of mixed micelles in surfactant solutions which
contain two or more surfactant components can be significantly
affected by the structures of the surfactants involved. The
observed critical micelle concentration (cmc) is often significantly lower than would be expected based on the cmc's of the
pure surfactants. This clearly demonstrates that interactions
between different surfactant components in the mixed micelles
are taking place.

0097-6156/84/0253-0141$06.00/0
© 1984 American Chemical Society

Nonideal mixed micelle models based on the pseudo-phase separation approach and a regular solution approximation have been developed (1-4) to describe this behavior. Here, the details of micellar structure are ignored and the interactions between two different surfactant components are accounted for by a single generalized parameter which represents an excess heat of mixing. This approach has been successfully applied (2,3,5) to a considerable variety of binary nonideal surfactant mixtures including mixed nonionic and ionic surfactants even though the model neglects effects due to the bonding of counterions. In the description of nonideal mixed ternary surfactant systems (3) no adjustable parameters beyond those for the binary mixtures are required to obtain good results. Together these results show that the net interaction parameters (β) obtained for binary surfactant mixtures can provide a useful measure of nonideal mixing in micelles. However, it is not as clear how well this approach (using the regular solution approximation) actually reflects the thermodynamics of mixed micellization.

Calorimetric measurements represent a promising way of gaining thermodynamic information about mixed micellization. Of the possible types of measurements, heats of micellar mixing obtained from the mixing of pure surfactant solutions are perhaps of the greatest interest. Also of interest is the titration (dilution) of mixed micellar solutions to obtain mixed cmc's. While calorimetric measurements have been applied in studies of pure surfactants (6,7) and their interaction with polymers (8), to our knowledge, applications of calorimetry to problems of nonideal mixed micellization have not been previously reported in the literature.

Among the purposes of this paper is to report the results of calorimetric measurements of the heats of micellar mixing in some nonideal surfactant systems. Here, attention is focused on interactions of alkyl ethoxylate nonionics with alkyl sulfate and alkyl ethoxylate sulfate surfactants. The use of calorimetry as an alternative technique for the determination of the cmc's of mixed surfactant systems is also demonstrated. Besides providing a direct measurement of the effect of the surfactant structure on the heats of micellar mixing, calorimetric results can also be compared with nonideal mixing theory. This allows the appropriateness of the regular solution approximation used in models of mixed micellization to be assessed.

Theory

The derivation of a pseudo-phase separation model for treating nonideal mixed micellization is given in detail in reference 3. This leads to the generalized result

$$\frac{1}{C^*} = \sum_{i=1}^{n} \frac{\alpha_i}{f_i C_i} \qquad (1)$$

which can be used to relate the mixed cmc (C^*) of a nonideal surfactant system to the activity coefficients (f_i) of the surfactant components in the mixed micelles, their mole fractions (α_i) and pure cmc's (C_i). In the case where the activity coefficients equal unity this expression reduces in form to that previous derived for ideal mixed micelles (1,9).

Consideration of the thermodynamics of nonideal mixing provides a way to determine the appropriate form for the activity coefficients and establish a relationship between the measured enthalpies of mixing and the regular solution approximation. For example, the excess free energy of mixing for a binary mixture can be written as

$$G^E = RT(x_1 \ln f_1 + (1 - x_1) \ln f_2) \qquad (2)$$

Taking the partial derivative with respect to the mole fractions in the micelle (x_i) and using the Gibbs-Duhem relation to eliminate some of the resulting terms gives

$$\left(\frac{\partial G^E}{\partial x_1}\right)_{T,P} = RT (\ln f_1 - \ln f_2) \qquad (3)$$

Combining these relationships then allows the activity coefficients to be expressed in terms of the excess free energy of mixing

$$\ln f_1 = \frac{1}{RT} \left(G^E + (1 - x_1) \frac{\partial G^E}{\partial x_1} \right) \qquad (4)$$

$$\ln f_2 = \frac{1}{RT} \left(G^E - x_1 \frac{\partial G^E}{\partial x_1} \right) \qquad (5)$$

The regular solution approximation is introduced by assuming (by definition) that the excess entropy of mixing is zero. This requires that the excess free energy equal the excess enthalpy of mixing. For binary mixtures the excess enthalpy of mixing is ordinarily represented by a function of the form

$$H^E = \beta x_1 (1 - x_1) RT \qquad (6)$$

where β times RT represents a difference in interaction energy between the mixed and unmixed systems. This corresponds to the leading term in the lattice model description of liquid mixtures (10). When this expression is substituted for G^E in equations 4 and 5 the resulting binary activity coefficients (in the regular solution approximation) take the form

$$f_1 = \exp \beta (1 - x_1)^2 \tag{7}$$

$$f_2 = \exp \beta x_1^2 \tag{8}$$

with β providing a measure of the nonideality of mixing in the system.

The β parameters in the above expressions are determined from the experimental mixed cmc's of binary systems. This requires solving iteratively for x_1 (at the cmc) using a relationship such as

$$x_1^2 \ln \left[\frac{\alpha_1 C^*}{x_1 C_1} \right] = (1 - x_1)^2 \ln \left[\frac{\alpha_2 C^*}{(1-x_1) C_2} \right] \tag{9}$$

followed by substitution into the expression

$$\beta = \frac{\ln \left[\frac{\alpha_1 C^*}{x_1 C_1} \right]}{(1 - x_1)^2} \tag{10}$$

While this treatment is strictly developed for nonionic surfactant mixtures it can often be applied empirically to mixtures containing ionic surfactants with success. Some examples of the types of surfactant mixtures to which this model has been successfully applied and the corresponding β parameters are given in Table I. It is readily seen that negative values are typically seen for deviation from ideal behavior. These would correspond to exothermic heats of mixing in the micelles. Assuming the validity of the regular solution approximation, it should then be possible to directly relate heats of mixing in the micelles (as a function of mole fraction) to the value of the β parameter via equation 6.

Table I. Values of the Net Interaction Parameter β for Some Binary Surfactant Mixtures (from Ref. 3)

β	Binary Mixture	Conditions
-3.7	$C_{10}(CH_3)_2PO/C_{12}OSO_3 \cdot Na^+$	1mM Na_2CO_3(24°C)
-2.4	$C_{10}(CH_3)SO/C_{12}OSO_3^- \cdot Na^+$	" " "
0	$C_{10}(CH_3)_2PO/C_{10}(CH_3)SO$	" " "
-4.4	$C_{12}(CH_3)_2NO/C_{12}OSO_3^- \cdot Na^+$	0.5mM Na_2CO_3(23°C)
-3.6	$C_{10}(OCH_2CH_2)_4OH/C_{12}OSO_3^- \cdot Na^+$	" " "
-0.8	$C_{12}(CH_3)_2NO/C_{10}(OCH_2CH_2)_4OH$	" " "
-13.2	$C_{10}N^+(CH_3)_3 \cdot Br^-/C_{10}OSO_3^- \cdot Na^+$	.05M NaBr (23°C)
-4.1	$C_8(OCH_2CH_2)_4OH/C_{10}OSO_3^- \cdot Na^+$	" " "
-1.8	$C_{10}N^+(CH_3)_3 \cdot Br^-/C_8(OCH_2CH_2)_4OH$	" " "

Experimental Section

Isoperibol calorimetric measurements were carried out using a
Tronac, Inc. Model 550 Calorimeter interfaced with a digital
voltmeter and micro-computer for data aquisition. In these
experiments a Dewar flask reaction vessel was stirred at a
constant rate while immersed in a constant temperature bath (at
25°C) with mixing initiated by either injection or titration of
another solution held at bath temperature into the vessel. The
experiments were initiated when the bath and reaction vessel
temperatures were equal and the changes in temperature were
monitored by a thermistor referenced to the bath. A slightly
modified version of a software package developed by Grime et.al.
(11) was used for data acquisition and reduction (i.e.,
correction for stirring heat, etc.). The compounds used in this
study were pure single specie surfactants determined to be
greater than 98% purity by thin-layer or gas chromatography.
     Heats of mixing for micellar solutions were determined by
mixing various ratios of pure equimolar surfactant solutions by
injection. These measurements were carried out for pentaoxyethylene glycol monodecyl ether ($C_{10}E_5$) with sodium dodecyl
sulfate (SDS), sodium dodecyl dioxyethylene sulfate ($C_{12}E_2S$) and
sodium decyl pentaoxyethylene sulfate ($C_{10}E_5S$), respectively at
25°C. In order to subtract out effects due to additional micellization of pure surfactant monomers on mixing, the experiments
were carried out at two concentrations above the cmc (.02M and
.08M) and the difference in the results taken. Under these
conditions the composition of the mixed micelles and solution
should be approximately equal since the concentrations are on the
order of 20 to 100 times the mixed cmc in each case. The final
differenced results should then correspond to the mixing of pure
micelles to form mixed micelles, and therefore to good approximation, to the excess heats of micellar mixing as a function of
mole fraction.
     The mixed cmc's of micellar solutions of tetraoxyethylene
glycol monooctyl ether (C_8E_4) with SDS and $C_{12}E_2S$, respectively
were also determined by calorimetry. Mixtures with C_8E_4 were
chosen due to the higher cmc and consequent larger and more
adequate total enthalpies of demicellization compared to mixtures
with $C_{10}E_5$. The measurements were carried out by the titration
(dilution) of mixed surfactant solutions with varying ratios of
components and .12M total concentration into distilled water.
Under these conditions the heat of demicellization is observed as
a function of titrant added, with a sharp break occurring once
the cmc is reached in the reaction vessel. Values for the cmc
endpoint were taken from the intersection of least squares fits
of line segments immediately before and after the break in the
titration results (corrected for stirring heat and volume
changes). Cmc's were determined for the pure surfactants in a
similar manner except for that of SDS. Here, the heat of

micellization is quite small and dilution effects are significant making an accurate determination of the cmc by calorimetry diffi-cult. Therefore the cmc of SDS in distilled water was determined by surface tension measurements using a tensiometer with du Nouy ring, as in previous work (3).

Results and Discussion

Results from the heats of micellar mixing experiments are shown in Figure 1. Here, the heat of mixing per mole (after differencing to correct for monomer contributions) is plotted versus the mole fraction of $C_{10}E_5$ in the mixed micellar system. It is clearly seen that the interaction of SDS with $C_{10}E_5$ is significantly stronger than that of the alkyl ethoxylate sulfate surfactants with $C_{10}E_5$ In addition, the symmetry in the heat of mixing curves is strikingly different with those for $C_{12}E_2S$ and $C_{10}E_5S$ showing an asymmetric maximum at about a 1:2 mole ratio with $C_{10}E_5$, while the SDS results are symmetric about a 1:1 ratio. These observations clearly demonstrate that the presence of ethoxylation in the structure of the sulfate surfactants has a pronounced effect on their heats of mixing.

Titration results for the mixed cmc's of the SDS/C_8E_4 and $C_{12}E_2S$/C_8E_4 systems as a function of their relative mole fraction in solution are shown in Figures 2 and 3, respectively. Here, the experimentally determined points are compared with calculated results from the nonideal mixed micelle model (solid line) and the ideal mixed micelle model (dashed line). Good agreement with the nonideal model is seen in each case. Figure 2 shows that the binary SDS/C_8E_4 system deviates sig-nificantly from ideality with a β value of −3.3. This result is comparable to the parameters found for other alkyl sulfate/alkylethoxylate nonionic systems (see Table I). In the case of the $C_{12}E_2S$/C_8E_4 system (see Figure 3) a significantly smaller deviation from ideality is observed, giving a value of −1.6 for β. This seems to be consistent with the smaller heats of mixing observed for $C_{12}E_2S$ compared to SDS in Figure 1.

Together these results show that the nonideal behavior of sulfate surfactants can be significantly affected by the presence or absence of ethoxylation in their structures. Here, both the heats of mixing and deviations of the cmc from ideality are smaller for alkyl ethoxylate sulfates than alkyl sulfates when mixed with alkyl ethoxylate nonionics. This is presumably due to improved screening or separation of charge in pure alkyl ethoxylate sulfate micelles compared to alkyl sulfate micelles. This interpretation would be compatible with the "charge separation" effect previously used in explaining the mixed cmc data of unethoxylated sulfate/alkyl ethoxylate non-ionic surfactant mixtures (5,12). The surprising differences

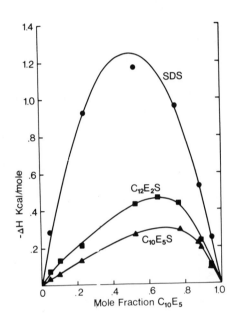

Figure 1. Heats of micellar mixing after correction for monomer contributions (see text) as a function of composition for SDS, $C_{12}E_2S$ and $C_{10}E_5S$ with $C_{10}E_5$ (at 25°C).

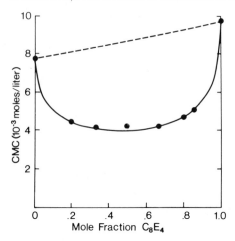

Figure 2.   Cmc's of mixtures of SDS and C8E4 in distilled water
(at 25°C). The plotted points are experimental data, the solid
line is the result for the nonideal mixed micelle model with
β = -3.3, and the dashed line is the result for ideal mixing.

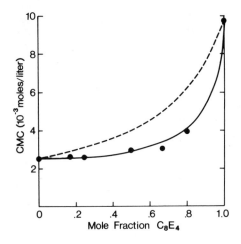

Figure 3.   Cmc's of mixtures $C_{12}E_2OSO_3^-.Na^+$ and C8E4 in
distilled water (at 25°C). The plotted points are experimental
data, the solid line is the result for the nonideal mixed
micelle model with β = -1.6, and the dashed line is the result
for ideal mixing.

observed in the symmetry of the micellar mixing heats with composition (Figure 1) are more difficult to understand. Here, it may be possible that differences in the ratio where the maximum in the heat of mixing occurs arise from differences between the preferred minimum energy packing geometry of the alkyl ethoxylate sulfate and alkyl sulfate surfactants with alkyl ethoxylates in the micelle.

The information obtained about the heats of mixing in nonideal mixed micelles can also be used to examine the appropriateness of the regular solution approximation used in nonideal mixed micelle models. Here, equation 6 allows a comparison between the observed calorimetric results and both the symmetry and magnitude of the excess heats of mixing that would be predicted by the regular solution approxima- tion used in the models. Referring to Figure 1, the $SDS/C_{10}E_5$ results show a good fit to the form of the regular solution approximation (solid line) calculated from equation 6 with $\beta=-8.4$. While it is seen that the predicted symmetry of the regular solution approximation is reproduced quite well, the magnitude of the β parameter necessary to obtain the fit is significantly outside the range of β values normally found when applying the nonideal micelle model to mixtures of alkyl sulfates and alkyl ethoxylate nonionics (see Figure 2 and Table 1). This indicates that the regular solution approxima- tion does not adequately describe mixing in this system. An examination of the heats of mixing for $C_{12}E_2S$ and $C_{10}E_5S$ with $C_{10}E_5$ clearly shows that the predicted symmetry of the regular solution approximation used in the mixed micelle models is not observed. Together these results clearly demonstrate that in spite of the success of the nonideal mixed micelle models in describing nonideal behavior, the regular solution approximation does not properly account for the heats of mixing in these systems.

This conclusion implies that the excess entropy of mixing is non-zero and that the mixed micelles presumably acquire more internal order than they would by random mixing. An examination of the magnitude of the deviations from the regular solution approximation shows that there must be a large TS^E contribution to the excess free energy of mixing. A similar situation is commonly observed in mixtures of liquids, where the regular solution approach often gives good results for the excess free energies, but poor results for the heats of mixing (13). Unfortunately, the excess entropies cannot be easily extracted from calorimetric mea- surements of micellar mixing heats such as those in Figure 1 and measurements of the mixed cmcs to obtain G^E. This is because the composition of the micelles at the cmc are not known and may vary significantly from the composition of the overall surfactant mixture.

The finding that the assumptions of the regular solution
approximation do not hold for the mixed micellar systems
investigated here suggests a re-examination of how the thermodynamics of mixing enter the nonideal mixed micelle model.
For example, it is readily seen (equations 4 and 5) that
when the functional form of equation 6 is substituted for
G^E, the form of the activity coefficients in equations 7 and
8 results. It is clear then that for cases where the excess
entropy is not zero, empirically assuming the form

$$G^E = \beta x_1 (1 - x_1)RT \tag{11}$$

for the excess free energy of mixing (as in reference 4),
will give the same form for the activity coefficients. Under
these circumstances, the β value from the nonideal mixed
micelle model should be interpreted as a dimensionless excess
free energy parameter, rather than an excess heat of mixing
parameter.

Conclusions

Calorimetric measurements can be used to obtain heats of mixing
between different surfactant components in nonideal mixed
micelles and assess the effects of surfactant structure on the
thermodynamics of mixed micellization. Calorimetry can also be
successfully applied in measuring the cmc's of nonideal mixed
surfactant systems. The results of such measurements show that
alkyl ethoxylate sulfate surfactants exhibit smaller deviations
from ideality and interact significantly less strongly with
alkyl ethoxylate nonionics than alkyl sulfates.
     The mixed cmc behavior of these (and many other) mixed
surfactant systems can be adequately described by a nonideal
mixed micelle model based on the psuedo-phase separation
approach and a regular solution approximation with a single net
interaction parameter β. However, the heats of micellar
mixing measured by calorimetry show that the assumptions of the
regular solution approximation do not hold for the systems
investigated in this paper. This suggests that in these cases
the net interaction parameter in the nonideal mixed micelle
model should be interpreted as an excess free energy parameter.

Acknowledgments

The author wishes to thank Mr. R. P. Burwinkel and Mr. D. F.
Etson for experimental work in the author's laboratory, Dr. J.
B. Kasting for allowing his Calvet isothermal calorimeter to be
used for some preliminary measurements and Dr. D. N. Rubingh for
helpful discussions.

Legend of Symbols

f_i activity coefficient of surfactant i in mixed micelles

x_i mole fraction of surfactant i in mixed micelles

C_i cmc of pure surfactant i

α_i mole fraction of surfactant i in total mixed solute

C^* cmc of mixed system

β dimensionless net interaction parameter

G^E excess free energy of mixing

H^E excess enthalpy of mixing

S^E excess entropy of mixing

R gas constant

T absolute temperature

Literature Cited

1. Lange, H.; Beck, K. H. Kolloid Z.Z. Polym. 1973, 251, 424.
2. Rubingh, D. N., in "Solution Chemistry of Surfactants"; Vol. 1, Mittal, L., Ed.; Plenum Press: New York, 1979; p. 337.
3. Holland, P. M.; Rubingh, D. N. J. Phys. Chem. 1983, 87, 1984.
4. Kamrath, R. F.; Franses, E. I. Ind. Eng. Chem. Fundam. 1983, 22, 230.
5. Scamehorn, J. F.; Schechter, R. S.; Wade, W. H. J. Disp. Sci. Tech. 1982, 3, 261.
6. Benjamin, L. J. Phys. Chem. 1964, 68, 3575.
7. Mazer, N.A.; Olofsson, G. J. Phys. Chem. 1982, 86, 4584.
8. Kresheck, G. C.; Hargraves, W. A. J. Colloid Interface Sci. 1981, 83, 1.
9. Clint, J. J. Chem. Soc. 1975, 71, 1327.
10. Münster, A. "Statistical Thermodynamics"; Springer-Verlag: Berlin-Heidelberg-New York, 1974, Vol. 2, p. 650.
11. Grime, J. K.; Staab, R. A.; Wernery, J. D., SAC '83 International Triennial Conference and Exhibition on Analytical Chemistry, Univ. of Edinburg July 1983.
12. Schick, M. J.; Manning, D. J. J. Amer. Oil Chem. 1966, 43, 133.
13. Hildebrand, J. H.; Prausnitz, J. M.; Scott, R. L. "Regular and Related Solutions"; Van Nostrand: New York, 1970; Chap. 7.

RECEIVED January 10, 1984

Influence of Structure and Chain Length of Surfactant on the Nature and Structure of Microemulsions

TH. F. TADROS

ICI Plant Protection Division, Jealott's Hill Research Station, Bracknell, Berkshire, RG12 6EY, England

The theories of microemulsion formation and stability have been reviewed. Three main approaches, namely mixed film, solubilisation and thermodynamic theories, have been briefly discussed. This is then followed by a section on factors determining w/o versus o/w microemulsion formation. The influence of surfactant and cosurfactant structure and chain length on the structure of microemulsions was described. In particular the cosurfactant chain length and structure has a considerable effect on the structure of the microemulsion. With short chain alcohols ($<C_6$) and/or surfactants, there is no marked separation into hydrophobic and hydrophilic domains and the structure is best described by a bicontinuous solution with easily deformable and flexible interfaces. With long chain alcohols ($>C_6$), well defined "cores" may be distinguished with a more pronounced separation into hydrophobic and hydrophilic regions. It was also concluded that microemulsions can be formed by a single surfactant provided this has the right geometry for packing at the interface and is capable of reducing the interfacial tension to low values. Addition of a cosurfactant is necessary in some cases to ensure packing of the surfactant molecule and to lower the interfacial tension.

0097–6156/84/0253–0153$06.25/0

The term microemulsion was first introduced by Hoar and Schulman (1) to describe the "transparent" or "translucent" systems, formed spontaneously when oil and water were mixed with relatively large amounts of an anionic surfactant (such as potassium oleate) and a cosurfactant (medium chain alcohol such as pentanol or hexanol). These systems are dispersions of very small drops (radius of the order of 5-50 nm) of water in oil (w/o) or oil in water (o/w). Microemulsions differ from ordinary emulsions (sometimes referred to as macroemulsions) in two main respects, namely their lack of turbidity and thermodynamic stability. For small particles, light scattering is proportional to the square of the volume of particles (2) and hence these systems with small droplets scatter little light and are not turbid. The thermodynamic stability of such systems is the consequence of the zero or negative free energy of their formation (see below), since the interfacial tension is so low that the remaining free energy of the interface is over compensated by the entropy of the dispersion of the droplets in the medium.

Inspite of the advances made in recent years in the theoretical basis of explaining the physics and chemistry of microemulsions (3), the science of their formation has not reached the point of their exact formulation (4). The mechanics of microemulsion formation differ somewhat from those used in making macroemulsions. The most significant difference lies in the fact that putting work into a macroemulsion or increasing the surfactant concentration usually improves their stability; but this is not the case with microemulsions which appear to be dependent for their formation on specific interactions among the constituent molecules and with the interface. If these interactions are not realised, neither the work input nor increasing the surfactant concentration will produce a microemulsion (4). On the other hand, once the conditions are right, spontaneous formation occurs and no mechanical work is required. Thus, the crucial step in formulating the required microemulsion lies in the choice of emulsifiers. In other words, the nature (whether w/o or o/w) and structure of microemulsions depend to a large extent on the structure and chain length of the surfactant and co-surfactant used in their preparation. This is the subject of the present review.

Before describing how microemulsion nature and structure are determined by the structure and chain length of surfactant and cosurfactant, it is necessary first to briefly review the theories of microemulsion formation and stability. These theories will highlight the important factors required for microemulsion formation. This constitutes the first part of this review. The second part describes the factors that determine whether a w/o or o/w microemulsion is formed. This is then followed by a section on the effect of structure and chain length of surfactant and cosurfactant on the structure of the microemulsion formed. Finally, the question of whether a microemulsion can be formulated with a single surfactant will be raised.

Theories of Microemulsion Formation and Stability

As discussed before (4) it is perhaps convenient to classify these theories into three main categories : interfacial or mixed film theories, solubilisation theories and thermodynamic theories. Below a brief description of each of these classes will be given with particular emphasis on the role of surfactant nature and structure.

Mixed film theories (4-8) The essential feature of the mixed film theories is to consider the film as a liquid, two dimensional third phase in equilibrium with both oil and water, implying that such a monolayer could be a duplex film, i.e., one giving different properties on the water side than on the oil side (4). According to these theories, the interfacial tension γ is given by the expression,

$$\gamma = (\gamma_{o/w})_a - \pi \tag{1}$$

where $(\gamma_{o/w})_a$ is the o/w interfacial tension, which is reduced by the presence of the alcohol cosurfactant (hence the subscript a), and π is the two dimensional spreading pressure of the mixed film. Contributions to π were considered to be the crowding of the surfactant and cosurfactant molecules and penetration of the oil phase into the hydrocarbon part of the interface. According to equation (1) if $\pi > (\gamma_{o/w})_a$, γ_T becomes negative leading to expansion of the interface until γ reaches zero.

The above concept of duplex film can be used to explain both the stability of microemulsions and the bending of the interface. Considering that initially the flat duplex film has different tensions (i.e., different π values) on either side of it, then the deriving force for film curvature is the stress of the tension gradient which tends to make the pressure or tension in both sides of the curved film the same. This is schematically shown in Figure 1.   For example if $\pi'_w > \pi'_o$ on the flat film, then the film has to be expanded much more at the water side than at the oil side (which indeed contracts as a result of the curvature effect) until the surface presssures become equal on both sides of the duplex film (i.e. $\pi_w = \pi_o = 1/2 \ (\gamma_{o/w})_a$). This means that an o/w microemulsion results in this case.   On the other hand, if $\pi_o > \pi_w$, then the film expands at the oil side and contracts at the water side of the interface resulting in the formation of a w/o microemulsion.

The above simple theory can be applied to predict the nature of the microemulsion. In a duplex film, the surface pressures at the oil and water sides of the interface depend on the interactions of the hydrophobic and hydrophilic portions of the surfactant at both sides respectively. For example, if the hydrophobic portions are bulky in nature relative to the hydrophilic groups, then for a flat film, such hydrophobic portions tend to crowd forming a higher surface pressure at the oil side of the film. In this case bending occurs to expand the oil side forming a w/o microemulsion. On the other hand, with a surfactant molecule with a relatively bulky hydrophilic group, crowding occurs at the water side of the interface, tending to form an o/w microemulsion.

A quantitative theory based on the lateral stress gradient resulting from the difference in swelling of the heads and tails across the interface was developed by Robbins (9). This stress gradient was expressed in terms of physically measureable quantitites, namely, surfactant molecular volume, interfacial tension and interfacial compressibility. Relating the pressure difference across a curved interface to the activity of water in a w/o microemulsion, Robbins (9) established criteria for spontaneous water uptake without postulating a negative interfacial tension. It should be mentioned, however, that any duplex film theory has the draw back that two interfacial tensions must be defined at the oil side and water side of the interface (this is certainly difficult to define in a thermodynamic sense). Moreover, there is no way that such interfacial tensions can be measured and, therefore, the mixed film theory must only be regarded as approximate and is of only historial interest.

Solubilisation Theories (10-13) The solubilisation concept introduced by Shinoda and coworkers (10-13) who preferred to treat microemulsions as swollen micellar systems, thus relating them directly to the phase diagrams of the components. For example, the phase diagram of a three component system of water, ionic surfactant and alcohol usually displays one isotropic aqueous liquid region L_1, from the water corner and one isoptropic liquid region L_2, from the alcohol corner (reverse micelles). The latter can dissolve a large amount of a hydrocarbon oil. Alternatively, such inverse micelles may be produced if the alcohol is dissolved in the oil followed by the addition of water and surfactant. Since the final solution is isotropic and no phase separation takes place when going from the pure hydrocarbon state to the microemulsion state, Shinoda and coworkers (10-13) preferred to describe these systems as swollen micelles.

Solubilisation can best be illustrated by considering the phase diagrams of non-ionic surfactants containing poly(oxyethylene oxide) head groups. Such surfactants do not generally need a cosurfactant for microemulsion formation. At low temperatures, the ethoxylated surfactant is soluble in water and at a given concentration is capable of solubilising a given amount of oil. The oil solubilisation increases rapidly with increase of temperature near the cloud point of surfactant. This is clearly illustrated in Figure 2a which shows both the solubilisation curve and cloud point curve of the surfactant. Between the two curves, an isotropic region of o/w solubilised system exists. At any given temperature any increases in the oil weight fraction above the solubilisation limit results in oil separation i.e. o/w solubilised + oil, whereas at any given sufactant concentration any increase in temperature above the cloud point of surfactant results in separation into oil water and surfactant.

On the other hand, if one starts from the oil phase with dissolved surfactant and add water, solubilisation of the latter takes place and hence solubilisation increases rapidly with reduction of temperature near the haze point of the surfactant. This is illustrated in Figure 2b which shows both the haze point and solubilisation curve. Between the two curves, an isotropic region of w/o solubilised system exists. At any given temperature, any increase in water weight fraction above the solubilisation limit results in water separation, i.e. w/o solubilised + water, whereas at a given surfactant concentration, any decrease in temperature below the haze point result in separation into water, oil and surfactant.

Thus, with nonionic surfactants, both types of micremulsions can be formed depending on the conditions. As shown above, with such systems temperature is the most crucial factor since the solubility of the surfactant in water or in oil depends on the temperature. One should remember that in aqueous solutions, the solubility of nonionic ethoxylated surfactants decreases with increase of temperature, whereas the reverse is true with oil solutions.

Thermodynamic Theories (14-18) Two main treatments have been considered, namely by Ruckenstein et al (14-16) and Overbeek et al (17,18). The treatments follow roughly the same procedure, but vary somewhat in detail (3). Ruckenstein et al (14,15) considered the free energy of formation of microemulsions ΔG_m, to consist of three main contributions ΔG_1 an interfacial energy term, ΔG_2 an energy of interaction between the droplets term and ΔG_3 an entropy term accounting for the disperion of droplets into the continuous medium. The interfacial free energy term ΔG_1 was considered to consist of two contributions due to the creation of an uncharged surface (given by the product of area created and specific surface free energy of the interface) and a contribution due to the formation of electrical double layers (which is given by the product of the interfacial area and the specific surface free energy due to creation of an electrical

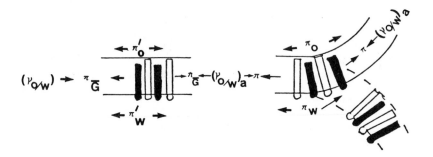

Figure 1. Schematic representation of film bending

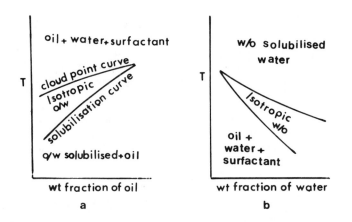

Figure 2. Schematic representation of solubilisation
           (a) oil solubilised in a nonionic surfactant
           solution ; (b) water solubilised in an oil
           solution of nonionic surfactant

double layer). The double layer contribution was calculated
using the Debye-Huckel approximation. For the calculation of
ΔG_2 a pairwise additivity of interaction potentials was
assumed, whereas for the calculation of the entropy contribution
term ΔG_3, a lattice model was used to calculate the number of
configurations of droplets in the continuos medium. From the
variation of ΔG_m with droplet radius R, various states could be
distinguished which illustrated the transition from instability
kinetic stability thermolynamic stability. This transition may
be obtained by reducing the value of ΔG_1, i.e. reducing the
specific surface energy, f_s. The reduction of f_s to
sufficiently small values was accounted for by Ruckenstein (15)
in terms of the so called "dilution effect". Accumulation of
surfactant and cosurfactant at the interface not only causes
significant reduction in the interfacial tension, but also
results in reduction of the chemical potential of surfactant and
cosurfactant in bulk solution. The latter reduction may exceed
the positive free energy caused by the total interfacial tension
and hence the overall ΔG of the system may become negative.
Further analysis by Ruckenstein and Krishnan (16) have showed
that micelle formation encountered with water soluble surfactants
reduces the dilution effect as a result of the association of the
the surfactants molecules. However, if a cosurfactant is added,
it can reduce the interfacial tension by further adsorption and
introduces a dilution effect. The treatment of Ruckenstein and
Krishnan (16) also highlighted the role of interfacial tension in
the formation of microemulsions. When the contribution of
surfactant and cosurfactant adsorption is taken into account, the
entropy of the drops becomes negligible and the interfacial
tension does not need to attain ultralow values before stable
microemulsions form.
   In Overbeek theory (17,18) the free energy of microemulsion
formation was also considered to consist of three main
contributions; ΔG_1 due to mixing of surfactant with water and
co-surfactant with oil; ΔG_2 due to the interfacial area in
forming the droplets and ΔG_3 due to the mixing of droplets
into the continuous phase. ΔG_1 is simply given by the sum of
the product of number of moles of each component and its chemical
potential with reference to the standard state. The interfacial
area term ΔG_2 is given by the final interfacial tension and
final area of the interface plus a chemical potential term due to
adsorption of surfactant and co-surfactant. Finally, the free
energy term due to mixing of droplets into the continuous medium
have been obtained using the hand-sphere model of Percus Yevick
(19) and Carnahan and Starling (20) which was originally used by
Agterof et al (21) to describe the non-ideal behaviour of
microemulsions.

Taking the above three contributions into account, Overbeek (17) derived the following expression for the free energy dG_m for microemulsion formation,

$$dG_m = dA \left[\gamma_{uncharged} + \int \psi_o d\sigma + \frac{A^2 kT}{12 (n_w v_w)^2} (\ln \phi - 1 + \phi \frac{4-3\phi}{(1-\phi)^2} + \ln \frac{v_o}{v_{hs}}) \right] \quad (2)$$

where dA is the change in interfacial area. The first term between the square brackets is the interfacial tension term, $\gamma_{uncharged}$ that is obtained if no double layers had been formed $\psi_o d\sigma$ is the electrical contribution to the interfacial tension as a result of formation of an electrical double layer, with ψ_o being the surface potential and σ the surface charge density. The third term is the osmotic contribution due to mixing of microemulsion droplets (treated as hard spheres) and V_{hs} is their molar volume. The number of hard sphere droplets is given by,

$$n_{hs} = \frac{A^3 \quad N_{av}}{36 \pi (n_w v_w)^2} \quad (3)$$

where A is the area, n_w is the number of water molecules and V_w is their molar volume.

The three terms of equation (2) are unequal in magnitude. The last term is always negative. For example, for a high ratio of surfactant to water $n_s/n_w = 0.04$ (corresponding to a mass ratio of 0.7) and $\phi = 0.5$, this term is ~ 0.2 m Nm^{-1}. For $\phi = 0.1$, the value would be ~ 0.5 m Nm^{-1} and to make this term $- 1.0$ m Nm^{-1}, ϕ has to be below 10^{-5}, i.e. a very dilute microemulsion.

On the other hand, the electrical free energy per unit area of double layer (second term) is high and positive even for relatively low surface potential. The contribution of this term could be tens of m Nm^{-1}. This requires $\gamma_{uncharged}$ to have a high negative value to reach the condition $dG = 0$. The conclusion so far reached from this analysis is that in a microemulsion the interfacial tension, including the electrical term must have very low but positive value. The small variations in the total interfacial tension required to balance the variation in the free energy of mixing (osmotic term) can be easily obtained by small variations in the amount of surfactant and cosurfactant at the interface leading to variations in $\gamma_{uncharged}$.

Factors Determining w/o versus o/w Microemulsion Formation

Severed factors play a role in determining whether a w/o or o/w
microemulsion is formed. These factors may be considered in the
light of the theories described in section 2. For example, the
duplex film theory predicts that the nature of the microemulsion
formed depends on the raltive packing of the hydrophobic and
hydrophilic portions of the surfactant molecule, which determines
the bending of the interface. For example, a surfactant molecule
such as Aerosol oT, having the structure shown below, is
favourable for formation of w/o microemulsion, without the need
of adding a cosurfactant. As a result of the presence of a
stumpy head group and large volume to length (V/l) for the non-
polar group, the interface tends to bend with the head groups
facing inwards thus forming a w/o microemulsion. This geometric
constraint for the Aerosol OT molecule was considered in detail
by Oakenfull (22) who showed that the molecule has a V/l >0.7.,
which was considered to be necessary for microemulsion formation.

With ionic surfactants for which V/l <0.7, microemulsion
formation needs the presence of a cosurfactant. The latter has
the effect of increasing V without affecting l (if the chain
length of the cosurfactant does not exceed that of the
surfactant). These cosurfactant molecules act as "padding"
separating the head groups.
    The importance of geometric packing on the nature of the
microemulsion has also been considered by Mitchell and Ninham
(23). According to these authors the nature of the aggregate
unit depends on the packing ratio $v/a_o l_c$ where v is the
partial molecular volume of the surfactant, a_o is the head
group area of a surfactant molecule and l_c is the maximum chain
length. Thus, this packing ratio provides a measure of the
hydrophobic-hydrophilic balance (HLB). For values of
$v/a_o l_c < 1$, normal i.e., convex aggregates are predicted,
whereas for $v/a_o l_c > 1$ inverse drops are expected. The packing
ratio is affected by many factors including hydrophobicity of
head group, ionic strength of solution, pH, temperature and the

addition of lipophilic compounds such as cosurfactants. Addition
of electrolyte or increase of temperature causes a change in the
area per head group and hence effects the packing ratio. With
the Aerosol OT molecule $v/a_o l_c$ is greater than 1 since both
a_o and l_c are small. Thus, this molecule is favourable for
formation of w/o microemulsion.

The packing ratio also explains the nature of microemulsion
formed by using nonionic surfactants. If $v/a_o l_c$ increases
with increase of temperature (as a result of reduction of a_o),
one would expect the solubilisation of hydrocarbons in nonionic
surfactact to increase with temperature as observed, until
$v/a_o l_c$ reaches the value of 1 where phase inversion would be
expected. At higher temperatures, $v/a_o l_c > 1$ and water in oil
microemulsions would be expected and the solubilisation of water
would decrease as the temperature rises again as expected.

The influence of surfactant structure on the nature of the
microemulsion formed can also be predicted from the thermodynamic
theory by Overbeek (17,18). According to this theory, the most
stable microemulsion would be that in which the phase with the
smaller volume fraction forms the droplets, since the osmotic
term increases with increasing ϕ. For w/o microemulsion prepared
using an ionic surfactant, the hard sphere volume is only
slightly larger than the water volume, since the hydrocarbon
tails of the surfactant may interpenetrate to a certain extent,
when two droplets come close together. For an oil in water
microemulsion, on the other hand, the double layer may extend to
a considerable extent, depending on the electrolyte concentration
(the thickness of the double layer $1/\kappa$ is of the order of 100 nm
for 10^{-5} ml dm^{-3} and 10nm for 10^{-3} mol dm^{-3} 1:1
electrolyte). Thus, the hard sphere radius can be increased by 5
nm or more unless the electrolyte concentration is high
(say 10^{-1} mol dm^{-3} where $1/\kappa \sim$ 1nm). Thus this factor
works in favour of the formation of w/o microemulsions especially
for small droplets. Furthermore, establishing a curvature of the
adsorbed layer at a given adsorption is easier with water as the
disperse phase, since the hydrocarbon chains will have more
freedom around than if they were inside the droplet.

Influence of Surfactant and Cosurfactant Structure and Chain Length on the Structure of Microemulsions

Both the structure and chain length of surfactants and
cosurfactants have a striking influence on the structure of the
micremulsion formed. The most systematic studies have been on
the influence of the cosurfactant chain length and structure on
the nature of the microemulsion region. Two main studies have
been carried out to elucidate the difference obtained, namely
electrical conductivity and NMR investigations. As we will see

later, the results of such investigations led to the
classification of microemulsions into two main sytems, namely
those with well defined "cores" with pronounced separation into
hydrophobic and hydrophilic regions and those systems in which
there is no marked separation into hydrophobic and hydrophilic
domains and the structure is best described by a bicontinuous
solution with easily deformable and flexible interface.

One of the best examples of the influence of chain length
and structure of the cosurfactant on the nature of the
microemulsion region is that obtained by Clausse and coworkers
(24,25). Figure 3 shows the phase diagrams of the system
water/sodium dodecylsulphate/alkanols benzene at various chain
length of the alkanol from C_2 to C_7 (i.e. ethanol to
heptanol). In such systems, the molar ratio of surfactant to
alcohol was kept constant (1:2) at the temperature was 25°C. The
pase diagrams may be classified into two distinct systems. In
the first case, with cosurfactant chain length C_1 to C_4
(Figure 3a-d), the transparent domain (sometimes referred to as
the Winsor IV domain) consists of a unique area that has the
shape of a curvilinear triangle leaning against a large portion
of the surfactant (S) - water (W) side of the phase diagram.
These systems were referred to by Clausse et al (21) as Type U
systems. On the other hand, with cofurfactants with chain length
C_5 to C_7 (Figur 3 e-g), the Winsor IV domain is split into
two disjointed areas that are separated by a composition zone
over which viscous turbid and birifringent media are encountered.
This second class of systems was referred as Type S systems
(24). It can also be seen that the Winsor IV domain reaches its
maximum extension at C_4 reducing in size below and above C_4.
Moreover, at C_5, one observes a small monophasic region near
the W apex (probably o/w microemulsion of the Schulman's type)
which vanishes as the alcohol chain length is increased to C_6.

The influence of cosurfactant structure is best illustrated
by using various isomeric alcohols with the same chain length.
This is shown in Figure 4 for the system water, benzene,
potassium oleate and amylic alcohols (25). The mass ratio of
potassion oleate to COH is 3:5 and the temperature was 22°C.
This figure shows the progressive geometrical deformation
undergone by the monophasic transparent domain upon substitution
of a C_5 alcohol for one of its isomers. It appears that the
substitution of one isomeric pentanol for another induces a
transition from systems displaying a unique Winsor IV domain to
systems displaying a Winsor IV domain split into two disjointed
areas. Thus, the substitution of an isomeric pentanol for
another one, induces a progressive transition between the case of
Type U systems and that of Type S systems, in contrast with the
sharp transition observed upon substituting a longer straight
alkanol for a shorter one (Figure 3).

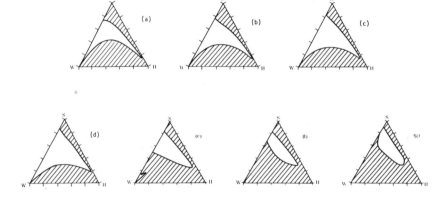

Figure 3.   Phase diagrams of the system water Sodium
dodecyl sulphate/alkanols benzene (a)
ethanol ; (b) 2-propanol; (c) 1-propanol; (d)
1-butanol; (e) 1-pentanol; (f) 1-hexanol; (g)
1-heptanol.

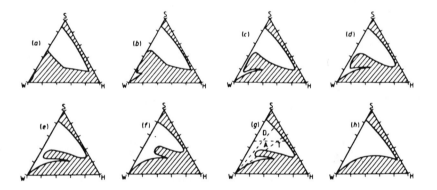

Figure 4.   Phase diagrams of the system water benzene
potassium oleate/amylic alcohols (a)
1-pentanol (1P); (b) 3-methyl-1 butanol
(3M1B); (c) 2-methyl-1-butanol (2M
1B) (d) 2,2-dimethyl-1-proptanol (22M1P);
(e) 2-pentanol (2P);(f) 3-pentanol (3P);
(g) 3 methyl-2-butanol (3M2B);(h) 2-methyl-
2-butanol (2M2B).

The above differences observed with the various cosurfactants are reflected in the conductivity - water volume fraction (ϕ_w). This is shown in Figures 5 and 6. It can be seen that the conduction behaviour is strongly influenced by the nature of the alcohol. For example Figure 5 shows that for the shorter alcohols ethanol, 1-propanol and 1-butanol (which give Type U systems) the conductivity increases rapidly as the water content increases above a certain ϕ_w value, which is smaller the shorter the alcohol. In contrast with the larger alcohol, 1-pentanol, the conductivity does not take high values and its variation with ϕ_w is more smooth. With even higher chain length alcohols, namely 1-hexanol and 1-heptanal (which give Type S system), the conductivity is very low (<0.05 Sm^{-1}) and hence appears close to zero on the scale of Figure 5. However, if the scale is expanded, the conductivity of such S-Type shows a maximum. This is illustrated in Figure 7 for the system water/hexadecane/potassium oleate/hexanol which shows the trend. Similar results have been recently obtained for the system xylene/water/sodium dodecyl benzene sulphonate/hexanol (26). Thus, the difference in conductive behaviour between the two systems must reflect a difference between the structural units involved.

With the U-Type systems (i.e. with the low chain alcohols) the trends in the conductivity - ϕ_w curve are consistent with percolative conduction originally proposed to explain the behaviour of conductance of conductor-insulator composite materials (27). In the latter model, the effective conductivity is practically zero as long as the conductive volume fraction is smaller than a critical value ϕ^P, called the percolation threshold, beyond which κ suddenly takes a non-zero value and rapidly increases with increase of ϕ_w. Under these conditions,

$$\kappa \sim (\phi_w - \phi_w^P)^{8/5} \qquad \text{when } \phi_w > \phi_w^P \qquad (4)$$

$$\text{and } \kappa \sim (\phi_w^C - \phi_w)^{-0.7} \qquad \text{when } \phi_w < \phi_w^P \qquad (5)$$

By fitting the conductivity data to the above equations, one usually finds a ϕ^P value for droplets (which should also include surfactant) near to the theoretical limit of 0.29. At this volume fraction, charge transfer between w/o globular micelles submitted to attractive interactions take place. Moreover, as we will see later, such systems contain easily deformable and flexible interfaces.

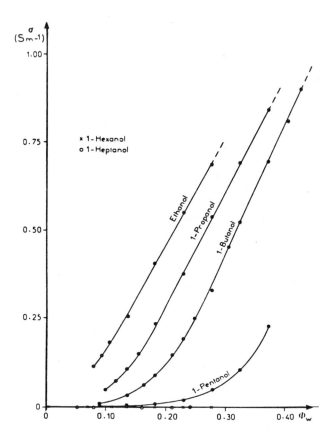

Figure 5.   Influence of alcohol chain length on the
            conductivity - water volume fraction curves

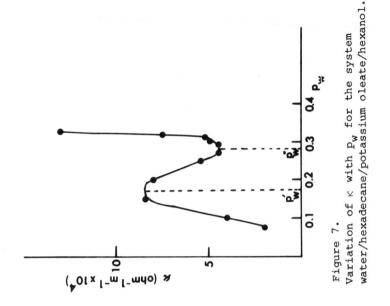

Figure 7.
Variation of κ with p_w for the system water/hexadecane/potassium oleate/hexanol.

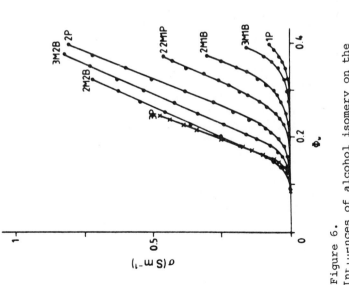

Figure 6.
Influences of alcohol isomery on the conductivity - water volume fraction curve: (symbols are the same as those for Figure 4).

With the S-type system, the trend in conductivity - p_w (which p_w is the water weight fraction) (Figure 7) is more complex showing a number of transitions indicating more subtle changes in the conduction units as the water concentration is increased. It seems that in the low water concentration region i.e. below the maximum, inverted swollen micelles or microdroplets do not exist (premicellar region) and the increase in conductivity with p_w in this region is likely to be due to the increase in the dissociation of the surfactant on the addition of water (26). The region beyond the maximum, i.e. where the conductivity decreases with increase of P_w, is the region of the existence of w/o microemulsion of the Schulman type. The sharp increase at the second p_w region (i.e. p_w) may be associated with a "facilitated" path for ion transport. This can be ascribed to the formation of non spherical particles resulting from swollen micelle clustering and subsequent cluster interlinking, a process that is indicative of system stability breakdown and of crystalline structure organisation (26). Thus, these systems represent those of "true"microemulsions, i.e., with definite water cores and therefore are non-percolating.

The maximum in the conductivity - ϕ_{H_2O} curves for the Type S systems can be explained in terms of the model proposed by Eicke and Denss (28). These authors considered charge production to be caused by two processes. The first by dissociation of the surfactant molecules inside the micellar entities, a process thought to be independent of the medium in which the aggregates are embedded. The second process is a transfer of charges, i.e., of an anion which stems from the dissociation of the surfactant inside the droplet, to another (originally neutral) micellar aggregate. This latter process is considered to represent a phase transfer between the droplet pseudo phase (built-up by the ensemble of microemulsion droplets) and the solution phase consisting of a very dilute solution of charged micellar aggregates in the non-polar solvent. The charge transfer will actually proceed in two steps : ion (droplet phase) $\rightarrow$ ion (non-polar solvent) and ion (non-polar solvent) $\rightarrow$ ion (resolubilised in micellar aggregate).

Further information on the dependence of structure of microemulsions formed on the alcohol chain length was obtained from measurement of self diffusion coefficient of all the constitutents using NMR techniques (29-34). For microemulsions consisting of water, hydrocarbon, an anionic surfactant and a short chain alcohol (C_4 and C_5) the self diffusion coefficient of both water, cosurfactant and hydrocarbon was quite high (of the order of 10^{-9} m^2 s^{-1}) being two orders of magnitude higher than the value to be expected for a discontinuous medium (10^{-11} m^2 s^{-1} and below). This can be

attributed to three alternative effects or a combination of all
of them: (i) bincontinuous solutions, i.e., both oil and water
continuous, (ii) easily deformable and flexible interface, or
(iii) absence of large aggregates. Lang et al (35) have shown
that addition of short chain alcohol to a micellar solution
results in a considerable reduction of the life time of
surfactant monomer in the micelle. This result is consistent
with the picture of disorganised and flexible interfaces. Thus,
the picture of microemulsion based on an anionic surfactant and
short chain alcohols is that of very small aggregates with any
internal interface with very limited spatial extension or very
dynamic and flexible structure which breaks up and reform at very
high speed, or both.

On the other hand, with microemulsions based on an anionic
surfactant and a long chain alcohol, D_w was fairly low for
certain concentrations, indicating that distinct water droplets
in a hydrophobic medium may form. The system investigated by
Lindman et al (29-34) was based on octanoic acid - decanol -
octane-water. This means that the anionic "surfactant" used
contains only seven carbon atoms in the alkyl chain which is
fairly short. With longer chain surfactants, one would expect
well defined "water cores" provided the alcohol is also long-
chain. Such well defined "water cores" have also been confirmed
by Lindman et al (34) for the Aerosol OT - hydrocarbon system.
In that case the self diffusion coefficient - concentration curve
shows a behaviour distinctly different from the cosurfactant
microemulsions. D_w has a quite low value throughout the
extension of the isotropic solution phase up to the highest water
content. This implies that a model with closed droplets
surrounded by surfactant anions in a hydrocarbon medium gives an
adequate description of these solutions. However, since D_w was
found to be significantly higher than D_s, the authors came to
the conclusion that a non-negligible amount of water must exist
between the emulsion droplets.

Thus, in summary, self diffusion measurements by Lindman
et al (29-34) have clearly indicated that the structure of
microemulsions depends to a large extent on the chain length of
the cosurfactant (alcohol), the surfactant and the type of
system. With short chain alcohols ($<C_6$) and or surfactants
there is no marked separation into hydrophobic and hydrophilic
domains and the structure is best described by a bicontinuous
solution with easily deformable and flexible interfaces. This
picture is consistent with the percolative behaviour observed
when the conductivity is measured as a function of water volume
fraction (see above). With long chain alcohols ($> C_6$) on the
other hand, well defined "cores" may be distinguished with a more
pronounced separation into hydrophobic and hydrophilic regions.

The same is true with other specific systems such as that based on Aerosol OT without the presence of a cosurfactant. This structure is also consistent with the non-percolative nature of such systems (see above).

Thus it can be concluded that the structure of microemulsions depends on the structure of surfactant and cosurfactant. Moreover, this structure also determines the amount of solubilisation of oil and or water in microemulsions. In recent studies by Gracia et al (36) and Baraket et al (37) the criteria for structuring surfactants to maximise solubilisation of oil and water were investigated. Using a number of ethoxylated alkyl phenols with the same HLB, but different molecular weight i.e. in which both the hydrophilic and hydrophobic moities of the surfactant were increased simultaneously, Gracia et al (36) found that solubilisation increases with increase of molecular weight. On the other hand, Baraket et al (37) showed that branching of the surfactant hydrocarbon tail has a significant effect on solubilisation; the less the branching the higher the solubilisation.

Microemulsions Based on a Single Surfactant and the role of the cosurfactant. It is interesting now to raise the following question. Can microemulsions be prepared using a single surfactant?. This question can be answered in the light of the criteria needed for the formation microemulsion discussed above. Two main criteria are important: surfactant geometry and low interfacial tension. As mentioned above the relative packing of the hydrophobic and hydrophilic portion of the surfactant molecule, plays a major role. Such packing is favourable with molecules such as Aerosol OT with a stumpy head group and a large volume to length ratio for the non-polar group, which tend to bend with the head groups facing inwards. Moreover such a molecule when adsorbed at the w/o interface lowers the interfacial tension sufficiently such that a w/o microemulsion can be formed without the need of a cosurfactant.

As mentioned before, nonionic surfactants also form microemulsions without the need of adding a cosurfactant. This is still the case with "pure" nonionic surfactants which consist of a single chain length. Under conditions whereby a microemulsion is formed, nonionic surfactants satisfy the above mentioned criteria, namely packing and low interfacial tension. For example, in the homogeneous isotropic region between the solubilisation and cloud point curve (Figure 2a), geometric packing of the molecules favours the formation of oil in water microemulsions and in this region the interfacial tension also falls to very small values. Similarly, between the haze point and solubilisation curve (Figure 2b) water in oil microemulsion form since geometric packing favours bending with the larger polyethylene oxide head group solubilising the water. In this region, the interfacial tension also falls to very small values.

The cosurfactant plays two major roles; firstly it effects the packing of the surfactant molecule at the interface and secondly it brings the necessary reduction in interfacial tension, if this could not be produced by the surfactant molecules alone. As mentioned before, the packing of the surfactant molecules depends on the ratio of the volume to the length of the non-polar group. Addition of a cosurfactant increases the volume without affecting the length and hence it enables the geometrical packing of the molecules at the interface. Moreover, such molecules act as "padding" separating the head groups.

The role of the cosurfactant in reducing the interfacial tension can be understood from application of the Gibbs adsorption equation in the form (14).

$$\left(\frac{\partial \gamma \partial}{\partial \mu^i}\right)_{T,p,} \text{ all } n_j \text{ except } n_i, \ A = -\left(\frac{\partial n^i}{\partial A}\right)_{T,p,} \text{ this all } n_j \text{ except } n_i,$$

$$= \Gamma i \qquad\qquad (6)$$

By using equation (6) twice for the cosurfactant (i = Co) and for the surfactant (i = Sa), the following equation can be derived,

$$\gamma = \gamma_o - [\int_{C(Co=o)}^{C(Co)} Co \ \mu_{Co}] \ T,P,^{C(sa) = 0]}$$

$$-[\int_{C(Sa=0}^{C(Sa)} Sa \ d\mu_{Sa}] \ T,P,n(Co), \ n \ oil, \ n_w \qquad (7)$$

It is clear from equation (7) that the addition of a second surfactant results in further decrease in γ; the essential requirements being a not too small adsorption of the second surfactant. Whether it replaces the first surfactant or is adsorbed in addition to it is immaterial, just as it is not essential for the two surfactants to form a complex. If the two surfactants are of the same type e.g. both water soluble anionic surfactants, they will form mixed micelles and this will lower the activity of the second surfactant added and decrease both its Γ and dμ. However, if the two surfactants are different in nature, e.g. one predominantly water soluble and the other oil soluble, they will only slightly affect each other's activity and their combined effect on the interfactial tension may be large enough to bring γ to zero at finite concentrations.

Literature Cited

1. Hoar, T.P. and Schulman, J.H., Nature (London) 152, 102 (1943)
2. Kerker, M., "The Scattering of Light" Academic Press, New York (1969) pp. 30-39.
3. Tadros, Th. T., "Microemulsions - An Overview"., in "Solution Properties of Surfactants" Editor Miltal, K.L., Plenum Publishing Corporation (in Press).
4. Prince, L. M, "Microemulsions, Theory and Practice" Academic Press, New York (1977).
5. Bowcott, J.E. and Schulman, J.H., J. Electrochem. 54, 283, (1955)
6. Schulman, J.H., Stoeckenhuis, W. and Prince, L. M., J. Phys. Chem. 63, 1677 (1959)
7. Prince, L.M., J Colloid Interface Sci 23, 165 (1967)
8. Prince, L.M., J. Cosmet Chem. 21, 183 (1970)
9. Robbins, M. L., "Theory for the Phase Behaviour of Microemulsions", Paper No 5839, presented at the Improved Oil Recovery Symposium of the Society of Petroleum Engineers of AIME, Tulsa, Oklahoma, March 22-24 (1976).
10. Shinoda, K and Friberg, S., Adv. Colloid Interface Sci. 4, 281 (1975)
11. Saito, H. and Shinoda, K., J Colloid Interface Sci 24, 10 (1967); 26; 70 (1968)
12. Shinoda, K. and Kunieda, H., J Colloid Interface Sci. 42, 381 (1973)
13. Ahmed, S.I., Shinoda, K and Friberg, S., J. Colloid Interface Sci. 47, 32 (1974)
14. Ruckenstein, E. and Chi, J.C., J. Chem. Soc. Faraday Trans. 71, 1690 (1975)
15. Ruckenstein, E, J Colloid Interface Sci 66, 369 (1978)
16. Ruckenstein, E and Krishnan, R, J Colloid Interface Sci 71, 321 (1979); 75, 476 (1980) ; 76, 188, 201 (1980)
17. Overbeek, J. Th. G., Farraday Disc. Chem. Soc 65, 7 (1978)
18. Overbeek, J. TH. G., de Bruyn. P.L. and Verhoeckx, F., in "Surfactants" Editor Tadros, Th. F., Academic Press (in Press)
19 Percus, J. K. and Yevick, G. J., Phys. Rev. 110, 1 (1958)
20. Carnahan, N. F. and Starling, K. E., J Chem Phys. 51, 635 (1969).
21. Agterof, W. G. M., Van Zameren, J. A. J. and Vrij, A., chem Phys. Letters, 43, 363 (1976)
22. Oakenfull, D., J. Chem Soc. Faradey Trans I, 76, 1875 (1980)
23. Mitchell, D. J. and Ninham, B. W., J Chem Soc. Farradey Trans II, 77, 601 (1981)

24. Clausse, M., Peyrelasse, J., Boned, C., Heil, J., Nicolas-Margantine, L. and Zradba, A., in "solution properties of surfactants" Editor Mittal, K. L., Plenum Corporation (in Press)
25. Peyrelasse, J., Boned, C., Heil, J. and Clausse, M., J Phys C: Solid State Phys., 15, 7099 (1982)
26. Baker, R. C., Florence, A. T. and Tadros, Th. F., J. Colloid Interface Sci, submitted.
27. Kilpatrick, S., Mod. Phys. 45, 574 (1983)
28. Eicke, H. F. and Denss, A., in "Solution Chemistry of Surfactants, K. L. Mittal, editor, Vol 2 Plenum Press, New York (1979) p607
29. Lindman, B., and Wennerstrom, H., in "Topics in current chemistry" (F.L. Boschke Ed) 87, 1-83, Springer-Verlag, Heidelberg (1980).
30. Winnerstrom, H. and Lindman, B., Phys. Rep 52, 1 (1970)
31. Lindman, B., Kamenka, N., Kalhopoulis, T.M., Brumand, B, and Nilsson, P. G., J. Phys Chem 84, 1485 (1980)
32. Fabre, H., Kamenka, N., Lindman, B., J. Phys. chem. 85, 3493 (1981)
33. Stilbs, P., Mosely, M. E. and Lindman, B., J. Magn. Reson. Sci. 401 (1981)
34. Lindman. B., Stilbs, P. and Moseley, M.E., J. Colloid Inteface Sci. 83, 569 (1981)
35 Lang, J., Djavanbakht, A, and Zana, R. in "Microemulsions", Rubb, I.D., Editor, Plenum Press, New York (1982) pp 233-255.
36 Gracia, A., Fortney, L. N., Schechter, R. S., Wade, W. H. and Yiv, S., Soc, Pet. Eng. SPE 9815 (1980)
37 Barakat, Y., Fortney, L. N., Schechther, R. S., Wade, W. H., and Yiv, S. H., J. Colloid Inteface Sci 92, 561 (1983)

RECEIVED January 20, 1984

Reaction of N-Dodecyl-3-carbamoyl Pyridinium Ion with Cyanide in Oil–Water Microemulsions

LEONA DAMASZEWSKI and R. A. MACKAY[1]

Department of Chemistry, Drexel University, Philadelphia, PA 19104

The rate constants for the reaction of a pyridinium ion with cyanide have been measured in both a cationic and nonionic oil in water microemulsion as a function of water content. There is no effect of added salt on the reaction rate in the cationic system, but a substantial effect of ionic strength on the rate as observed in the nonionic system. Estimates of the ionic strength in the "Stern layer" of the cationic microemulsion have been employed to correct the rate constants in the nonionic system and calculate effective surface potentials. The ion-exchange (IE) model, which assumes that reaction occurs in the Stern layer and that the nucleophile concentration is determined by an ion-exchange equilibrium with the surfactant counterion, has been applied to the data. The results, although not definitive because of the ionic strength dependence, indicate that the IE model may not provide the best description of this reaction system.

A pseudophase ion exchange model has been applied to reactions in micellar systems with varying success ($\underline{1}$-$\underline{7}$). According to this model, the distribution of nucleophile is considered to depend on the ion-exchange equilibrium between the nucleophile and the surfactant counterion at the micelle surface. This leads to a dependence on the ion-exchange constant (K_{IE}) as well as on the degree of dissociation (α) of the surfactant counterion. The ion exchange (IE) model has recently been extended to oil in water microemulsions ($\underline{8}$).

[1]Current address: Chemical Branch, Research Division, Chemical Research and Development Center, Aberdeen Proving Ground, Aberdeen, MD 21010

0097–6156/84/0253–0175$06.00/0

Ion exchange constants have been calculated from kinetic data for the reaction of p-nitrophenyl diphenyl phosphate with hydroxide and fluoride in a cationic microemulsion, and from acid-base equilibrium data in an anionic microemulsion (8-9) using the ion exchange model for microemulsions. Although the kinetic data can be fit to the IE model, they can also be explained by means of the effective surface potential (ESP) model which considers the local nucleophile concentration to be a function only of α . Thus, if the ESP model is valid, attempts to fit kinetic data to the IE model should produce similar values of the ion-exchange constant (K_{IE}) for any nucleophile. This was in fact observed for the above reaction. However, hydroxide and fluoride have similar ion exchange constants with respect to bromide. Therefore, a reaction involving a nucleophile having a significantly different ion exchange constant should provide a more definitive test of the model. Unfortunately, attempts to react nucleophiles such as cyanide or thiophenoxide with the phosphate ester used in the above studies were unsuccessful. Thus, it was decided to employ a different substrate, and to that end we have followed the reaction of interphase-bound N-dodecyl 3-carbamoyl pyridinium bromide with cyanide ion at pH $\geq$ 11.8 (equation 1) in a cationic and in a nonionic microemulsion.

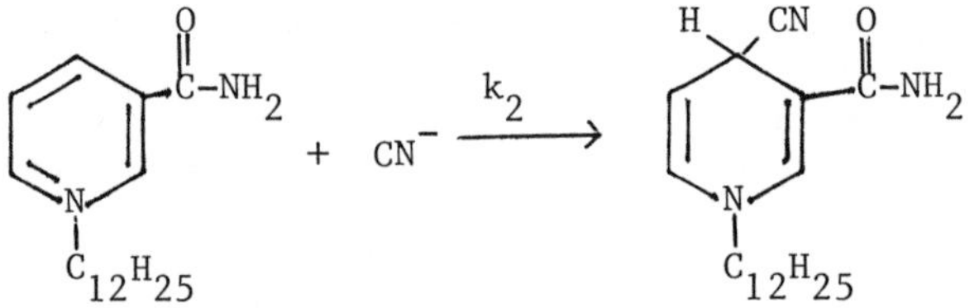

$$ \tag{1} $$

This reaction has been used as a test of the ion-exchange model in micelles (2), and the value of K_{IE} for cyanide differs from that of fluoride or hydroxide by about an order of magnitude.

Experimental

Systems and materials. The reaction was carried out at several compositions in an ionic and in a nonionic system. The ionic system consisted of an emulsifier (49.6 wt % cetyltrimethyl ammonium bromide (CTAB)/50.4% n-butanol), hexadecane, and water. The nonionic emulsifier consisted of 65.7% polyoxyethylene (10) oleyl ether (Brij 96) and 34.4% n-butanol, again with hexadecane and water. In both systems, microemulsion (µE) compositions used were obtained by diluting an initial 90 weight percent (%) emulsifier/10% oil mixture with varying amounts of water. Microemulsion samples thus obtained had final compositions of 30 to 80% water. Phase maps describing these systems have been published (10-11).

     N-dodecyl 3-carbamoyl pyridinium bromide (I) was prepared
from vacuum-distilled n-dodecyl bromide (K&K) and nicotinamide
(Kodak) according to a published procedure (15) and identified
by IR spectrum and melting point (218-220°C). The molar extinction coefficient of (I) at 264.5 nm was determined as 4120 $m^{-1} CM^{-1}$
(methanol), 4250 (water), and 4220 (60% water CTAB μE). Fisher
technical grade CTAB was recrystallized before use (16). Brij 96,
obtained from Sigma, was neutralized and deionized before use
(17) to remove buffering impurities. Aldrich reagent grade KCN,
Aldrich 99% hexadecane, Aldrich gold label n-butanol (99 + %),
and Fisher certified 0.200 N and 0.0200 N NAOH were used as
received.

Kinetics. The reaction of N-dodecyl 3-carbamoyl pyridinium
bromide (I) with cyanide ion in the microemulsions was observed
by following the 340 nm absorption maximum of the 4-cyano adduct
(II). See equation (1) . Following the work of Bunton, Romsted
and Thamavit in micelles (2), a 5/1 mole ratio of KCN to NaOH
was employed to prevent cyanide hydrolysis. The pH of each
reaction mixture was measured on a Coleman 38A Extended Range pH
meter to insure that the system was sufficiently basic to allow
essentially complete ionization of the cyanide. The appropriate
amounts of cyanide and hydroxide were added to the microemulsion
sample within 10 minutes of running a reaction. Cyanide concentration varied between 0.02 and 0.08 M with respect to the water
content. Substrate was injected via a Unimetrics model 1050
syringe directly into a known volume of the μE-nucleophile
mixture in a 1.0 cm UV quartz cell. Absorbance at 340 nm was
followed as a function of time on a Perkin-Elmer model 320
spectrophotometer at 25.0 $\pm$ 0.3°C. Since the initial bulk concentration of substrate was 10^{-4}M, cyanide was always present in
considerable excess.

     The reaction is reversible (12-13). However, it was possible
to monitor approximately 60 percent of the reaction without any
significant contribution from the reverse reaction. At most μE
compositions, infinite-time absorbance readings could be obtained
by the addition of small amounts of solid KCN. For those
compositions which could not be forced to completion, the average
value of the available final (infinite time) absorbance measurements in the μE system was used with only a small decrease in the
accuracy of the results. It is known that rearrangement (12-13)
or hydrolysis (14) of the product (II) may occur in various media.
In our experiments, however, these secondary reactions did not
interfere with observation of the reaction of interest over the
time span involved. Also, hydroxide was not found to react
appreciably with the substrate (I) at the concentrations employed.

     As a check on the method, the reaction was run in aqueous CTAB
micelles. The results obtained were in agreement ($\pm$ 5%) with

those previously reported ($\underline{2}$, $\underline{15}$). It should also be noted that the molar extinction coefficient which corresponds to our infinite-time absorption measurements in the above CTAB (7100 $M^{-1}cm^{-1}$) and Brij (6900 $M^{-1}cm^{-1}$) μE systems corresponds closely to the extinction coefficient (7120 $M^{-1}cm^{-1}$) obtained for the hexadecyl analog of ($\underline{II}$) in CTAB micelles ($\underline{15}$).

At each composition, at least three separate runs were performed. In most cases, the reported rate constant is the average of four to five runs and up to eight runs were carried out at some compositions. In all cases, the (n-1) standard deviation was less than 15%, and most often 10% or less.

Effect of Ionic Strength. Both μE systems were examined for ionic strength effects. Microemulsion compositions were prepared at 70% water, with a cyanide concentration of 0.032 M with respect to the water content. Potassium bromide was used to vary the ionic strength of the reaction mixtures. Ionic strength in the CTAB μE was varied from 0.04 to 0.34. Since the Brij μE tolerated a much higher salt concentration without phase separation, ionic strength in that system was varied between 0.04 and 1.80. As will be seen, the Brij system exhibits a salt effect, while the CTAB μE does not. Rate constants obtained for reaction (1) in the Brij μE were therefore corrected to take into account the effect of ionic strength in that system (vide infra).

Results

Kinetics. With nucleophile present in large excess, pseudo-first-order rate constants were obtained for reaction (1) in both μE systems employed. As already noted, the reverse and the possible successive reactions did not interfere with observation of the reaction of interest over the observed time span. The levels of hydroxide ion employed held the pH in the reaction mixtures at 11.8 or greater, thus insuring essentially complete ionization without itself producing any significant reaction.

The reaction is second order overall. The second order rate constant corrected for phase distribution of cyanide ion, $k_{2\phi}$, is given by $k_1/(CN^-)_w$, where k_1 is the observed pseudo-first-order rate constant and $(CN^-)_w$ is the concentration of cyanide ion with respect to the water content. Polarographic data on the N-dodecyl-3-carbamoyl pyridinium ion in μE shows that the substrate is completely associated with the hydrophobic region ($\underline{18}$), making it unnecessary to consider the effect of phase distribution of substrate on the measured reaction rate. The kinetic results are summarized in Table I. The microemulsion composition is expressed as phase volume $\phi = 1 - wg$, where w is the weight fraction of water and g is the specific gravity of the μE.

Table I.   Phase volume corrected second order rate constants
($k_{2\phi}$) as a function of phase volume (ϕ) for the
reaction of cyanide with N-dodecyl-3-carbamoyl-
pyridinium ion in CTAB and Brij 96 microemulsions

ϕ	CTAB[a]	BRIJ[b]
0.73	0.067	1.10
0.63	0.077	0.67
0.53	0.094	0.67
0.43	0.13	0.55
0.32	0.18	0.54
0.23	0.29	0.53

a. $M^{-1}s^{-1}$. Average standard deviation $\pm$ 10%.

b. $M^{-1}s^{-1}$. Average standard deviation $\pm$ 12%.

As expected, $k_{2\phi}$ increases by a factor of 4 with decreasing
phase volume in the CTAB μE over the composition range studied
(Figure 1). In the Brij system $k_{2\phi}$ quickly becomes approximately
linear (within experimental accuracy) at the various compositions.
An anomalously high value in the Brij μE occurs at ϕ = 0.72. In
this high-emulsifier region, however, the μE may have a different
structure.

The average rate constant in the nonionic μE is, however,
considerably higher than all the rate constants in the cationic
system, rather than lower. As discussed above, this is due to
the effect of ionic strength on the reaction rate and a correction
must be made for it if a meaningful comparison of the data is to
be made. The dependence of the rate constant $k_{2\phi}$ on the ionic
strength (I) is shown in figure 2. Within experimental error,
$-\log k_{2\phi}$ = 0.78 I + 0.16. In order to estimate the value of I
to employ for the correction, it was assumed that the reaction
took place in an outer shell (e.g. Stern Layer) of thickness s
using a droplet radius of 34A and corrected for the fraction of
free counterion α. The value of I thus obtained varied with
water content since α varies with ϕ. However, the variation was
not significant, and an average value of I of 3.0 (s = 4A) and
6.0 (s = 2A) was used. The corrected values of $k_{2\phi}$ for Brij were
0.03 and 0.01 $M^{-1}s^{-1}$ respectively (Figure 1).

Reaction models. The effective surface potentials (ψ_ϕ) as defined
by equation (2) are given in Table II, along with the values of α.
The values of ψ_ϕ for the reaction of p-nitrophenyl diphenyl

$$k_{2\phi} \text{ (CTAB)}/k_{2\phi} \text{ (Brij, corrected)} = \exp (e\psi_\phi/kT) \qquad (2)$$

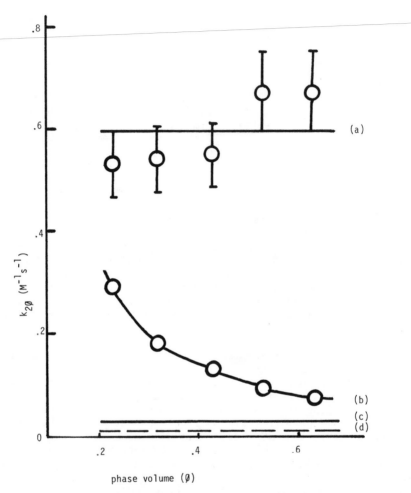

Figure 1. Phase volume corrected rate constant ($k_{2\phi}$) vs. phase volume (ϕ) for the N-dodecylnicotinamide-cyanide reaction in (a) Brij μE and (b) CTAB μE. Curves (c) and (d) are the ionic strength corrected Brij rate constants for Stern layer thicknesses of 4A and 2A, respectively (vide text).

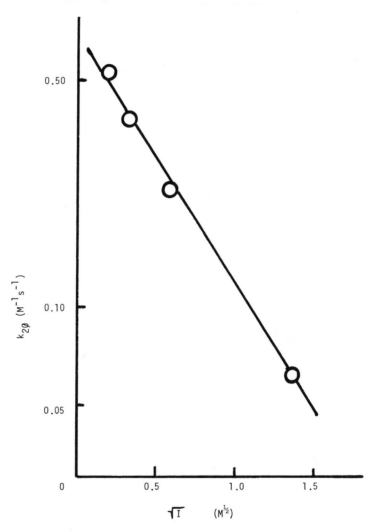

<u>Figure 2.</u> Dependence of the rate constant ($k_{2\phi}$) for the cyanide
reaction (c.f. figure 1) in Brij μE at a phase volume $\phi = 0.32$
on ionic strength (I) adjusted by means of added potassium
bromide.

phosphate (PNDP) are also presented for purposes of comparison. It is immediately evident that the values of ψ_ϕ using s = 4A for

Table II. Effective surface potential (ψ_ϕ) and degree of dissociation (α) as a function of phase volume (ϕ)

ϕ	α^a	$(CN^-)^{b,c}$	$(CN^-)^{b,d}$	$(OH^-)^e$	$(F^-)^e$
.63	.10	48	20	26	30
.53	.12	53	25	30	33
.43	.14	66	38	34	36
.32	.16	76	48	41	43
.23	.18	88	60	49	54

a. reference 18.
b. substrate N-dodecyl-3-carbamoyl pyridinium bromide. (ψ_ϕ in mV. $\pm$ 22%).
c. s $\doteq$ 2A°.
d. s = 4A°.
e. substrate PNDP (ψ_ϕ in mV, $\pm$ 20%).

cyanide are comparable to those for fluoride and hydroxide, and thus consistent with the ESP model. The values for s = 2A are higher, as would be expected based on the IE model. Additional insight may be obtained from an attempt to fit the data to equation (3), obtained from the microemulsion IE model (8).

$$\psi_\phi = \frac{kT}{e}\left\{\ln\left[\left(\frac{1-\phi}{\phi}\right)\left(\frac{K_\phi}{1+K_\phi}\right)\right]+ \ln\frac{r}{3s}\right\} \qquad (3)$$

Here, $K_\phi = K_{IE}(1-\alpha)/\alpha$, kT/e = 25.6 mV at 25°C, and all other symbols have their previously defined meanings. A plot of ψ_ϕ vs $\ln\left[(1-\phi)/\phi\right]\left[K_\phi/(1+K_\phi)\right]$ using the data in Table II is shown in Figure 3 for three different values of K_{IE}. The values of the slope and the value of s calculated from the intercept using the theoretical slope of 25.6 mV and a radius of 34A are given in Table III.

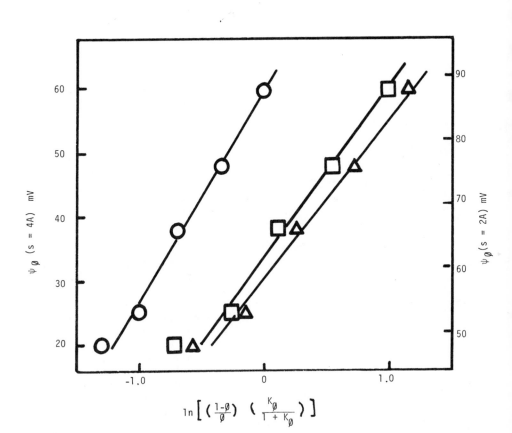

$$\ln \left[\left(\frac{1-\phi}{\phi} \right) \left(\frac{K_\phi}{1 + K_\phi} \right) \right]$$

Figure 3.   Ion-exchange model plot (vide text). The data are for values of K_{IE} of 0.1 (circles), 1.0 (squares), and 10 (triangles).

Table III. Values of the slope and reaction region thickness (s)
for the values of K_{IE} from figure 3.

K_{IE}[a]	Slope (mv)	Thickness (A)[b]	
		S = 2A	S = 4A
0.1	34	0.4	1.1
1.0	26	1.0	3.0
10	25	1.2	3.5

a. $K_{IE} = (CN^-)_{free} \; (Br^-)_{Stern} / (CN^-)_{Stern} \; (Br^-)_{free}$

b. Determined using values of ψ_ϕ obtained from the ionic
strength corrected Brij $k_{2\phi}$ values, where I was estimated
assuming a thickness s = 2A or 4A.

The values of s obtained from the higher set of ψ_ϕ values
are unrealistically small. The values of the slope and s for
K_{IE} = 1 using the lower ψ_ϕ values are acceptable, but again this
set of ψ_ϕ values is also in accord with the ESP model.

The major difficulty here is of course the ionic strength
effect. It is clear that a reasonably definitive test of the
model will require a reaction which (a) is relatively insensitive
to ionic strength and (b) can employ the same oil soluble substrate with reactant ions of widely varying K_{IE}, preferably over
two orders of magnitude. It should be noted that the pyridinium
ion substrate is more water soluble than the ester, and in addition has a positive charge. This may result in the protrusion of
part of the head group out of the Stern layer, which in turn
means that the nucleophile concentration should be governed by the
ESP model. There is precedent for this type of behavior in the
reaction of an alkylated pyridine derivative with hydroxide in an
anionic system (19). A neutral, oil soluble substrate such as
the ester may be more likely to react in the Stern layer. This
is illustrated in Figure 4. There may also be substrate –
nucleophile systems which are distributed such that they exhibit
behavior intermediate between the two extremes of the IE and ESP
models.

Conclusion

The rate constants for the reaction of N-dodecyl-3-carbamoyl-
pyridinium ion with cyanide in both cationic and nonionic o/w
microemulsions have been measured as a function of phase volume.
Added salt has no effect in the cationic system, but the rate
constants in the nonionic system depend upon ionic strength as
would be expected for a reaction between two ions. In order to
compare the two microemulsions, the ionic strength in the reaction
region has been estimated using thicknesses of 2-4A. The former
produces values of the effective surface potential which yield

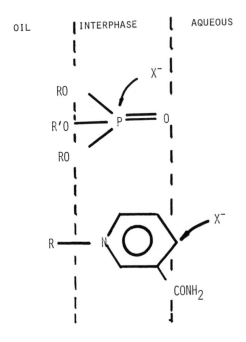

Figure 4. Schematic diagram illustrating the attack of an anionic nucleophile on a substrate located in an interface with a net positive charge, where the attack occurs (a) in the Stern layer or (b) in the double layer.

small s values, while the latter yield reasonable s values but
are also in accord with the ESP model. It must be concluded that
at present there is no compelling evidence in favor of the IE
model. Definitive experiments must utilize the employment of
substrate – nucleophile combinations which meet the criteria
discussed above.

Acknowledgment. The authors wish to thank Dr. N. S. Dixit for
valuable technical assistance, and the US Army Research Office
for financial support.

Literature Cited
1. Chaimovich, H.; Bonilha, J.B.S.; Politi, M.J.; Quina, F.H.
    J. Phys. Chem. 1979, 83, 1851.
2. Bunton, C.A.; Romsted, L.S.; Thamavit, C. J. Am. Chem. Soc.
    1980, 102, 3900.
3. Quina, F.H.; Politi, M.J.; Cuccovia, I.M.; Baumgarten, E.;
    Martins-Franchetti, S.M.; Chaimovich, H. J. Phys. Chem. 1980,
    84, 361.
4. Bunton, C.A.; Romsted, L.S.; Sepulveda, L. ibid, 1980, 2611.
5. Bunton, C.A.; Romsted, L.S.; Savelli, G. J. Am. Chem. Soc.
    1979, 101, 1253.
6. Bunton, C.A.; Frankson, J.; Romsted, L.S. J. Phys. Chem.
    1980, 84, 2607.
7. Rupert, L.A.M.; Engberts, J.B.F.N. J. Org. Chem. 1982, 47,
    5015.
8. Mackay, R.A. J. Phys. Chem. 1982, 86, 4756.
9. Mackay, R.A., in proceedings of 4th International Symposium
    on Surfactants in Solution, Lund, Sweden, 1982. Plenum Press,
    Mittal, K.L., ed, in press.
10. Hermansky, C.; Mackay, R.A., in "Solution Chemistry of Sur-
    factants", Mital, K.L., ed, Plenum Press, NY 1979.
11. Hermanski, C.; Mackay, R.A. J. Coll. Interface Sci., 1980,
    73, 324.
12. Bruice, T.C.; Benkovic, S.J. "Bioorganic Mechanisms",
    Chapter 9, W.A. Benjamin, New York, 1966.
13. Lindquist, R.N.; Cordes, E.H. J. Am. Chem. Soc. 1968, 90,
    1269.
14. Rappaport, Z., in "Chemistry of the Cyano Group", Patai, S.,
    ed., Interscience, New York, 1970.
15. Baumrucker, J.; Calzadilla, M.; Centeno, M.; Lehrmann, G.;
    Urdaneta, M.; Lindquist, P.; Dunham, D.; Price, W.; Sears, B.;
    and Cordes, E.H. J. Am. Chem. Soc., 1972, 94, 8164.
16. Mukerjee, P.; Mysels, K.L. J. Org. Chem. 1955, 77, 2937.
17. Mackay, R.A.; Hermansky, C. J. Phys. Chem. 1981, 85, 739.
18. Mackay, R.A., in "Microemulsions", Robb, I.D. ed., Plenum
    Press, London, 1982.
19. Mackay, R.A.; Dixit, N.S. and Agarwal, R. "Inorganic
    Reactions in Microemulsions", ACS Symposium Series No. 177,
    Holt, S.L. ed., 1982, 179-194.

RECEIVED January 20, 1984

EFFECT OF STRUCTURE
ON PERFORMANCE
IN VARIOUS APPLICATIONS

Interactions of Nonionic Polyoxyethylene Alkyl and Aryl Ethers with Membranes and Other Biological Systems

ALEXANDER T. FLORENCE, IAN G. TUCKER[1], and KENNETH A. WALTERS[2]

Department of Pharmacy, University of Strathclyde, Glasgow G1 1XW, Scotland

Many nonionic surfactants of the poly(oxyethylene) alkyl and aryl ether class interact with biological membranes increasing their permeability and causing increased trans-membrane solute transport. The mechanisms of such effects are not fully understood. Studies of the interaction of homologous series of nonionic surfactants (varying either in hydrophobic or hydrophilic chain length) with a variety of biological substrates, reviewed here, indicate there to be an optimal lipophilicity for maximal membrane activity. Problems with the hydrophile-lipophile balance (HLB) as an index of lipophilicity result from the structural non-specificity of this value. The biological effects of surfactants are concentration dependent and structure dependent. In many cases C_{12} hydrocarbon chain compounds appear to exert maximal effects and in all series a parabolic relationship between membrane activity and lipophilicity is observed. The effects are complex resulting from penetration of the membrane, its fluidization and, at high surfactant concentrations, solubilization of structural components.

Many formulated food and pharmaceutical products contain nonionic surfactants as emulsifiers, suspending, wetting or solubilizing agents (1). As nonionic surface active agents may have intrinsic biological activity or may influence the bioligical activity of other molecules the optimisation of formulations cannot be carried out without consideration of the biological

[1]Current address: C.S.I.R.O. Division of Animal Production, Box 239, Blacktown 2148, Australia

[2]Current address: Fisons Pharmaceuticals plc, Loughborough, England

0097–6156/84/0253–0189$06.00/0

properties of the surfactants and indeed the whole formulation.

It has been recognised for some time (see for example reference 1), that surfactants can increase the rate and extent of transport of solute molecules through biological membranes by fluidisation of the membrane. It is only recently, however, that sufficient work has been carried out to allow some analysis of structure-action relationships. In this overview an attempt is made, by reference to our own work and to work in the literature, to define those structural features in polyoxyethylene alkyl and aryl ethers which give rise to biological activity, especially as it is manifested in interactions with biomembranes and subsequent increase in the transport of drug molecules. This paper is largely confined to the alkyl and aryl polyoxyethylene ethers, as these form series variable in both hydrophobic and hydrophilic chain length, and hence HLB, with minimal structural variation.

Structure-activity relationships are generally applied in the pharmaceutical sciences to drug molecules. The value of any structure-activity correlation is determined by the precision of the biological data. So it is with studies of the interaction of nonionic surfactants and biomembranes. Analysis of results is complicated by the difficulty in obtaining data in which one can discern small differences in the activity of closely related compounds, due to i) biological variability in tissues and animals, ii) potential differential metabolism of the surfactants in a homologous series (2), iii) kinetic and dynamic factors such as different rates of absorption of members of the surfactant homologous series (2) and iv) the typically biphasic concentration dependency of nonionic surfactant action (3). Few, if any, of the studies of the biological properties of nonionic surfactants have been carried out with purified non-ionic poly(oxyethylene) ethers, rendering comparison of work from different laboratories somewhat difficult. Nevertheless it is clear from the work reviewed here that in homologous series of surfactants with constant ethylene oxide chain length, there will be found an optimal hydrocarbon chain length for biological activity, most often C_{12}, and given a constant hydrophobic chain length, an optimal hydrophilic chain length will be observed. (Figure 1a,b). As in most work on structure-activity relationships of drug series a parabolic activity curve is also obtained with nonionic surfactant series, (Figure 1c), so it appears that data on surfactant activity fits conventional wisdom. However, with drug molecules increasing biological activity is always found with increasing lipophilicity (log P) until beyond the maximum (log P_0) a decrease in activity is brought about by limitations in solubility, protein binding, drug association etc. Not all of the surfactants considered in this paper are freely soluble, e.g. the E_2 derivatives are dispersible at the 0.1% level, and $C_{16}E_{10}$ and $C_{18}E_{10}$ have solubilities below 1%.

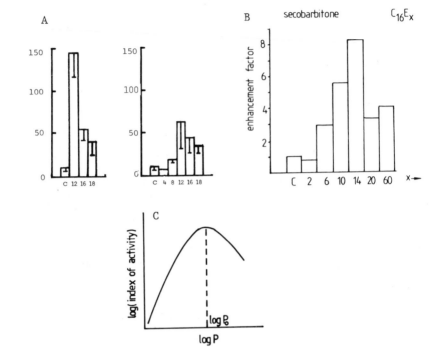

Figure 1.

a) The effect of hydrocarbon chain length of (left) 1% C_xE_{10} and (right) 1% C_xE_{20} on paraquat transport through isolated rabbit gastric mucosa. (Walters, Dugard and Florence, 1981).

b) The increase in the absorption of secobarbital in goldfish in the presence of 0.1% $C_{16}E_x$ alkyl ethers as a function of ethylene oxide chain length, x. (Walters, Florence and Dugard, 1982a).

c) An idealised log (activity) - log P profile showing optimal biological activity at a partition coefficient of P_o.

However the reduction in biological activity with increasing
hydrocarbon chain length beyond the optimum (Figure 1a) and the
increase in biological activity on initially increasing the
hydrophilic nature of the surfactants (Figure 1b) requires
elucidation, especially at lower concentrations.

Studies with homologous series of alkyl and aryl polyoxyethylene
ethers

In this section several recently published studies on the
interaction of nonionic surfactants with a variety of biological
systems, including enzymes, bacteria, erythrocytes, leukocytes,
membrane proteins, low density lipoproteins and membranes
controlling absorption from the gastrointestinal tract, nasal
and rectal cavities, will be assessed. This is a selective
account, work having been reviewed that throws light on structure-
activity relationships and on mechanisms of surfactant action.
        Reference will be made to the hydrophile-lipophile balance
(HLB) of the surfactants studied as a convenient index of
hydrophilicity. Early attempts to observe correlations between
activity and HLB include those by Marsh and Maurice (4) and
Florence and Gillan (5, 6).   Marsh and Maurice found that the
compounds most efficient in effecting increased penetration of
fluorescein into the anterior chamber of the eye had HLB values
in the range 16-17, but several surfactants with HLB numbers
in that range were inactive or poorly active. Florence and
Gillan (5)   proposed from work on the goldfish (Carassius
auratus) gill membrane that "bulky" surfactants had low membrane
activity and suggested that there might be a physical blocking
mechanism as a result of the adsorption at the membrane surface
of long chain ethylene oxide derivatives.
        The difficulty with HLB as an index of physicochemical
properties is that it is not a unique value, as the data of
Zaslavsky et al. (7) on the haemolytic activity of three
alkyl mercaptan polyoxyethylene derivatives clearly show in
Table 1. Nevertheless data on promotion of the absorption of
drugs by series of nonionic surfactants, when plotted as a
function of HLB do show patterns of behaviour which can assist
in pin-pointing the necessary lipophilicity required for optimal
biological activity. It is evident however, that structural
specificity plays a part in interactions of nonionic surfactants
with biomembranes as shown in Table 1. It is reasonable to
assume that membranes with different lipophilicities will "require"
surfactants of different HLB to achieve penetration and
fluidization; one of the difficulties in discerning this optimal
value of HLB resides in the problems of analysis of data in the
literature. For example, Hirai et al. (8) examined the
effect of a large series of alkyl polyoxyethylene ethers (C_4, C_8,
C_{12} and C_{12} series) on the absorption of insulin through the nasal
mucosa of rats. Some results are shown in Table II.

Table I.   Haemolytic activity of some nonionic surfactants

$R_mS(OCH_2CH_2)_nOH$			
m	n	HLB	C*
8	8	12.6	10.7
10	10	12.6	30.0
12	12	12.6	2.9

from B. Yu Zaslavsky <u>et al</u>. (1978) <u>Biochim.Biophys.Acta,507</u>,1.

*concentration required to produce 50% haemolysis compared with concentration of standard $R_{12}S(OCH_2CH_2)_{7.8}OH$ required to do the same.

Table II.   Effect of 1% nonionic surfactants on nasal absorption of insulin in rats

Surfactant	HLB	n	D%*
Control	–		6.0 ± 0.9
$C_{12}E_5$	8.6	5	59.3 ± 3.0
$C_{12}E_9$	11.5	9	60.9 ± 2.9
$C_{12}E_{10}$	12.1	4	63.3 ± 2.5
$C_{12}E_{20}$	15.5	5	50.8 ± 2.3
$C_{16}E_5$	7.2	4	13.5 ± 3.6
$C_{16}E_{10}$	10.6	4	64.0 ± 1.3
$C_{16}E_{20}$	14.1	4	54.6 ± 2.4

from S. Hirai <u>et al</u>. <u>Int. J. Pharmaceutics</u>, 1981, <u>9</u>, 165.
*D is the percentage decrease in glucose levels. n = number of experiments.

In the C_{12} series optimum activity is seen with the $C_{12}E_{10}$ compound (HLB 12.1), whereas in the C_{16} series the peak activity is achieved with $C_{16}E_{10}$ whose HLB is only 10.6. This might be explained by the fact that the $C_{16}E_{14}$ compound with an HLB of 12.8 was not included in the study. These data when combined produces impressive corroboration of the previously suggested parabolic behaviour   (see Figure 2).

Results with some esters are included in Figure 2. The activity of the esters is generally lower possibly due to their rapid metabolism (hydrolysis) in the mucosal layer. Also it is evident on close examination of the data that (as with the data

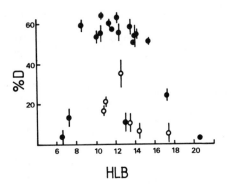

Figure 2.

Relationship between HLB values of nonionic surfactant

ethers and esters and the nasal absorption of insulin

(10U/kg) in rats measured as a percentage reduction (D)

in glucose levels from 0-4h.   Surfactant applied at a

concentration of 1%.   ● ethers, O esters.   Data

redrawn from Hirai et al. (1982a) as mean datum points

± SE n = 4-9.

of Marsh and Maurice (4) some apparent anomalies occur in the
ether series. The low activity of two surfactants with HLB
values of 13.7 and 16.8 represent C_4 compounds with short
ethylene oxide chains; this is in agreement with the previously
demonstrated lack of activity of short hydrocarbon chain analogues.
(Figure 1).

The promotion of insulin absorption from the rectal cavity
has been studied by Touitou et al. (9). Figure 3 illustrates
the marked influence of ethylene oxide chain length in the C_{12}
series and of alkyl chain length in the C_xE_9 series. This
striking demonstration of the structural dependence of activity
is indicated with the loss of activity when the $C_{12}E_{25}$ and $C_{12}E_{40}$
compounds are used. Here $C_{12}E_9$ appears to be the most active
of the series in promoting insulin absorption, although time
dependent effects make it appear (see Figure 3) that another
compound ($C_{16}E_9$) is more active.   This may point to
differences in the rate of absorption and interaction.

Several studies indicate the special nature of the dodecyl
chain in interactions with a variety of biomembranes: Zaslavsky's
(7) data on the haemolysis caused by alkyl ethers, our own
data (10) on the absorption of paraquat by gastric mucosa,
Walters and Olejnik's (11) data on methyl nicotinate transfer
through hairless mouse skin. All indicate maximal activity
residing with the C_{12} ether. Not only do the latter two studies
indicate a falling off in effectiveness as the chain length
is increased to C_{16} and C_{18} but they also indicate that the
oleyl (unsaturated chain) ethers are more active than their
saturated analogues.

In some biological systems nonionic surfactants have an
intrinsic biological activity; the C_{12} alkyl ethers were too
toxic to be used in the experiments of drug absorption with
goldfish. The activity of the C_{12} ethers was quantified by
measurement of the fish turnover time, T. When the reciprocal
of the turnover time is plotted against alkyl chain length for
the series C_x and E_{10} and C_{12} compound is distinguished by its
marked effect. (12).

In all of these data what requires explanation is the
effectiveness of the dodecyl chain, the decreasing activity
with increasing ethylene oxide chain length above E_{10-14} the
increase in activity when we move from a very hydrophobic
surfactant with short ethylene oxide chain to the optimum,
and the decrease in activity with increasing lipophilicity of
compounds with alkyl chain lengths greater than C_{12}.

Penetration of the biomembrane is undoubtedly essential for
most membrane activity.   Araki and Rifkind (13) obtained esr
spectra of stearic acid spin labelled erythrocyte membranes in the
presence of diverse compounds including Triton X100, chlorpromazine
and glutaraldehyde. The two surfactants chlorpromazine and Triton
X100 both increase the rate of haemolysis and are shown to increase
membrane fluidity. Glutaraldehyde as expected decreases fluidity and
decreases the rate of haemolysis.

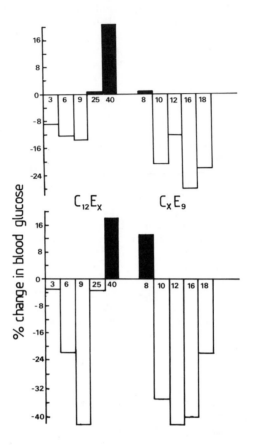

Figure 3.

The influence of hydrophilic chain length and alkyl

chain length of a series of alkyl ethers ($C_{12}E_x$ and C_xE_9

respectively) on the percentage change in blood glucose

on administration rectally of insulin with the surfactants.

The upper plot shows the results 30 minutes after

administration and the lower plot results after 60 minutes.

(Redrawn from data in Touitou et al, 1978).

Surfactant lipophilicity

Penetration of the surfactant demands a certain minimum
lipophilicity. Partitioning of nonionic surfactants into
simple organic phases has been measured for a limited range of
nonionics. Crook et al. (14) and Harusawa et al. (15, 16)
have studied the partition coefficients (P) of homogeneous octyl
phenyl ethers and nonyl phenyl ethers, the former between isooctane
and water at 25° and the latter between cyclohexane and water.
Harusawa et al. (16)   observe, assuming a linear relationship
between log P and ethylene oxide chain length (n):

$$\log P = 5.19 - 0.418n$$

Using this relationship one can calculate log P values of the
nonyl phenyl ethoxylates used by Cserhati et al. (17) in their
study on the effect of the surfactants on the growth of Coronilla
rhizobium and Bacillus subtilis var niger and interaction with
dipalmitoyl phosphatidylcholine liposomes containing tracer
amounts of $42KCl$.   Results on growth inhibition of the organisms
and efflux of $42K$ from the liposomes are plotted as a function of
log P in figure 4.
     These results are obtained with a constant hydrophobic head
group, whereas most activity - log P plots in biological (drug)
systems are derived by variation of this head group. The nature
of the membrane interacting with the surfactant will play a role
in determining the optimal lipophilicity, log P_0; peak activity
is noted in C. rhizobium at log P_0 = 1.0, in the liposomes at
log P_0 = 1.55 and in B. subtilis at 1.90. While such shifts
can indicate the differing degrees of hydrophobic nature of the
target membrane they can also reflect differences in the structure
of the barrier membrane. (18).

Concentration dependence of effects

Fig. 5 shows the concentration dependence of the effect of three
alkyl polyoxyethylene ethers, $C_{16}E_2$, $C_{16}E_{10}$ and $C_{16}E_{20}$, on the
absorption of secobarbitone by the goldfish. The latter two
compounds are active at very low temperatures. No obvious change
in activity occurs at the critical micelle concentrations of the
surfactants. Other data appended to the graph are the areas/
molecule of the surfactant at the air water interface (27,110
and 148Å2 respectively) and calculated log P values using π values
from Hansch (19).   Calculated values are only valid below the
CMC as Harusawa's (16) data shows.   It must be assumed,
therefore, that because of the lack of effect of $C_{16}E_2$ on this
system at any concentration in spite of its extreme lipophilicity,
a certain degree of amphipathicity is required for activity. This
may be because the site of action of the surfactant is not directly

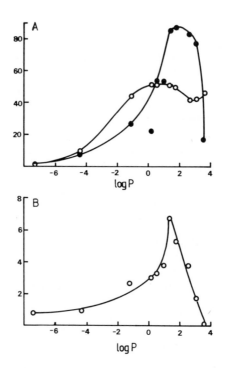

Figure 4.

Results of Cserhati et al. (1982) plotted as a function
of the calculated log P of the surfactants used A) in
determining the % reduction in growth of ● B. subtilis
var niger and O C. rhizobium and B) the change in efflux
of ^{42}K from liposomes prepared from dipalmitoyl
phosphatidyl choline.

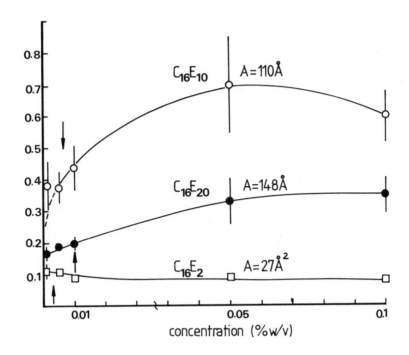

Figure 5.

Concentration dependent effects of surfactants $C_{16}E_2$, $C_{16}E_{10}$ and $C_{16}E_{20}$ on the reciprocal turnover times (T^{-1}, min^{-1}) of goldfish in the presence of secobarbital showing the areas/molecule of the surfactants at the air/water interface and calculated log P values for the surfactant molecules. Data from Walters, Florence and Dugard, (1982b).

on the lipid bilayer but may be mediated by adsorption and interaction with membrane proteins. $C_{16}E_2$ is however not freely
soluble and its dispersion rather than solution might complicate
interpretation. Tiddy has, in fact, suggested (20) that the
existence of lamellar mesophases in short ethylene oxide chain
species might account for the low biological activity. Mitchell
et al. (21) have studied the phase behaviour of a range of pure
C_mE_n compounds ranging from C_8 to C_{16}. Whereas normally they
appear only at high concentrations, mesomorphous L_α phases exist
over a large concentration range for $C_{16}E_4$. Extensive L_α phases
occur only with surfactants with less than 5 ethylene oxide units.
The L_α phase is extremely sensitive to ethylene oxide chain
length as $C_{12}E_4$ has an extensive lamellar phase at 37° while
$C_{12}E_5$ does not. The melting points of the L_α phase are shown
in Figure 6 as a function of hydrocarbon chain length and
hydrophilic chain length; there is an interesting structural
dependence here which might well throw light on the interactions
of the molecules with the lipid bilayer.

Evidence from studies on the penetration of cholesterol
monolayers by nonionic surfactants of two Brij series suggests
penetration occurs at extremely low concentrations
(22), the C_{12} compounds interacting at lower concentrations
than C_{18} compounds. (Figure 7).

Haemolytic activity and the permeability-enhancing activity
of nonionic alkyl ethers correlate well (23).
Ponder in his classic work on haemolysis (24) has
proposed that haemolysis by surfactants involves various stages
viz, i) approach of the surfactant, ii) contact with surface i.e.
adsorption, iii) reorientation of the molecule to allow film
penetration and iv) "complex" formation in the membrane bilayer.
This complex may either remain part of the altered surface
ultrastructure or may become detached and solubilized in the
external phase. Adsorption and membrane changes are essential
but do not necessarily follow. Dahlgren et al. (25) used a
series of fatty acid (C_2-C_{18}) esters of polyoxyethylene glycol
6000 to modulate the activity of polymorphonuclear leukocytes.
Binding first becomes apparent with the C_{12} derivative but as
expected increases with increasing hydrocarbon chain thereafter,
yet the C_{14} derivative was most effective in causing the release
of the superoxide anion, a measure of the activation of cell
metabolism. Superoxide production was negligible below C_{12}.

The question of optimal hydrocarbon chain length

In many of the experiments described C_{12} compounds have generally
proved to be most active when a series of surfactants with
identical hydrophilic tails are studied. Schott (26)
suggested that the particular effectiveness of the lauryl
surfactants is due to the balance of two properties. As the
homologous series is ascended the lipophilicity of the compounds

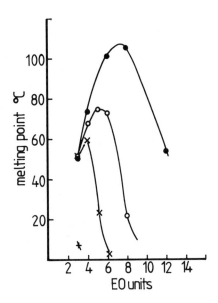

Figure 6.

The melting point of the L mesomorphic phase of alkyl ether nonionic surfactants taken from the results of Mitchell et al. (1983) showing the influence of alkyl and hydrophilic (EO) chain lengths. ● C_{16}; O C_{12}; × C_{10}; + C_8.

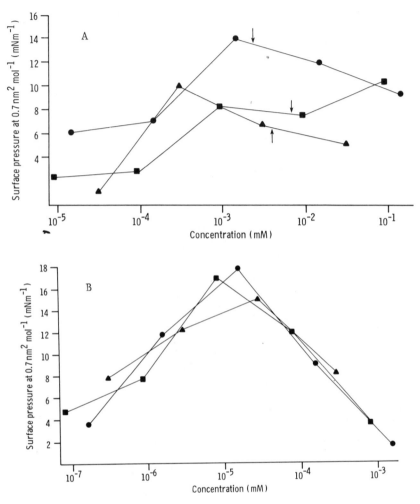

Figure 7.

a) Plot of surface pressure at constant molecular area
   against surfactant concentration. ▲, Brij 72 ($C_{18}E_2$);
   ● Brij 76 ($C_{18}E_{10}$), ■ Brij 78 ($C_{18}E_{20}$). Arrows
   denote the CMC for each surfactant.

b) Plot of surface pressure at constant molecular area
   against surfactant concentration. ▲ , Brij 30 ($C_{12}E_4$)
   ●, Brij 36T ($C_{12}E_{10}$, ■ , Brij 35 ($C_{12}E_{23}$).

increases, but the CMC decreases, thus limiting the concentration
of monomers which can exist in the aqueous phase in the case of
higher members of the series. From C_8 to C_{12} as the partition
coefficient increases there is an increased opportunity for the
surfactants to enter the biophase, whereas from C_{12} to C_{18}, while
the thermodynamic tendency to partition into non-aqueous
environments increases the decreasing concentration of monomers
(the active species of surfactant) may elicit a smaller response.
Adsorption of surfactant onto the high available surface areas
of membrane components can also influence behaviour. (27).
        Some workers have suggested that the lauryl chain
is of intrinsic biological importance in relation to its ability
to disrupt lipid bilayers, having the optimal physical properties
of lipophilicity and size, but as C_{12} compounds are also maximally
irritant to the skin   (28) where simple lipoidal barrier
membranes are probably not involved, other factors are no doubt
implicated. Dominguez et al. (29) have considered Schott's
(26)   approach to the biological uniqueness of the dodecyl chain,
but have postulated that its properties of skin penetration are
related to the conformation of the chain, especially when adsorbed
to or interacting with protein. Dominguez et al. postulate that
by adopting a compact configuration the dodecyl chain can migrate
deeper into skin structure and thereby be more active than more
lipophilic compounds. This is very speculative and requires
more experimental and theoretical study.
    Schott's arguments should apply to a homologous series in
which the hydrophilic chain length alters too. In Figure 8
experimentally determined CMC's of a series of nonionic surfactants
are employed with idealised diagrams of monomer concentrations as
a function of total surfactant concentration. When measurements
of surfactant activity in a biological system are made at 0.01%
concentration levels, the monomer concentrations will approximate
the CMC. It seems likely that at such low concentrations the
activity of the hydrophobic members of the series is limited by
micellisation, and in the case of $C_{16}E_2$, by low solubility.
Hydrophilic members of the series have very low values of P and
are unlikely to have much affinity for the membrane, in addition
to having a larger surface area.   If it is assumed that the
monomer is the active species, a maximum in activity is readily
shown by choice of partition coefficients such as those shown
for members of the homologous series. (3). But
these assumed values are very different from thos calculated
using the Hansch approach. It is likely then that partitioning
of the complete surfactant molecule into the membrane is not
necessary for activity.
    It could well be that many of the results of increased
transport rates are determined by membrane damage, rather than
by reversible physical effects when surfactants are used above
their CMC. Studies on the solubilization of protein have in
several cases demonstrated a remarkable parallel between
solubilizing ability and membrane action (30).

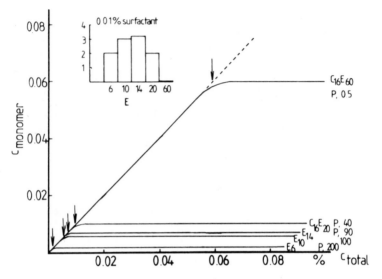

Figure 8.

Monomer concentrations as a function of total surfactant

$(C_{16}E_x)$ concentration. Experimental CMC's are marked

by arrows.   At 0.01% total surfactant levels, the monomer

concentrations will be approximately given by the CMC

values except for $C_{16}E_{60}$ whose CMC is 0.06..   Inset

shows absorption profile if values of partition coefficient

(P) shown apply.   From Florence (1982).

Model membranes: surfactant in low density lipoprotein interactions

According to Helenius and Simons (30) solubilization of the
membrane is preceded by saturation of the bilayer with surfactant.
Certainly critical surfactant/phospholipid ratios must be attained
before membrane disruption occurs.

Photon correlation spectroscopy was used to study the effects
of a series of nonionic surfactants on the Stokes radius (R) of
low density lipoprotein (LDL$_2$) particles (31, 32). LDL$_2$
interacted with surfactants in a manner similar to membranes.
That is, nonionic surfactants caused an increase in the Stokes
radius (R) of the particles due to penetration of the phospholipid surface layer and unfolding of apoprotein B molecules
leading to particle assymetry at molar ratios of surfactant LDL$_2$
of ca. 1000/1. At higher molar ratios, corresponding to 1-2
moles surfactant per mole of phospholipid, ionic surfactants and
nonionics with HLB values < 14.6 caused rapid decreases in the
Stokes radius due to breakdown of LDL$_2$ into the lipid surfactant
and protein surfactant micelles.

The data suggest that surfactant HLB is important in
determining ability to delipidate LDL$_2$ rapidly. Nonionic
surfactants with HLB values < 14.6 caused a fall in R which
can be interpreted as due to delipidation of LDL$_2$. Umbreit
and Strominger (33) found that surfactants with HLB values of
12.5 - 14.5 are most effective in solubilizing membranes. Thus
LDL$_2$ is responding like a biological membrane and this is
supported by the fact that the fall in R starts at approximately
one mole of surfactant per mole of phospholipid, a typical value
for membranes (30). However, kinetic data suggest that
surfactants with higher HLB values (e.g. C$_{12}$E$_{23}$, HLB 16.9)
might delipidate membranes over a prolonged period (20 hours).
The reason for these differences is not well known but it is
interesting to speculate. The lower HLB surfactants tested
are also those with the shortest hydrophilic chains, and so
the smallest molecular areas (10). They should therefore have
the highest concentrations at the LDL$_2$ surface at saturation
causing disruption and solubilization. The slow action of
C$_{12}$E$_{23}$ could be due to slow replacement of C$_{12}$E$_{23}$ molecules by
lower HLB impurities in the surfactant samples.

Conclusions

A parabolic relationship between membrane activity and lipophilicity of nonionic surfactants is clearly established in series of
surfactants in which either the hydrocarbon chain length or
ethylene oxide chain length is varied. Activity at low and high
concentrations should be considered separately.

membranes can occur at very low surfactant concentrations; structural specificity is demonstrated by the superior action of compounds with the C_{12} moiety. Depth of penetration of the hydrocarbon chain may not be equal in any homologous series varying in ethylene oxide chain length, hence even in the $C_{12}E_x$ series biological activity varies considerably. When x is below the optimum this is probably due to the low CMC which limits monomer activity and perhaps is also due to penetration of a portion of the EO chain into the membrane. When x is above the optimum number, increasing ethylene oxide chain length leads to increasing molecular area, decreasing surface activity and a likely less efficient penetration of the membrane by the hydrocarbon chain. If, as seems likely, the molecular dimensions of the C_{12} chain intercalate most efficiently with the bilayer structure than any alteration in the depth of penetration will change its activity. There is no direct evidence, however, to support this concept, and further detailed work is necessary to determine mechanisms of action at sub-solubilizing surfactant concentrations.

At concentrations well above the CMC, these effects are complicated by solubilization, but while it seems clear that protein solubilizing effectiveness and membrane activity correlate well, the structural basis for protein solubilization has not yet been clarified either.

Literature Cited

1. Attwood, D: Florence, A.T. in "Surfactant Systems: their Chemistry, Pharmacy and Biology"; Chapman and Hall, London. 1983.
2. Drotman, R.B. Toxicol. Appl. Pharmacol. 1980, 52, 38-44.
3. Florence, A.T. Pure and Applied Chem. 1982, 53, 2057-2068.
4. Marsh, R.J.; Maurice, D.M. Exp. Eye Res. 1971, 11, 43-48.
5. Florence, A.T; Gillan, J.M.N. Pesticide Sci. 1975, 6, 429-439.
6. Florence, A.T.; Gillan, J.M.N. J. Pharm. Pharmacol. 1975, 27, 152-159.
7. Zaslavsky, B. Yu. et al. Biochim. Biophys. Acta. 1978, 507, 1.
8. Hirai, S.; Yashiki, T.; Mima, H. Int. J. Pharmaceutics, 1981, 9, 175-182.
9. Touitou, M.; Donbrow, M.; Azaz, E. J. Pharm. Pharmacol. 1978, 30, 662-663.
10. Walters, K.A.; Dugard, P.H.; Florence, A.T. J. Pharm. Pharmacol. 1981, 33, 207-213.
11. Walters, K.A.; Olejnik, O. J. Pharm. Pharmacol. 1983, 35, Suppl. 81P.
12. Florence, A.T.; Walters, K.A.; Dugard, P.H. J. Pharm. Pharmacol. 1979, 30, Suppl. 29P.
13. Araki, J.; Rifkind, J.M. Biochim Biophys. Acta, 1981, 645, 81-90.

14. Crook, E.H.; Fordyce, D.B.; Trebbi, G.F. J. Colloid Sci. 1965, 20, 191-204.
15. Harusawa, F.; Saito, T.; Nakajima, H.; Fukushima, S. J. Colloid Interface Sci. 1980, 74, 435-440.
16. Harusawa, F.; Nakajima, H.; Tanaka, M. J. Soc. Cosmet. Chem. 1982, 33, 115-129.
17. Cserhati, T.; Szogyi, M.; Bordas, B. Gen. Physiol. Biophys. 1982, 1, 225-231.
18. Kubinyi, H. in "Drug Research"; Jucker, E., Ed.; Birkhauser: Basel, 1979; 23, 97-198.
19. Hansch, C. in "Drug Design"; Ariens., Ed.; Academic: New York, 1971: Vol. 1.
20. Tiddy, G.J.T.; Personal Communication, 1983.
21. Mitchell, D.J.; Tiddy, G.J.T.; Waring L.; Bostock, T.; McDonald, M.P. J.C.S. Faraday Trans. I, 1983, 79, 975-1000.
22. Walters, K.A.; Florence, A.T.; Dugard, P.H. J. Colloid Interface Sci. 1982a, 89, 584-587.
    Walters, K.A.; Florence, A.T.; Dugard, P.H. Int. J. Pharmaceutics, 1982b, 10, 153-163.
23. Hirai, S.; Tashiki, T.; Mima, H.M. Int. J. Pharmaceutics, 1981b, 9, 1973.
24. Ponder, E. in "Haemolysis and Related Phenomena", Grume and Stratton: New York, 1984, 138 et. seq.
25. Dahlgren, C.; Rundquist, J.; Stendahl, O.; Magnusson, K.E. Cell Biophysics, 1980, 2, 253-267.
26. Schott, H. J. Pharm. Sci. 1973, 62, 341-343.
27. Kirkpatrick, F.H.; Gordesky, S.E.; Marinetti, G.V. Biochim. Biophys. Acta, 1974, 345, 154-161.
28. Ferguson, T.F.M.; Prottey, C. Fd. Cosmetic Toxicol. 1976, 14, 431-434.
29. Dominguez, J.C.; Parra, J.L.; Infante, M.R. et al. J. Cosmetic Chem. 1977, 28, 165-182.
30. Helenius, A.; Simons, K. Biochim. Biophys. Acta. 1975, 415, 29-79.
31. Tucker, I.G.; Florence, A.T.; Stuart, J.F.B. J. Pharm. Pharmacol. 1982, 34, 19P.
32. Tucker, I.G.;Florence, A.T. J. Pharm. Pharmacol. 1983, 35, 705-711.
33. Umbreit, J.N.; Strominger, J.L. Proc. Natl. Acad. Sci. U.S.A. 1973, 70, 2997-3001.

RECEIVED March 6, 1984

Modification by Surfactants of Soil Water Absorption

RAYMOND G. BISTLINE, JR., and WARNER M. LINFIELD

Agricultural Research Service, U.S. Department of Agriculture, Philadelphia, PA 19118

This work describes the application of a
previous study which dealt primarily with
organic synthesis and physical properties
of reaction products of pure fatty acids
with DETA. Derivatives to amplify the
previous study were prepared from various
industrial fatty materials. The reaction
product, from 1 mole diethylenetriamine
(DETA) with 2 moles fatty acid, was thought
to be the primary amine, $RCON(CH_2CH_2NH_2)$-
CH_2CH_2NHCOR, rather than the secondary amine,
as cited in the literature. The amine
was readily dehydrated to the imidazoline,
$R\underline{C} = NCH_2CH_2NCH_2CH_2NHCOR$. The imidazolines in
the presence of moisture hydrolyzed upon standing,
to the open chain derivatives. These cationic
surfactants were examined as water repellents
for soil. Water repellency was evaluated by
contact angle measurements and water infiltration through sand, sandy soil, and soil containing 30% clay. A large number of derivatives
made clay soil hydrophobic, whereas only a few
caused this effect on sandy soils. The following
factors influenced soil water repellency. Open
chain derivatives were more hydrophobic than
the corresponding imidazolines. Hydrophobicity
intensified with increasing molecular weight of
the saturated fatty acids. Unsaturation, as in
the oleic acid derivatives, enhances hydrophilicity. Hydrocarbon branching in the fatty acid

also reduces water repellency. The soil hydro-
phobing agents in treated soils greatly restrict
seed germination.

Farming, particularly in arid areas, requires efficient
utilization of water. Over the years several methods have
been developed to harvest water. Water harvesting is the
process of collecting water from plots that have been made
water repellent so that the runoff from these plots may be
employed in agriculture. Physical waterproofing methods,
such as coating the ground with nylon sheets, asphalt, fuel
oil, and paraffin wax (1) have appeared in the literature;
however, references to chemical interactions with the soil
to produce hydrophobic surfaces are few (2, 3). A chemical
soil treatment would provide an alternative means of
harvesting water.

A search of the chemical literature suggested that
surfactants, containing amine groups, might impart water
repellent properties to soils. While studying the cation
exchange mechanism of large substituted ammonium ions on
clay, Gieseking (4) found that clays saturated with the
organic ammonium ions did not show the water absorption,
swelling, and dispersion characteristics of untreated clay.
Subsequently, Law and Kunze (5) examined the effects of
three different types of surfactants on clays. They reported
that commercial cationic surfactants were strongly adsorbed
on clay through ionic bonding, in amounts equal to or even
greater than the cation-exchange capacities of the clays.
According to these authors, the presence of surfactants on
clays significantly reduced hydration and water content at
high treatment rates. Furthermore, Greenland (6) observed
that cationic surfactants created a layer of increased
hydrophobic character beneath the soil's surface.

Various types of cationic surfactants were reviewed by
us to select one, relatively low in cost and readily prepared
on a large scale, for which the relationship of chemical
structure to surface active properties could be conveniently
studied.

One type of cationic surfactant was the fatty acid
derivatives of polyamines. The properties of the derivatives
of fatty acids and ethylenediamine have been described in
the literature (7-9). It appeared from these reports that
the 2-alkyl-2-imidazolines would not impart sufficient
hydrophobicity to soils. However, the analogous series of
homologous compounds from the fatty acids and diethylenetriamine (DETA) appeared likely to do so because of their
higher molecular weight.

In 1940, Ackley (10) reacted fatty acids with DETA to form precursors for fabric softeners. However, it was not until recently (11) that the course of the reaction between fatty acids and DETA, and properties of the reaction products were studied. The reaction appears to proceed, according to the following scheme:

$$RCO_2H + HN(CH_2CH_2NH_2)_2 \xrightarrow{90°C} \begin{array}{c} Amine \\ Soap \end{array} \Big] \quad 130°C$$

$$\begin{array}{c} \rule{0.3cm}{0.4pt} RCO_2H + RCON(CH_2CH_2NH_2)_2 \longleftarrow \end{array}$$

$$\Big\downarrow \begin{array}{c} 150°C \\ 6 \ hr \end{array}$$

$$RCON(CH_2CH_2NH_2)CH_2CH_2NHCOR$$

$$\Big\downarrow \begin{array}{c} 150°C, \ 2 \ hr \\ Vacuum \end{array}$$

$$R-C=N-CH_2CH_2-N-CH_2CH_2NHCOR + H_2O$$

Initially, one molecule of fatty acid reacted with the secondary amine of DETA, followed by a second fatty acid molecule, reacting with one of the two primary amines. The product formed was the N-(2-aminoethyl) diamide, which will be referred to as the diamide. The diamides could be cyclized by heating under vacuum to form imidazolines, which were unstable to hydrolysis. In the presence of water, or upon exposure to atmospheric moisture, they revert back to the diamide.

DETA derivatives of C_9-C_{22} saturated fatty acids, as well as the C_{18} unsaturated acids, oleic and elaidic, were prepared and evaluated in the previous publication (11). Hydrophobicity determination, via contact angle measurements, proved to be nondiscriminatory and, therefore, a more meaningful test, the sand penetration test was devised. The results of this test demonstrated that the diamides of the C_{14} and higher saturated fatty acids were water repellents. On the other hand, the unsaturated oleic acid derivatives enhanced hydrophilicity.

Materials and Methods

Apparatus. Infrared spectroscopy, Perkin-Elmer Model 257 infrared spectrophometer, Perkin-Elmer Corporation (Norwalk, CT); ultraviolet spectroscopy, Perkin-Elmer Model 559 UV-vis spectrophotometer, Perkin-Elmer Corporation, Coleman

Instruments Div. (Oak Brook, IL); Fisher-Johns melting
point apparatus, Fisher Scientific Company (Pittsburgh,
PA); contact angles, Gaertner goniometer, Gaertner Scientific Corporation (Chicago, IL).

Reagents. The following chemicals were used: diethylenetriamine, Aldrich Chemical Company (Milwaukee, WI); C_{12}, C_{14},
C_{16}, C_{18} saturated fatty acids, Armak Industrial Chemicals
Division (Chicago, IL); oleic acid, A. Gross & Company
(Newark, NJ); elaidic acid, laboratory preparation (12);
tallow, Corenco Corporation (Philadelphia, PA) (titer, t =
42°, iodine no. 40); tallow fatty acid (T - 22) and partially
hydrogenated tallow fatty acid (T - 11), Proctor & Gamble,
Industrial Chemicals Division (Cincinnati, OH) (T - 22, t =
40°, I. no. = 60) (T - 11, t = 46°, I. no. = 41); hydrogenated
tallow fatty acids, Acme-Hardesty Company, Incorporated
(Jenkintown, PA), (t = 62°, I. no. = ——); tall oil fatty
acids, Arizona Chemical Company (Wayne, NJ) (t = 24°,
I. no. = 75); MO-5 fatty acid, Union Camp Corporation
(Jacksonville, FL) (t = 33°, I. no. 75); isostearic acid,
Emery Industries Incorporated (Cincinnati, OH) (t = liquid,
I. no. = 12).
        Drierite, W. A. Hammond Drierite Company (Xenia, OH);
Ottawa Sand ASTM, Arthur H. Thomas Company (Philadelphia,
PA); SYLON-CT Silylating Reagent, Supelco, Incorporated
(Bellefonte, PA).

Soils. Granite Reef soil is a sandy loam soil, supplied by
the U.S. Water Conservation Laboratory, USDA, Phoenix, AZ.
Walla Walla soil is a soil which contains 30% clay and
organic matter, supplied by the Columbia Plateau Conservation
Research Center, USDA, Pendleton, OR.

Experimental

Preparation of commercial fatty acid-DETA derivatives: the
previously described procedure (11) was slightly modified,
using a 5% excess fatty acid to ascertain that all DETA was
consumed. Amine analysis was conducted, according to AOCS
test method (13). The reaction products were used without
purification.
        Procedures for infrared and ultraviolet absorption and
contact angle measurements are reported in the previous
publication (11).

Infiltration Tests

Sand Penetration Test

Ten grams of indicating Drierite was placed in a 120-mL silylated jar and covered with 80 g treated ASTM Ottawa sand, prepared according to the previous report (11). The depth of the soil was 35 mm. Distilled water, 8 mL, was then placed on the surface of the sand. The time required to change the indicating Drierite completely from blue to pink was recorded as the infiltration time.

Soil Infiltration Test

A soil infiltration test was devised to screen a large number of compounds within a limited time span. The amounts used are far in excess of quantities used in field application. A 5% diamide solution in isopropanol, 15 mL, was added to 50 g soil, air dried overnight, and then placed in a vacuum oven at 50° for 1 hr to remove traces of isopropanol. The treated soil, 10 g, was placed in a 25 X 500 mm glass chromatographic column with a coarse porosity fritted disc on top of a detachable adapter base. The soil was tapped down lightly with a wooden dowel to a depth of 12 mm in order to prevent channeling. Forty-five cm of water covered the soil. The period required for 200 mL distilled water to penetrate through 10 g of treated soil was recorded as the infiltration time. The test was arbitrarily discontinued after 2 weeks.

Glycerine Effect on Hydrophobicity. The partially hydrogenated tallow fatty acid-DETA reaction product, 10 g, was weighed into each of five 60-mL bottles. The following amounts of glycerine, expressed as percent, were weighed, using an analytical balance, into the bottles: 0.36, 1.02, 2.60, 5.10, 11.11. Mixtures were then melted and stirred. Contact angles were measured, according to the procedure in the previous publication (11).

Soil Extraction. Granite Reef soil, 100 g, was treated with 30 mL of a 5% isopropanol solution of the partially hydrogenated tallow fatty acid-DETA reaction product and then air dried overnight and finally in a vacuum oven at 50°C for 1 hr to remove residual isopropanol. On the following day the treated soil was extracted with ethanol 8 hr in the Soxhlet apparatus. The extracted solvent was evaporated recovering 1 g residue. As a control experiment, untreated soil, 100 g, was also extracted 8 hr with ethanol,

recovering 0.2 g material. The dried extracted soil was
placed in the chromatographic column and the infiltration
test with 200 mL distilled water repeated.

Plant Growth Effect Test. Potting soil, 3.4 kg, was weighed
into a 55 X 27.5 X 7.5 cm flat, treated with 1.2 L of a 5%
solution of the partially hydrogenated tallow fatty acid-DETA
diamide in isopropanol, and air dried for 2 days until the
isopropanol had evaporated. Untreated potting soil, 3.4
kg, was weighed into another flat to be used as a control.

Soybeans

Fifty soybean seeds each were planted in the diamide-treated
soil as well as in a flat with an untreated control soil.
Each flat was watered weekly with 2 L tap water. Water
absorption and plant growth were recorded.

Corn

Since the nitrogen requirement of corn plants is greater
than the nitrogen content found in potting soil, additional
nitrogen fertilizer, 15 g 20:20:20/2 L water, was added to
the soil, prior to treatment with water repellent chemicals.
Fifty corn kernels were planted in the treated soil and an
untreated control flat. The flats were watered weekly as
described above, and growth recorded. After several weeks
the plants were harvested, weighed, and dried in a vacuum
oven and weighed again.

Results and Discussion

This work describes the application to soil of compounds of
a previous study (11) which dealt primarily with organic
synthesis and physical properties of reaction products of
pure fatty acids with DETA. In this study derivatives were
prepared from various industrial fatty materials. In
addition, water infiltration studies on sand, sandy soil,
and clay soils were carried out on the previously prepared
and new compounds. Finally, an investigation was initiated
to determine the biological effects of one water-repelling
chemical, the partially hydrogenated tallow-fatty acid-DETA
reaction product, on seed germination and plant growth.
    The DETA derivatives from industrial fatty materials
for soils application were of special interest because of
their ready availability, low cost, and greater solubility

than the DETA derivatives from the purified fatty acids.
Diamides were prepared from tallow, tallow fatty acids,
partially hydrogenated tallow fatty acids, hydrogenated
tallow fatty acids, tall oil fatty acids, MO-5 acids (a
mixture of oleic, isooleic, stearic, and elaidic acids), and
isostearic acid (a mixture of branched chain isomeric
acids).

Melting points and contact angle measurements of the
fatty acid-DETA reaction products are recorded on Table I.
The melting points of the pure saturated fatty acid-DETA
derivatives were over 100°C and possessed low solubility in
most organic solvents. The unsaturated fatty acid-diamides
had much lower melting points. The industrial fatty-DETA
derivatives with the exception of the product from hydrogenated tallow fatty acids melted at 60°C or below.

Contact Angle Measurements. It was planned to establish
the hydrophilic or hydrophobic nature of the diamides by
contact angle determination (14). The angles for the
lauric diamides and higher saturated homologs were all
above 90°. Thus, the contact angle measurements did not
discriminate between members of a homologous series. The
unsaturated fatty acid-DETA derivatives were predictably
more hydrophilic. Among the industrial fatty acid-DETA
derivatives, only three possessed contact angles over 90°.
These were the DETA derivatives of partially and fully
hydrogenated tallow fatty acids and the isostearic acid.
Although a 50:50 mixture of DETA derivatives of isostearic
and partially hydrogenated tallow fatty acids produced a
compound with the desired melting point, the contact angle
of this mixture fell to 73°. The data are reported in
Table I.

Sand Penetration Tests. Since contact angle measurements
appeared fairly insensitive to structure differences, a
sand penetration test was devised. Water penetration rates
through beds of treated sand were determined for the derivatives of the individual pure fatty acids, as well as for
those from the industrial fatty materials as shown in the
first column of Table II. Water passed quickly through the
C_{12} diamide-treated sand, whereas a period of 7 or more
days was required for the C_{14} and higher molecular weight-
saturated diamides. While water penetrated the oleic
diamide-coated sand within 5 min, 1-day was required for
penetration of the elaidic diamide-coated sand, showing the
superiority of the trans isomer. In the cis configuration,
the molecule is bent back upon itself, resulting in properties

Table I.   Physical Properties of
Diethylenetriamine Reaction Products

R-COOH	Melting Point, C	Contact Angle (°)
$C_{11} H_{23}$	110-111	99
$C_{13} H_{27}$	112-113	99
$C_{15} H_{31}$	116-117	96
$C_{17} H_{35}$	118-119	98
$\Delta^9 C_{17} H_{33}$, cis	52-53	50
$\Delta^9 C_{17} H_{33}$, trans	91-92	62

Derivatives of Industrial Fatty
Materials

	Melting Point, C	Contact Angle (°)
Tallow	50-60	37
Tallow Fatty Acid[a]	45-55	63
Tallow Fatty Acid[b]	55-65	97
Hyd. Tallow F.A.	90-95	94
Tall Oil F.A.	35-45	59
MO-5	30-40	56
Isostearic	30-40	91
Isostearic + Tallow Fatty Acid[b] (50:50)	40-45	73

[a] = T-22 fatty acids, C_{18} cis = 36%,
    C_{18} trans = 7%

[b] = T-11 fatty acids C_{18} cis = 6%, C_{18}
    trans = 24%

Table II. Infiltration Properties of
Diethylenetriamine Reaction Products

DETA-Fatty Acid Derivative from	Sand Penetration	Granite Reef Soil Infiltration	Walla Walla Soil Infiltration
$C_{11}H_{23}$ COOH	1 hr	3 hr	5 hr
$C_{13}H_{27}$ COOH	7 d	5 d	>2 wk
$C_{15}H_{31}$ COOH	>7 d	5 d	>2 wk
$C_{17}H_{35}$ COOH	>7 d	>2 wk	>2 wk
$\Delta^9 C_{17}H_{33}$ cis COOH	5 min	3 hr	8 hr
$\Delta^9 C_{17}H_{33}$ trans COOH	1 d	>2 wk	2 wk
Tallow	i[c]	4 hr	8 hr
Tallow Fatty Acid[a]	i	1 d	>2 wk
Tallow Fatty Acid[b]	i	1 wk	>2 wk
Hyd. Tallow F.A.	>7 d	>2 wk	>2 wk
Tall Oil F.A.	i	4 hr	8 hr
MO-5	i	2 hr	8 hr
Isostearic	i	2 hr	7 hr
Isostearic + Tallow F.A.[b] (50:50)	i	3 hr	2 wk
Control	i	4 hr	1 d

[a] = T-22 fatty acids

[b] = T-11 fatty acids

i[c] = instantaneous

roughly analogous to those of a C_9 fatty acid derivative.
Although the elaidic diamide derivative is more linear in
configuration than the oleic diamide, the trans unsaturation
disarranges the molecules, resulting in surface properties
which are less hydrophobic compared to those of the saturated
stearic acid derivative. These observations help explain
the results obtained with the industrial fatty acid derivatives. Instantaneous penetration was observed for sand
treated with the DETA derivatives of tallow, tallow fatty
acids, tall oil fatty acids, and MO-5 acids as predicted by
their low contact angles. All of these are high in oleic
acid content. In contrast, sand treated with the completely
hydrogenated tallow fatty acid-DETA diamide which had a
high contact angle required more than 1 week for water
penetration. In these two sets of examples, correlation as
to degree of hydrophobicity was observed between the sand
penetration tests and the contact angle measurement.

However, with the other derivatives there was disagree-
ment. Water immediately penetrated sand coated with the
DETA derivatives of partially hydrogenated tallow fatty
acids and isostearic acid, despite their high contact
angles of greater than 90°. Water quickly passed through
sand treated with a 50:50 mixture of diamides of isostearic
acid and partially hydrogenated tallow fatty acids. While
these penetration results on standardized ASTM sand were
interesting, they did not forcast the results found with
sandy and clay soils.

Soil Column Tests. In the sand penetration test, a minimal
amount of water was used. No consideration was given to
the hydrostatic pressure which would occur in nature from a
body of surface water. A new soil infiltration test was
developed to take this into consideration. This test used
a maximum amount of water (200 mL) on a minimum amount of
treated soil (10 g) and was restricted only by the dimensions of the laboratory equipment. Our aim was to prepare
an hydrophobe for soil which would support water over an
extended period of time. Whereas water passed through soil
treated with hydrophilic compounds within 8 hr, 2 weeks or
more were required for penetration through an hydrophobe-
treated soil. In the latter case the water level dropped
6 mm or less each day, showing that the cationic surfactant
greatly hindered, but did not completely restrict the
passage of water. The tests were usually terminated after
2 weeks, due to the large number of samples to be tested.
The two soils were selected because of differences in

composition and properties. The results of the soil infiltration test are recorded in Table II.

Although the infiltration times differed slightly from
each other, due to differences in soil structure, water
passed quickly through soil treated with hydrophilic compounds, whereas a week or longer was required for passage
through hydrophobic-treated soil. Water penetrated soil
treated with the lauric acid diamide more rapidly than the
controls, demonstrating the hydrophilic nature of the C_{12}
derivative. A much longer time period was required for
penetration through soils, treated with the saturated C_{14}
and above diamides. Drastic differences were observed for
the infiltration times of soils treated with the cis unsaturated oleic acid diamide (hr) compared to the trans unsaturated elaidic acid diamide (wk).

The industrial fatty acid-DETA derivatives were evaluated
for application to soil. Water infiltrated soil coated
with the DETA derivative of tallow more rapidly than the
controls. This was due to the high unsaturation content
and also in part to the glycerine retained in the product
as discussed below. We were unable to find a solvent
system which would readily separate the glycerine, formed
from the triglyceride, from the DETA reaction product. If
the glycerine were removed, the infiltration rates for the
tallow-DETA derivative should be identical with the rates
obtained for tallow fatty acid-DETA reaction product.

The soil treated with the tallow fatty acid-DETA
reaction product retarded moisture infiltration on both
soils. The partially hydrogenated and completely hydrogenated tallow fatty acid-DETA derivatives also displayed
hydrophobic properties on both soils. Although the completely hydrogenated tallow fatty acid-DETA reaction product
(m.p. 90°C) had optimum hydrophobic properties on all three
soils, it was difficult to dissolve in most organic solvents.
Even the partially hydrogenated tallow fatty acid-DETA
reaction product, m.p. 45°-50°C, was not readily soluble.
To overcome this difficulty, we examined the DETA derivatives
of some other fatty materials, hoping to find an hydrophobe
with a lower melting point.

Water infiltration tests were conducted on soils
treated with the DETA derivatives of tall oil fatty acids,
MO-5 acids, and isostearic acid, all of which melted below
45°. Water infiltrated sandy soil treated with these
compounds within 4 hr and through clay containing soil
within 8 hr, confirming their hydrophilic nature. The
blending of these diamides into the tallow derivatives in
order to lower melting points and enhance solubility causes

a serious loss of hydrophobicity in sandy soil. For example
infiltration rates dropped from 1 week to 3 hr passing
water through Granite Reef soil treated with a 50:50 mixture
of DETA derivatives of isostearic acid and partially hydrogenated tallow fatty acids. The water infiltration rate on
Walla Walla soil, treated with the 50:50 mixture, remained
similar to that obtained with the partially hydrogenated
tallow fatty acid-DETA derivative alone.
    The soil properties observed with these chemicals
suggest that the chemical structural requirements for an
hydrophobic cationic surfactant appear to be:
    a.   Molecular weight > 500.
    b.   Prepared from saturated starting materials.
    c.   Contain free amino groups to attach to soil
       particles.

Glycerine Effect on Hydrophobicity. The tallow-DETA reaction
product, containing over 10% free glycerine, had a contact
angle of 37°. Glycerine in measured amounts was added to
the partially hydrogenated tallow fatty acid-DETA reaction
product and contact angles measured (See Figure 1). The
purpose was to determine how much glycerine a reaction
product could tolerate and yet remain hydrophobic. The
results show that glycerine amounts of as low as 1% render
the reaction product hydrophilic and suggest that tallow
fatty acids rather than tallow be used in the synthesis of
DETA reaction products.

Retention on Soil. An experiment was conducted to demonstrate
the retention of the cationic surfactant on the soil.
Granite Reef soil, treated with the partially hydrogenated
tallow fatty acid-DETA reaction product together with an
untreated soil sample was extracted in the Soxhlet apparatus.
The extracted soils were returned to the chromatographic
column and the water infiltration test repeated. It has
been reported previously (15) that short-chain alcohols
form double layer complexes on the surface of soil particles,
causing a modification of soil properties. Whereas water
penetrated the untreated, unextracted Granite Reef soil
control in 4 hr, 12 hr was now required for passage through
the untreated, alcohol extracted soil. Water penetration
through extracted treated soil was incomplete after 4 days,
showing that the water repelling agent remained on the
soil. Fripiat and coworkers (16) have reported that amines
can protonate in soil and replace inorganic cations from
the clay complex by ion exchange. Amines are adsorbed with
their hydrocarbon chains perpendicular or parallel to the

clay surface, depending on the concentration (16). The
surface of the soil particles has now become hydrophobic
and delays the passage of water molecules.

The material (1 g) recovered by the alcoholic extraction of treated soil in the Soxhlet apparatus was identified by U.V. spectroscopy at 202 nm as being unreacted
partially hydrogenated tallow fatty acid-DETA reaction
product. The alcohol solubles from the untreated soil
absorbed in the range of 230-220 nm with only a trace
absorption at 202 nm. The 60% recovery suggests that the
5% concentration was too high and further work is required
to determine the proper concentration.

Effects on Plant Growth. Experiments were initiated to
determine the effect of a soil hydrophobe on seed germination. Potting soil was treated with the partially hydrogenated tallow fatty acid-DETA reaction product. Although
no crust formed on the soil's surface after the addition, a
coarser texture of the potting soil was observed. In
reviewing plants which would germinate rapidly, soybeans
were selected. After the seeds were planted in the treated
soil, as well as an untreated control, the flats were
watered weekly. We were surprised to see water roll off
and down the sides of the treated flat. Whereas the control
flat took up 1.7 kg of water, the treated soil flat gained
only 110 g. The soybeans in the control flat germinated in
10 days and grew rapidly. After a period of 8 weeks, one
seed germinated in the hydrophobe-treated soil (See Figure 2).
This experiment was continued for another month, but no
further plants appeared. The seeds were not able to germinate
because the passage of water through the soil was blocked.

Corn was used to demonstrate the effects of soil
hydrophobes on the cereal grasses. After adding nitrogen
fertilizer, the potting soil was treated with the partially
hydrogenated tallow fatty acid-DETA reaction product. As a
result of this enrichment all seeds germinated in the
treated soil and in the control. However, the corn plants
grown in the treated soil soon developed a yellow cast and
appeared stunted in growth. After 6 weeks the corn plants
were harvested. The stalks planted in the treated soil
weighed 54% of those grown in the control. The stalks were
then dried in a vacuum oven to constant weight, and the
plants from the hydrophobe-treated soil now weighed 47% of
the control plants. These results show that the growth of
plants was greatly impeded by the surfactant.

Figure 1.   Effect of glycerine concentration on contact
            angle the partially hydrogenated tallow fatty
            acid-DETA reaction product.

Figure 2.   Effect of soil hydrophobe on soybean germination.

Acknowledgments

D. Brower and M. P. Thompson carried out the plant experiments.

Literature Cited

1. Fink, D. H.; Cooley, K. R.; Frasier, G. W. J. Range Manage. 1973, 26, 396.
2. Myers, L. E.; Frasier, G. W. J. Irrig. Drain. Div. Amer. Soc. Civil Engin. Pro. 1969, 95, 43.
3. Fink, D. H.; Frasier, G. W. Soil Sci. Soc. Am. Publication #7,1975,173. ASCE 1969, 95, 43.
4. Gieseking, J. E. Soil Sci. 1939, 47, 14.
5. Law, J. P., Jr.; Kunze, G. W. Soil Sci. Soc. Am. Proc. 1966, 30, 321.
6. Greenland, D. J. Soc. Chem. Ind. Monograph 1970, 37.
7. Shapiro, S. H in "Fatty Acids and Their Industrial Application"; Pattison, E. S., Ed.; Marcel Dekker, Inc.: New York, NY, 1968; p. 98.
8. Ferm, R. J.; Riebsomer, J. L. Chem. Rev. 1954, 54, 593.
9. Tornquist, J. Proc. of the 4th International Congress on Surface Active Substances, Vol. 1, Sect. A, Gordon and Breach Science Publishers, London, England, 1967, p. 387.
10. Ackley, R. R.; U.S. Patent 2,200,815, 1940.
11. Bistline, R. G., Jr.; Hampson, J. W.; Linfield, W. M. J. Am. Oil Chem. Soc. 1983, 60, 823.
12. Swern, D.; Scanlan, J. T. Biochem. Prep. 1953, 3, 118.
13. AOCS Test Method TF 1b-64 and TF 2b-64 "Quantitative Determination of Amines"; Am. Oil Chem. Soc.: Champaign, IL, 1979.
14. Rosen, M. J. in "Surfactants and Interfacial Phenomena"; John Wiley & Sons; New York, NY, 1978; p. 179.
15. German, W. L.; Harding, D. A. Clay Miner. 1971, 9, 167.
16. Fripiat, J. S.; Servias, A.; Leonard, A. Bull. Soc.: Chim., France 1962; p. 635.

RECEIVED January 10, 1984

15

Binding of Alkylpyridinium Cations by Anionic Polysaccharides

A. MALOVIKOVA[1] and KATUMITU HAYAKAWA[2]

Department of Chemistry, Dalhousie University, Halifax, Nova Scotia, Canada B3H 4J3

JAN C. T. KWAK

Département de chimie, Université de Sherbrooke, Sherbrooke, Québec, Canada J1K 2R1

Solid state electrodes selective for alkylpyridinium cations are used to study the binding of these surfactants cations, with C_{12}, C_{14} and C_{16} alkyl chainlengths, to a number of anionic polyelectrolytes. The electrodes are shown to be effective from very low surfactant concentrations to the cmc, and can be used for accurate cmc determinations in solutions of high ionic strength. Binding isotherms of the alkylpyridinium cations with polyacrylate, alginate, pectate and pectinates are presented. All isotherms are highly cooperative. The surfactant chainlength dependence of the overall binding constant is identical to the case of micelle formation of the free surfactant, but for a given surfactant the overall binding constant depends strongly on the charge density of the polyion.

The binding of ionic surfactants by polyions of opposite charge distinguishes itself from the much more widely studied case of binding by neutral water soluble polymers mainly because binding occurs at much lower free surfactant concentrations (1,2). Thus, while the binding of ionic surfactants by neutral polymers is normally studied at concentrations close to or above the cmc of the surfactant, in the case of oppositely charged polyions and surfactant ions the first binding may take place at concentrations orders of magnitude below the cmc of the surfactant. It is therefore not surprising that a detailed study of this binding process had to await the development of suitable analytical methods, notably surfactant selective electrodes (2-10). The pioneering paper by Satake and Yang (2) demonstrates the strongly cooperative

[1]Current address: Institute of Chemistry, Slovak Academy of Sciences, Bratislava, Czechoslovakia
[2]Current address: Department of Chemistry, Kagoshima University, Kagoshima, 890 Japan

character of the binding of decylsulfate anions by highly charged
cationic polypeptides. These authors interpret their binding isotherms in terms of a nearest neighbour interaction model, with
hydrophobic interactions between neighbouring bound surfactant
ions accounting for the increase of the apparent binding constant
with increased binding. They apply the theory of Zimm and Bragg,
developed to describe the helix coil transition in polypeptides
(11), to the case of surfactant binding by polymers, effectively
fitting the observed binding isotherm to two parameters, i.e. K,
the "intrinsic binding constant" between an isolated polymer
binding site and a single surfactant ion, and u, a "cooperativity
parameter" presumably determined by the hydrophobic interactions
between neighbouring surfactants. The Satake-Yang treatment parallels the theory of Schwarz developed to describe the binding of
anionic dyes to linear biopolymers (12), and in fact the binding
of long chain surfactants may be a much better test of such a
theory than the case of relatively bulky and stiff dye molecules.
     Since the work of Satake and Yang, a number of other studies
have appeared employing surfactant selective electrodes to study
the binding of anionic surfactants by cationic polymers (13-15) or
of cationic surfactants by anionic polymers (7,10,16-19), with
most of these studies relying on the theories of Schwarz and Satake and Yang to describe the observed highly cooperative binding
isotherms. Although in a number of cases special attention was
given to conformational changes of the polymer induced by surfactant adsorption (2,7,20,21) it is of interest to note the difference between the models used to describe these data at very low
surfactant and polymer concentrations, and the much more widely
studied case of binding measurements close to or past the surfactant cmc (1,22-28). Thus, whereas the generally accepted model
for micellar binding envisages a complex where the polymer enve-
lops many distinct, micellar-like surfactant aggregates (24,26) in
surfactant ion polyion binding studies at low surfactant concentrations it is generally assumed that the polyion maintains a well
defined solution conformation, or the polyion conformation is not
considered at all, with a description of the binding isotherm in
terms of a nearest neighbour model. Such a description leaves
open the question whether the hydrophobic part of the bound surfactant remains exposed to the aqueous phase, or whether after
binding the surfactants aggregate into micelle-like groups, with
the polymer surrounding the aggregates. It is important to note
that the binding between oppositely charged surfactant ions and
polyions initially takes place without phase separation, and is
fully reversible. This distinguishes the binding measurements at
low surfactant concentrations from the studies on precipitating
systems with or without subsequent redissolution (1,24).
     In previous studies we have described the binding of alkyl-
trimethylammonium ions to a variety of polyanions (16-19). As has
been observed by other authors, both the alkyl chainlength dependence and the ionic-strenght dependence of the binding process

were found to be remarkably similar to the case of micelle formation, apparently independent of whether the polyion has a well-
defined backbone configuration such as with DNA or pectic acid
(18,19), or may be presumed to be extremely flexible, e.g. dextransulfate or poly(styrenesulfonate) (16,17). On the other hand,
the binding constants for a given alkylammonium cation are found
to depend strongly on the polymer structure and charge density
(19). In this paper we extend these measurements to the case of
alkylpyridinium cations with dodecyl, tetradecyl, and hexadecyl
alkyl groups (to be abbreviated as $C_{12}Py^+$, $C_{14}Py^+$ and $C_{16}Py^+$ respectively). As polyanions we choose two polysaccharides of well de-
fined structure, i.e. alginic acid, a copolymer of mannuronic and
guluronic acid, and pectic acid, a linear polymer of D-galacturo-
nic acid (29). The influence of the charge density of the polyion
is studied by comparing pectate with pectinates with degrees of
esterification of the corresponding pectate varying from 20 to
70%. In addition, results are presented for binding to the sodium
salt of poly(acrylic acid) (PAA). Polymer structures are shown
in Figure 1.

Experimental

Surfactants. $C_{12}PyCl$ and $C_{16}PyBr$ were commercial products (Tokyo
Kasei Kogyi Co., Ltd and Eastman Kodak Co., respectively). These
products were purified by repeated recrystallization from acetone
and treatment with active charcoal. $C_{14}PyBr$ was synthesized by
reacting the corresponding alkylbromide (Eastman Kodak), purified
by fractional distillation, with a slight excess of pyridine (30).
The crude product was purified by extraction with diethylether,
followed by up to 6 recrystallizations from acetone and treatment
with active charcoal.
Polysaccharides. Na-alginate was isolated from Laminaria Hyperbo-
rea. The guluronic acid content was found to be 63.6%, corresponding to an M/G (mannuronic acid/guluronic acid) ratio of 0.57.
The total carboxylate concentration was determined by ion exchange
to the acid form followed by titration with NaOH. Pectins of various degrees of esterification, E, were prepared from a purified
citrus pectin (Genu Pectin, Kopenhagen, Denmark) by controlled
alkaline deesterification (31). Degree of esterification, intrinsic viscosity [η] and viscosity molecular weight were determined
using standard procedures (31,32). The concentration of free carboxyl groups in the initial (stock) solutions of potassium pectate and sodium pectinate were determined by precipitation with Cu^{2+}
(33,34).
        Analytical grade NaCl, polyvinylchloride (Aldrich, high mol.
wt.), bis(2-ethylhexyl)phtalate (GR, Aldrich), and tetrahydrofuran
(AR, BDH Chemicals) were used without further purification. Polyacrylic acid, mol. wt. 250,000 (Aldrich) was titrated with NaOH to
obtain a stock solution of NaPAA of known carboxylate concentration.

Potentiometry. Free surfactant concentrations were determined by
means of solid state membrane electrodes responding to the alkylpyridinium cations. The electrodes were made as before (16-19)
except that the carrier complex was prepared by reacting the required purified alkylpyridinium bromide with highly purified so-
diumdodecylsulfate and repeated recrystallization from acetone of
the resulting precipitate. Binding curves were determined by
means of a titration technique, where surfactant solution is added
to the polymer solution by means of a motorized piston buret. The
polymer concentration is kept constant by adding an equal volume
of polymer solution of double the initial concentration from a second piston buret (18). In a recent improvement of our experimental set-up, the electrometer output is now digitized and stored in
a microcomputer which also checks for constancy of the e.m.f. and
activates the piston burets. Thus complete binding curves can be
determined unattended. In all measurements the temperature was
constant at 30.0 ± 0.1^{o}C.

Results and Discussion

The surface tension of the pyridinium surfactants at 30°C as a
function of concentration were measured by means of a Du Nouy ring
tensiometer (Figure 2). No minima are apparent in the C_{14} and C_{16}
curves, but a small minimum in the C_{12} curve indicates the presence of a minor impurity in the commercial C_{12}PyCl used, even after
repeated recrystallizations. In spite of this, our result for the
cmc, 1.40 (± 0.04) x 10^{-2} m is in very reasonable agreement with
literature data reported as 1.46 x 10^{-2}, 1.48 x 10^{-2} and 1.78 x
10^{-2} from conductance (35-37) and 1.62 x 10^{-2} from surface tension
(37), all at 25°C. For C_{14}PyBr we find a cmc of 2.65 (± 0.05) x
10^{-3} m, typical literature values are given as 2.57 x 10^{-3} m from
surface tension (35) and 2.63 x 10^{-3} m from conductance (36). Finally, for C_{16}PyBr we obtain 6.2 (± 0.1) x 10^{-4} m, where literature values vary rather widely, i.e. 5.8 x 10^{-4} (35,38) at 25°C and
7.05 x 10^{-4} at 30°C (36), both values from conductance, and 6.6 x
10^{-4} at 25°C from surface tension (39). Note that Anacker (40)
has pointed out some difficulties in the determination of the cmc
of C_{16}PyCl by conductance.
    Typical electrode performances are shown in Figure 2 for C_{14}
and C_{16} pyridinium cations in the presence of a large excess of
NaCl. We note that the electrodes have an excellent selectivity
for the surfactant cation. Response is Nernstian from below 10^{-6}
m to the cmc for C_{16}Py^{+}, and from about 3 x 10^{-6} m to the cmc for
C_{14}Py^{+} and C_{12}Py^{+} (not shown). In fact, the electrodes provide for
a convenient and accurate method to determine the cmc in particu-
lar of the higher chainlength cationic surfactants in solutions of
high ionic strength, where other methods become increasingly more
difficult (40). An example of the possible application of these
electrodes in thermodynamic studies is shown in Figure 4, where
the cmc as obtained from data such as in Figure 3 is plotted vs
the total counterion concentration. When extrapolated to m_{NaCl} =

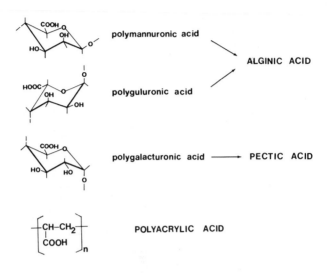

Figure 1.   Polymer structures.

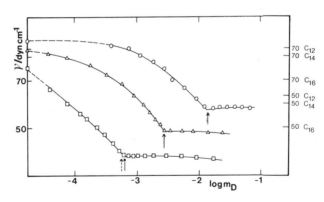

Figure 2.   Surface tension of alkylpyridinium halides at 30°C.
o $C_{12}PyCl$;   Δ $C_{14}PyBr$;   □ $C_{16}PyBr$.
Note shifts in vertical scale for C_{12} and C_{14}.

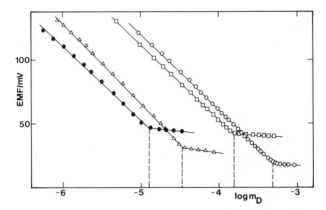

Figure 3.   Electrode response for $C_{14}Py^+$ and $C_{16}Py^+$ with NaCl. $C_{14}Py^+$: o 0.1 m NaCl, □ 0.5 m NaCl; $C_{16}Py^+$: Δ 0.1 m NaCl, o 0.5 m NaCl.

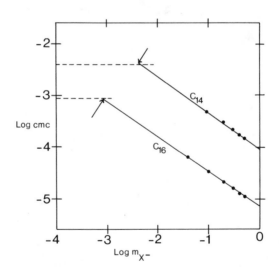

Figure 4.   Log cmc vs total counterion concentration in solutions of $C_{14}PyBr$ and $C_{16}PyBr$ with added NaCl. Broken lines: cmc of corresponding chlorides (refs. 35,36). Arrows: extrapolated cmc's.

0, we obtain for pure $C_{14}PyCl$ log cmc = -2.37 ± 0.03, and for
$C_{16}PyCl$ log cmc = -3.07 ± 0.02. Obviously better accuracy can be
obtained in a more complete study with data at lower NaCl concentrations, but even these numbers are in reaonable agreement with
literature data at 25°C for log cmc quoted as -2.40 (36) and -3.05
(35) for $C_{14}PyCl$ and $C_{16}PyCl$ respectively. Similarly, if we calculate the free energy of micellization, ΔG_m^o from the intercept at
log m_{NaCl} = 0 we obtain -29.9 kJ/mole for $C_{16}PyCl$ and -23.5 kJ/
mole for $C_{14}PyCl$, i.e. a contribution per CH_2 group of 3.2 kJ/mole
or 1.27 RT, and from the slope of the log cmc vs log m_{NaCl} curves
we calculate an apparent degree of counterion dissociation of
0.30 ± 0.02 in both cases. These numbers are in good agreement
with expected values (41), perhaps even surprisingly so given the
high ionic strength of the systems from which these values have
been calculated. For the moment they serve to underline the reliable performance of the electrodes in solutions of widely varying ionic strength, such as encountered in the present binding
studies, and we leave the application of the e.m.f. method to cmc
determinations especially of systems with low cmc and high concentration of various added electrolytes, including multivalent counterions, to the future.

   In previous publications (10,16) we have described the procedure used to obtain the degree of binding, β, defined as

$$\beta = (m_D - m_D^f)/m_p \qquad (1)$$

where m_D is the total surfactant concentration, m_D^f the free surfactant concentration and m_p the monomolar polyion concentration
(i.e. moles COO^-/kg H_2O), from the e.m.f. data. All our data will
be presented as "binding iostherms", where β is plotted vs log m_D^f.
As has been demonstrated before (19) the average linear charge separation on the polymer is the predominant factor in determining
the m_D^f region where cooperative binding is observed. This charge
separation on the polyion is often expressed in the form of a
charge density parameter ξ,

$$\xi = e^2/\epsilon bkT \qquad (2)$$

where e is the protonic charge, ε the dielectric constant, k the
Bolzmann constant and T the temperature, and b is the average linear charge separation on the polymer, i.e. the average distance
between charged groups on the fully extended polymer. Thus the
charge density parameter ξ of polyacrylate has the value 2.83 typical for vinylic polymer chains. For pectate ξ = 1.61 if we assume that the 10% neutral sugars in this polymer are not randomly
distributed. As is discussed in ref. 19, in the case of alginate
there are good reasons to assume that the charge density parameter
ξ is slightly larger than the value of 1.43 expected for a polyman-
nuronic acid chain. Both the presence of guluronic acid blocks,
and the larger flexibility of this polymer (42) would indicate a
value in between 1.43 and 1.61, the value for pectate (19). A typical example of the influence of the charge density is shown in

Figure 5, where we compare the binding of $C_{14}Py^+$ in the presence of
0.01 m NaCl to PAA, pectate and alginate. We note that the order
of the overall binding constant Ku, defined by (2,11,12,16) varies

$$Ku = (m_D^f)^{-1}_{0.5} \qquad (3)$$

in the order PAA >> alginate > pectate. This order of Ku is the
same as was observed for alkyltrimethylammonium ions (19), but in
the present case of alkylpyridinium ions the difference between
alginate and pectate is slightly more pronounced. A number of
other minor but typical characteristics can be observed in Figure
5. Binding to PAA reaches a second critical point around β = 0.7,
again similar to the case of the corresponding trimethylammonium
ions. All alkyl pyridinium binding curves give indications of a
two-step binding process, as may be deduced from the behaviour of
the binding isotherms below β = 0.5. The pectate binding curve
seems to level off above β = 1, possibly indicating that the alkylpyridinium ion can bind to the approximately 10% neutral sugars
present in pectate but not in alginate. All the pectate binding
curves exhibit a significantly lower cooperativity, i.e. the rise
in β with increasing m_D^f is less steep, than the corresponding alginate binding curve.
    In Figure 6 we compare the binding of $C_{12}Py^+$ and $C_{14}Py^+$ to PAA
to the case of the co-responding dodecyl- and tetradecyltrimethyl
ammonium ions (DTA$^+$ and TTA$^+$) both in the presence of 0.01 m NaCl,
and in Figure 7 a similar comparison is made for $C_{14}Py^+$ and $C_{16}Py^+$
binding to alginate and pectate. The remarkably consistent binding patterns of the various cations and polyions attest not only
to the reproducibility of the results, but also to the highly specific character of the binding process. In the case of post-mi-
cellar binding the polymer concentration is an important parameter
(24). In the present case the relation between the degree of binding, β, and the free surfactant concentration, m_D^f, is completely
independent of the equivalent polymer concentration. For instance, curves in Figures 5-8 represent polymer monomolal concentra-
tions of 10^{-4}, 5×10^{-4}, and 10^{-3} m, without any noticeable difference in the trends observed. Finally, in Figure 8 we show the
influence of the polymer backbone, varying the degree of esterification of the carboxyl group in pectinates derived from the same
polypectate. Relevant parameters characterizing the pectate and
pectinates used are given in Table I (31,43). All binding isotherms in Figure 8 are for an equivalent polyion concentration
(COO$^-$ concentration) of 1×10^{-3} m, note that therefore the actual
polymer concentration increases as the degree of esterification
increases. The influence of the charge density is evident. In
addition, the influence of the added NaCl concentration shows an
interesting pattern: the difference between the m_D^f values at the
half-bound point (β = 0.5) between m_{NaCl} = 0.01 and 0.02 is largest for pectate (and approximately as expected compared e.g. to
the case of dextransulfate (16)) and becomes progressively smaller
as the polyion charge density decreases.

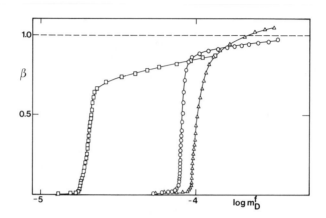

Figure 5. Binding isotherms for $C_{14}PyBr$. m_{NaCl} = 0.01 m. Polymers: ☐ NaPA (5 x 10^{-4} m); o Na-alginate, △ K-pectate (1 x 10^{-3} m).

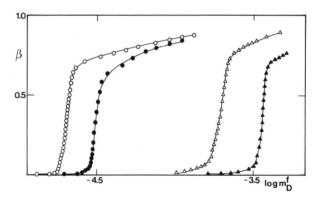

Figure 6. Binding isotherms with polyacrylate (5 x 10^{-4} m), m_{NaCl} = 0.01 m. △ $C_{12}PyCl$, o $C_{14}PyBr$; ▲ DTABr, ● TTABr (ref. 19).

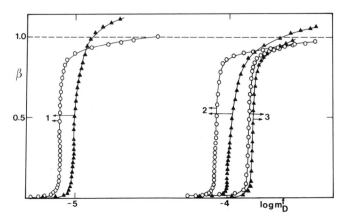

Figure 7.   Polyuronide binding isotherms. m_{NaCl} = 0.01 m.
o Na-alginate; ▲ K-pectate. Curves 1: $C_{16}PyBr$
(1 x 10^{-4} m uronide); 2: $C_{14}PyBr$, 3: $C_{12}PyCl$
(1 x 10^{-3} m uronide).

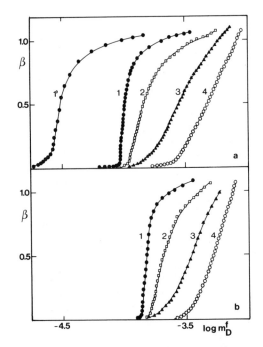

Figure 8.   Pectate and pectinate binding isotherms, $C_{14}PyBr$,
m_{COO^-} = 1 x 10^{-3}. esterification (E): 1 0%,
2 20.6%, 3 46.1%, 4 70%. a: 0.01 m NaCl.
b: 0.02 m NaCl. 1': 0%, salt free.

TABLE I.   Characterization of Pectate and Pectinates

Polyuronides	E^1	$[\eta]^2$	M_η^3	ξ
	%	%	m^3kg^{-1}	
K-pectate	84.9	0	0.133	29,000	1.61
Na-pectinate (2)	86.1	20.6	0.211	41,000	1.28
Na-pectinate (3)	83.8	46.1	0.327	56,000	0.87
Na-pectinate (4)	88.9	70.0	0.092	22,000	0.48

1 Degree of esterification
2 Intrinsic viscosity
3 Molecular weight calculated from viscosity (ref. 43).

All binding parameters derived from fitting the binding isotherms to the equations of Schwarz or Satake and Yang (2,16,19)
are collected in Table II. As has been stated before, the overall
binding constant Ku can be determined accurately (estimated at
± 2%), but the determination of K and u separately is much more
inaccurate. Generally we estimate the possible error in u at
± 20%. Even if the model considerations which equate u to a cooperativity parameter describing the aggregation of bound surfactant molecules, prove incorrect or inapplicable, from an experimental point of view u may be seen simply as a parameter indicating the slope of the binding isotherm in the cooperative region,
i.e. higher u values mean steeper binding isotherms. What is most
obvious from Table II is the identical surfactant chainlength dependence of the Ku values for all polymers, independent of the pyridinium of trimethylammonium head group of the surfactant and of
the presence of added salt. The difference per CH_2 group in ln Ku
for all cases presented in Table II averages $1.19_{\pm}$kT, very close
to the value of 1.23 kT found for the case of DTA^+ and TTA^+ binding to DNA with or without added NaCl (18). It is hard to find
any other explanation for this remarkable constancy than to assume
that this factor reflects only the difference in hydrophobic interactions between the C_{12}, C_{14}, or C_{16} alkyl groups, and that the
intrinsic binding between surfactant and polyion is unaffected by
the surfactant hydrophobic chainlength. Of course the similarity
between this hydrophobic effect in surfactant binding by polymers
and micelle formation has been pointed out many times, but it is
nevertheless satisfying to see this almost perfect correspondence
between such widely varying systems.
    Finally, we will consider the variation in Ku with surfactant
head group, polymer structure and charge density, and added salt
concentration. Generally, Ku decreases with increasing salt concentration, which is easily explained in terms of increased shielding of the polymer charges. It remains to be seen whether in
fact this is a correct interpretation, it might for instance be
argued that similar to the case of metal ion binding by polyelec-

TABLE II. Binding Constants Ku and Parameters u for Surfactant–Polyion Binding.

Polymer	m_{NaCl} (mol/kg)	$C_{12}Py^+$		$C_{14}Py^+$		$C_{16}Py^+$		DTA^+[2]		TTA^+[2]	
		log Ku	u[1]	log Ku	u[1]	log Ku	u[1]	log Ku	u[1]	log Ku	u[1]
PAA	0									5.39	20
	0.01	3.69	200	4.67	250			4.43	15	4.48	600
Alginate	0	3.58	100	4.08	2300			3.43	500	4.48	150
	0.01			4.52	770	5.08	4800	3.30	70	3.88	2000
Pectate	0	3.42	45	3.99	250					4.46	60
	0.02			3.82	320	5.00	1100	3.35	70	3.86	2000
Pectinate (20.6% E)	0.01			3.83	26						
	0.02			3.68	50						
Pectinate (46.1% E)	0.01			3.55	12						
	0.02			3.45	26						
Pectinate (70% E)	0.01			3.32	12						
	0.02			3.28	26						

1 Estimated precision ± 20%
2 reference 19.

trolytes it is the entropy gain of the released counterions (i.e.
Na^+) which should be considered ($\underline{44},\underline{45}$). Both approaches would
predict a decrease in Ku with increasing salt concentration for a
given polyion-surfactant system, and both approaches would also
predict a smaller dependence of Ku on the added salt concentration
the smaller the polyion charge density. This last effect is sen-
sitively demonstrated not only in our data for the pectinates, but
also in the comparison between the salt dependence of Ku for dextransulfate, polystyrene sulfate ($\underline{16}$) and polyacrylate (Table II),
all with charge density parameters of 2.8, and alginate and pectate.

The influence of the head group, i.e. pyridinium or trimethylammonium, on the overall binding constant Ku and the cooperativity parameter u is relatively small. It is perhaps surprising
that in all cases Ku for the pyridinium salt is larger than for
the trimethylammonium salt, as can be seen best e.g. by comparing
$C_{14}Py^+$ and TTA^+, just as the cmc for pyridinium salts is always lower than the cmc of the corresponding trimethylammonium salt.
This may be caused by two factors. First of all, in the case of
pyridinium salts there may be a contribution from the hydrophobic
interactions between neighbouring bound headgroups (an effect
which would not contribute to the free energy of micelle formation). Secondly, a steric hindrance effect may prevent the positive chrge on the trimethylammonium head group from approaching
close to the polyion charge.

In comparing binding data for the various polymers, it is
clear that indeed the charge density is the dominant factor governing Ku. The inversion in Ku values observed between alginate and
pectate should then be attributed to the larger flexibility of the
alginate polyion ($\underline{42}$), allowing it to bind and "envelop" the surfactant aggregates more efficiently. It is noteworthy that the Ku
values for DTA^+ at 0.01 m NaCl with dextransulfate and polyacryla-
te are virtually identical ($\underline{16},\underline{19}$) but that in the case of poly-
styrenesulfonate Ku for DTA^+ is much larger, and u very much lower.
These three polyions all have an identical charge density parameter of 2.8, and we conclude that only in the case of polystyrene-
sulfonate at least part of the surfactant alkyl group binds to the
hydrophobic polymer backbone, and does not contribute to the cooperative binding between neighbouring surfactants. Thus it seems
likely that, all other things being equal, more flexible polymers
are more efficient in binding surfactants, as is particularly
clear from the case of DNA ($\underline{18}$). We have pointed out the similarity between polyion surfactant ion interaction and micelle formation of free surfactants. This similarity can be seen from e.g.
the chainlength dependence of the overall binding constant as discussed above, or from a comparison of the thermodynamic parameters
describing both processes ($\underline{46}$). In the present context it now
seems likely that this similarity extends to the actual aggregation process of the bound surfactants, with the polyion enveloping
the micelle-like aggregates and neutralizing the charge of the

surfactant head groups. The remarkable fact remains that this charge neutralization allows aggregation to take place at free surfactant concentrations orders of magnitude below the cmc, dependent on the polyion charge density.

Acknowledgments

We are grateful to the Natural Sciences and Engineering Research Council of Canada and the Czechoslovak Academy of Sciences for the award of a scholarship under the auspices of the scientific exchanges agreement between the Council and the Academy, and the Killam Foundation for the award of a postdoctoral fellowship to one of the authors (A.M.). The authors are grateful to Drs. B. Larsen and O. Smidsrød, Institute of Marine Biochemistry, University of Trondheim, Norway for the donation of a fully characterized sample of Na-alginate, and to Dr. R. Kohn, Institute of Chemistry, Slovak Academy of Sciences, Bratislava, Czechoslovakia, for preparing and characterizing pectin samples with different degrees of esterification. This research is supported by the Natural Sciences and Engineering Research Council of Canada through grants to J.C.T.K.

Literature Cited

1. Goddard, D.E., Hannan, R.B. J. Colloid Interface Sci. 1976, 55, 73.
2. Satake, I., Yang, J.T. Biopolymers 1976, 15, 2263.
3. Gavach, C., Bertrand, C. Anal. Chim. Acta 1971, 55, 385.
4. Birch, B.J., Clarke, D.E. Anal. Chim. Acta 1973, 67, 387.
5. Cutler, S.G., Meares, P., Hall, D.G. J. Electroanal. Chem. 1977, 85, 145.
6. Yamauchi, A., Kunisaki, T., Minematsu, T., Tomokiyo, Y., Yamaguchi, T., Kimizuka, H. Bull. Chem. Soc. Jpn. 1978, 51, 2791.
7. Satake, I., Gondo, T., Kimizuka, H. Bull. Chem. Soc. Jpn. 1979, 52, 361.
8. Kale, K.M., Cussler, E.L., Evans, D.F. J. PHys. Chem. 1980, 84, 593.
9. Maeda, T., Ikeda, M., Shibaharu, M., Haruta, T., Satake, I. Bull. Chem. Soc. Jpn. 1981, 54, 94.
10. Hayakawa, K., Ayub, A.L., Kwak, J.C.T. Colloids Surf. 1982, 4, 389.
11. Zimm, B.H., Bragg, J.K. J. Chem. Phys. 1980, 84, 593.
12. Schwarz, G. Eur. J. Biochem. 1970, 12, 442.
13. Shirahama, K., Yuasa, H., Sugimoto, S. Bull. Chem. Soc. Jpn. 1981, 54, 375.
14. Fukushima, K., Murata, Y., Nishikido, N., Sugihara, G., Tanaka, M. Bull. Chem. Soc. Jpn. 1981, 54, 3122.
15. Fukishima, K., Murata, Y., Sugihara, G., Tanaka, M. Bull. Chem. Soc. Jpn. 1982, 55, 1376.

16. Hayakawa, K., Kwak, J.C.T. J. Phys. Chem. 1982, 86, 3866.
17. Hayakawa, K., Kwak, J.C.T. J. Phys. Chem. 1983, 87, 506.
18. Hayakawa, K., Santerre, J.P., Kwak, J.C.T. Biophys. Chem. 1983, 17, 175.
19. Hayakawa, K., Santerre, J.P., Kwak, J.C.T. Macromolecules 1983, 16, 1642.
20. Satake, I., Yang, J.T. Biopolymers 1975, 14, 1841.
21. Hayakawa, K., Ohara, K., Satake, I. Chem. Lett. (Jpn.) 1980, 647.
22. Robb, I.D. in "Surfactant Science Series", Vol. 11, E.H. Lucassen-Reynders, ed., Marcel Dekker, New York, 1981, p. 109.
23. Oteri, R., Dubin, P.L. Polymer Preprints 1982, 23, 45.
24. Dubin, P.L. Oteri, R. J. Coll. Interface Sci. 1983, 95, 453.
25. Murata, M., Arai, H. J. Coll. Interface Sci. 1973, 44, 475.
26. Gilanyi, T., Wolfram, E. Colloids Surf. 1981, 3, 181.
27. Nagarajan, R. Colloids Surf., in press.
28. Cabane, B., Colloids Surf., in press.
29. Kohn, R. Pure Appl. Chem. 1975, 42, 371.
30. Knight, A., Shaw, B.D. J. Chem. Soc. 682, 1938.
31. Kohn, R., Furda, I. Collect. Czech. Chem. Commun. 1967, 32, 1925.
32. Owens, H.S., Lotzkar, H., Schultze, T.H., Mackay, W.D., J. Amer. Chem. Soc. 1946, 68, 1628.
33. Tibensky, V., Rosik, J., Zitko, V. Nahrung 1963, 7, 321.
34. Kohn, R., Tibensky, V. Chem. Zvesti 1965, 19, 98.
35. Mukerjee, P., Mysels, K.J., "Critical Micelle Concentrations of Aqueous Solutions", NSRDS-NBS 36, USA, 1971.
36. Hoffman, H., Nagel, R., Platz, G., Ulbricht, W. Colloid Polymer Sci. 1976, 254, 812.
37. Rosen, M.J., Dahanayake, M., Cohen, A.W. Colloids Surf. 1982, 5, 159.
38. Evers, I.C., Kraus, C.A. J. Amer. Chem. Soc. 1948, 70, 3049.
39. Paluch, M. J. Coll. Interface Sci. 1978, 66, 582.
40. Anacker, E.W. J. Phys. Chem. 1959, 62, 41.
41. Anacker, E.W. in "Cationic Surfactants", Jungerman, E., Ed., Marcel Dekker, New York, 1970, p. 203.
42. Aspinall, G.O., "The Carbohydrates, Chemistry and Biochemistry", Academic Press; New York, 2nd ed., 1970, II, p. B515.
43. Kohn, R., Luknar, O. Collect. Czech. Chem. Commun. 1975, 40, 959.
44. Manning, G.S. Q. Rev. Biophys. 1978, 2, 179.
45. Mattai, J., Kwak, J.C.T. J. Phys. Chem. 1982, 86, 1026.
46. Santerre, J.P., Hayakawa, K., Kwak, J.C.T. Colloids Surf. in press.

RECEIVED March 6, 1984

Linear Sodium Alkylbenzene Sulfonate Homologs

Comparison of Detergency Performance with Experimental and Thermodynamic Wetting Theories

JAMES A. WINGRAVE[1]

Conoco, Inc., Chemical Research Division, Ponca City, OK 74603

Pure homologs of linear alkylbenzene sulfonate
sodium salts (LAS) were evaluated for detergency
performance. The surface tensions of the wash
liquors used for these detergencies were then
measured and used in conjunction with a wetting
model to calculate a theoretical detergency performance. The theoretical and experimental detergency results were compared. The molecular structure effects of the LAS homologs on detergency
performance were calculated by incorporating into
the detergency equation several molecular structure theories such as the cohesive energy ratio
concept, molar-attraction constants, internal
liquid pressure, and liquid thermal properties.
The assets and deficiencies of these approaches
are discussed.

The process of detergency involves the complete separation of
two substances by means of a detersive bath. The success of
such a process requires a knowledge of the chemistry between the
two substances to be separated (henceforth referred to as soil
and substrate). From this knowledge, a detergency bath can be
designed with chemical characteristics necessary to separate the
soil and substrate by overcoming the attractive forces between
them. However, stating the principles of detergency and achieving them in practice are different; the latter being far more
difficult, as noted by the voluminous literature on the subject
(1-10). It will, therefore, be the purpose of this paper to
develop a better understanding of how surfactant structure

[1]Current address: E. I. du Pont de Nemours, Inc., Chambers Works, Deepwater, NJ 08023

effects detergency performance by using a homologous series of
linear alkylbenzene sulfonate sodium salt (LAS) derivatives in
the detergency bath.

The kinetic effects of detergency will not be explored in
this study, but review articles on this topic can be found elsewhere (1-5). Instead, the agitation will be held constant (see
Experimental Section) so that the equilibrium (or near equilibrium) processes can be observed. Equilibrium was achieved for
similar soil/substrate systems within 5-10 minutes in previous
studies (3).

For this study, the soil will be a mineral oil constituted
as shown in Figure 1. The cloth used will be a cotton/polyester
blend with a polyethylene glycol fabric finish. Since the soil
will be a liquid, effective detergency will require a bath that
can overcome the physical forces between the soil and fabric
(i.e., the fabric finish). For physical forces, the problem becomes one of developing a model of the detergency system, then
combining this model with wetting theory in order to produce a
detergency equation for detergency performance.

Discussion

Soil-Fabric Morphology. In Figure 2, scanning electron micrographs of the fabric used in this study are shown. In the
threads of the fabric are fibers which run very nearly parallel.
When soiled with a liquid soil, a pendular drop of soil should
form between contiguous fibers as shown in cross section in
Figure 3. The bath/soil interface is shown planar and parallel
with the plane passing through the fiber centers (Figure 4a).
When one chooses a particular detergent bath/soil/fiber combination, the contact angle, ψ (measured through the soil
phase), at the line where all three phases meet will strive to
attain a specific value based on the properties of the three
phases, as shown in Figure 4b. Therefore, the triple-phase line
(TPL) will move to assume a value of ψ. This, in turn, will
cause a curvature J in the soil/bath interface. This movement
of the TPL is the process by which large soil drops can be
formed on/in the fabric as ψ approaches values of 90° and
greater, as shown in Figure 4c. This process is relatively
rapid, occurring within 10 minutes for most systems, and is generally referred to as the roll-up mechanism of soil removal.
From this description, it is obvious how agitation and buoyant
effects of the soil could speed up this mechanism (in fact,
roll up cannot occur at all unless buoyant or agitation forces
act on the soil), but soil-removal rate is a kinetic question
and will not be pursued further here.

Another soil-removal mechanism is also working, simultaneously, to remove soil albeit a much slower process (i.e., up
to several hours). When the bath/soil interface becomes curved
due to a change in ψ (see Figure 4d), its chemical potential is

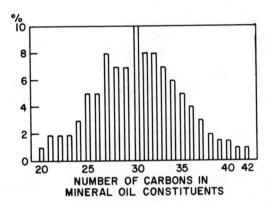

Figure 1.   Constitution of the Mineral Oil Soil.

Figure 2.   Scanning Electron Micrographs of Fabric Used for Detergency Studies.

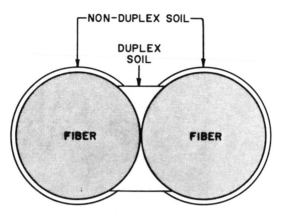

Figure 3.   Cross-Sectional View of Soil Drop Between Contiguous Fibers.

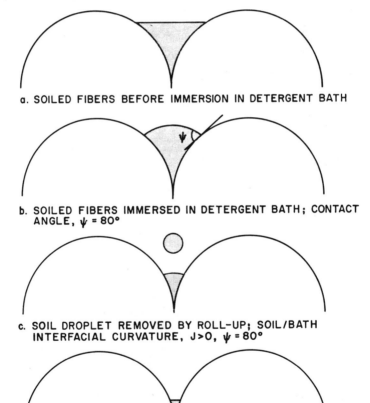

a. SOILED FIBERS BEFORE IMMERSION IN DETERGENT BATH

b. SOILED FIBERS IMMERSED IN DETERGENT BATH; CONTACT
   ANGLE, ψ = 80°

c. SOIL DROPLET REMOVED BY ROLL-UP; SOIL/BATH
   INTERFACIAL CURVATURE, J>0, ψ = 80°

d. SOIL REMOVAL THROUGH EQUILIBRATION; J=0, ψ = 80°

Figure 4.   Mechanism for Soiled Fabric Fiber Detergency.

greater than a flat, bath/soil interface. The instantaneous curvature J will change through soil dissolution until it achieves a smaller curvature (flatter interface) J_e. The soil removed in this way will either solubilize in the bath, collect as small macroscopic oil droplets at the bath/air interface, or emulsify if agitation is present. Hence, in a detergency system that has attained equilibrium, soil lenses or small (nominally 1 mm diameter) soil droplets are commonly seen floating at the air/bath interface. Under these circumstances, the curvature is very small (nominally 1 mm diameter), and when applied to the scale of pendular oil droplets between fibers (nominally 1 to 10 μm diameter) the bath/soil interface will be flat for all practical purposes. Therefore, the model in Figure 3, showing a flat bath/soil interface, should be valid for all detersive systems where bulk soil is observed and fabric fibers are small (<100 nm diameter).

From this model, the volume of soil can be determined for the geometry shown in Figure 5 and the Appendix. The equation for this volume of a partially soil-filled pore with a contact angle ψ compared to the total pore volume, V_R, is (11)

$$V_R = \frac{2 \cos \psi \ (2-\sin \psi) - (\pi - 2 \ \psi)}{(4-\pi)} \qquad (1)$$

Then since detergency is a soil-removal process, the relative detergency, D_R, can be written as a function of ψ only as

$$D_R = 1 - \frac{2 \cos \psi \ (2-\sin \psi) - (\pi - 2 \ \psi)}{(4-\pi)} \qquad (2)$$

The quantity D_R ignores the adsorbed soil on the fiber surfaces. However, this can be shown (11) to be an insignificant amount of soil compared to the pore-held soil unless the fiber diameters were <0.1 μm. Hence Equation 2 should be quite satisfactory for almost all cloth in which fiber diameters are generally ≥1.0 μm.

It is also noteworthy that the contact angle has been defined as being measured through the oil phase since that is the medium of interest. Furthermore, ψ varies directly with D_R by this definition (see Figure 6). Conventionally, contact angle is measured through the most dense medium (which would be the aqueous detergent bath) and given the symbol θ. Hence, the symbol ψ will be used in this work while it is clear that the two quantities are related as, $\psi = 180° - \theta$.

With this geometric model for D_R as a function of ψ, it is now necessary to develop a method to evaluate ψ from wetting or surface chemistry parameters.

Evaluation of the Soil-Fiber Morphology Equation. The most direct method of evaluating Equation 2 for D_R is to measure the contact angle ψ for the fiber-soil-bath system of interest.

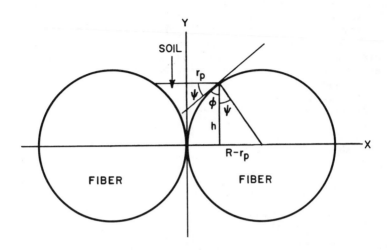

Figure 5.   Cross-Sectional View of Soil/Fiber Geometry.

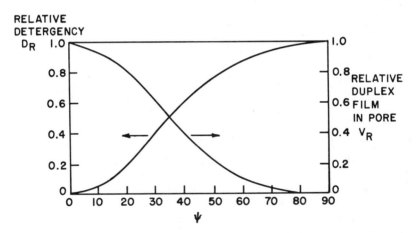

Figure 6.   Dependence of Detergency on Soil-Fiber-Water Contact Angle.

While direct, this method is the most difficult experimentally due to the diminutive nature of fiber diameters and the uncertainty involved with contact angle measurements and hysteresis. The value ψ can also be measured on flat sheets of the fiber material but due to fabric finishes and different surface properties incurred during manufacture, the surface energetics of the sheet and fiber may be very dissimilar. Therefore, the value of $\cos\psi$ was determined in the following manner from detergency data. The Kubelka-Munk Equation (12-13),

$$V_R(KM) = \frac{\alpha_{soil} - \alpha_{wash}}{\alpha_{soil} - \alpha_{clean}} \tag{3}$$

where
$$\alpha = \frac{K}{S} = \frac{(1-R_d)^2}{2R_d}$$

R_d = reflectance (see Experimental Section for more details)
K = coefficient of reflectivity (12-13)
S = coefficient of light scattering (12-13)
α_{wash} = K/S for soiled cloth after washing
α_{soil} = K/S for soiled cloth
α_{clean} = K/S for unsoiled cloth

was combined with Equations 1 and 3

$$V_R(KM) = \frac{\alpha_{soil} - \alpha_{wash}}{\alpha_{soil} - \alpha_{clean}} = \frac{2 \cos \psi \, (2-\sin \psi) - (\pi-2 \, \psi)}{(4-\pi)} \tag{4}$$

This yields an equation from which contact angle values, ψ, can be determined directly from reflectance readings (see Table 1).

Wettability Dependence of Detergency. The dependence of D_R on ψ was described above. The purpose of this section will be to develop an expression for ψ in terms of experimentally accessible interfacial tension quantities. To begin with, the TPL has a geometry as shown in Figure 7. For this system, the Young-Dupre' equation (14-16) can be written as

$$\gamma_{fw} = \gamma_{fs} + \gamma_{sw} \cos \psi \tag{5}$$

In a similar manner, an expression for the interfacial tension, γ_{fw} can be written using the geometric mean definition for work of adhesion, W_A (see Figure 8), and the Girifalco-Good interaction parameter (17-18), Φ as,

$$W_A = 2 \, \Phi_{fw}^* \sqrt{\gamma_{sw} \, \gamma_{fs}^*} \tag{6}$$

or

Table I. Kebulka-Munk Transformation of Detergency Reflectance Data to Wettability Data

Hardness (ppm)	LAS[a]	R_d	Detergency, D_R (%)	$\cos \psi$[b]
50	10	61.3	65.1	0.741
	11	63.7	70.7	0.702
	12	63.9	71.0	0.699
	13	64.9	73.2	0.682
	14	65.2	73.8	0.677
150	10	51.6	33.7	0.897
	11	58.1	56.4	0.793
	12	65.2	73.8	0.677
	13	65.8	75.0	0.677
	14	66.3	75.9	0.660
300	10	49.5	24.5	0.929
	11	62.2	67.2	0.727
	12	64.4	72.1	0.691
	13	61.6	65.7	0.737
	14	59.9	61.5	0.764

[a] value refers to the number of carbon atoms in the linear alkyl chain of the linear alkylbenzene sulfonate sodium salt.

[b] From Kebulka-Munk/detergency Equation 4.

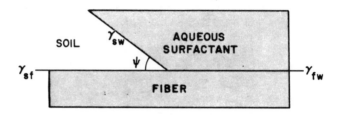

Figure 7.   Triple-Phase Line Geometry for Soil/Fiber System.

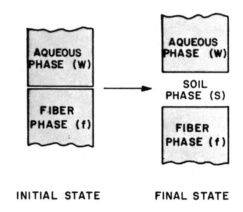

INITIAL STATE   FINAL STATE

Figure 8.
Process for Defining the Work of Adhesion Between Two Dense
Phases.

$$\gamma_{fw}^* = \gamma_{sw} + \gamma_{fs}^* - 2\ \Phi_{fw}^* \sqrt{\gamma_{sw}\ \gamma_{fs}^*} \tag{7}$$

where the asterisk signifies that γ_{fs}^* is measured in the absence of the detergent bath and γ_{fw}^* in the absence of the soil.

Notice that the asterisk designation is not required when the third phase is air, or the vapor of one or more of the components already present in the system, since interfacial interactions are not significantly affected by phase dissolution and adsorption of air or a common vapor. Hence the asterisk designation was not required in Girifalco and Good's original work (17-18). In a similar fashion, the asterisk is not used with γ_{sw} since the solid is assumed insoluble in both the soil and bath.

When Equations 5 and 7 are combined and rearranged, a Girifalco-Good-type equation results as[‡]

$$\cos\psi = 1 - 2\ \Phi_{fw}^* \sqrt{\frac{\gamma_{fs}^*}{\gamma_{sw}}} + \frac{\pi_{fs} - \pi_{fw}}{\gamma_{sw}} \tag{8}$$

where
$$\pi_{fs} = \gamma_{fs}^* - \gamma_{fs}$$
$$\pi_{fw} = \gamma_{fw}^* - \gamma_{fw}$$

If the substitution for Φ_{fw}^*

$$\Phi_{fw}^* = (\Phi_{fw}^* - 1) + 1 \tag{9}$$

is combined with Equation 8 and the resulting equation rearranged, then

$$\cos\psi = 1 - \frac{2\sqrt{\gamma_{fs}^*}}{\sqrt{\gamma_{sw}}} + \frac{\pi_{fs} - \pi_{fw} + 2\ (1 - \Phi_{fw}^*)\ \sqrt{\gamma_{sw}\ \gamma_{fs}^*}}{\gamma_{sw}} \tag{10}$$

Equation 10 is obviously a Girifalco-Good equation for a solid (f), liquid (s), liquid (w) system. If values of ψ and γ_{sw} are measured for a series of surfactant homolog solutions, a plot of $\cos\psi$ versus $1/\sqrt{\gamma_{sw}}$ will give a Girifalco-Good plot from which values of γ_{fs}^* and other surface energetic parameters can be determined under appropriate conditions. This evaluation will be the topic of the following sections.

Wettability Parameters From Detergency Data. If a series of surfactant homolog solutions is used to evaluate ψ and γ_{sw} for a given fiber and soil system, a Girifalco-Good plot can be constructed. Such data was gathered for LAS homologs at 50, 150, and 300 ppm hard ion concentration and plotted in Figure 9. From the form of Equation 10, it is clear that a plot of $\cos\psi$ versus $1/\sqrt{\gamma_{sw}}$ will give a linear plot which will fit a straight-line equation

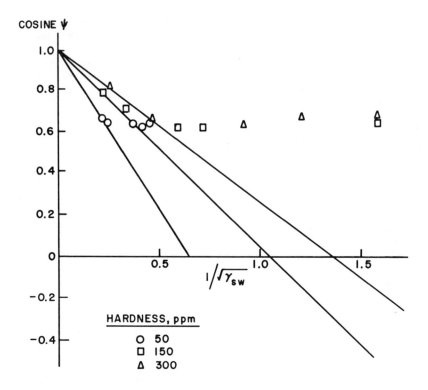

Figure 9.
Girifalco-Good Plot of Wettability Data From Detergency and
Interfacial Tension Data.

$$\cos \psi = - \frac{a}{\sqrt{\gamma_{sw}}} + 1 + \frac{b}{\gamma_{sw}} \tag{11}$$

where $a = 2\sqrt{\gamma_{fs}^*}$

$b = \pi_{fs} - \pi_{fw} + 2 (1 - \Phi_{fw}^*)\sqrt{\gamma_{sw}\gamma_{fs}^*}$

From the form of Equation 11, it is clear that the slope, a, will be independent of the aqueous surfactant solution. However, b in the intercept, $1 + (b/\gamma_{sw})$, will not be zero; in fact, it will not even be independent of the aqueous surfactant solution. Therefore, the conventional methods of Girifalco-Good plot analysis with an intercept of unity will not work for the detergency systems of interest in this work. The impediment to the Girifalco-Good analysis method is obvious from Figure 9, where no set of data points (3 or more) lies on a line passing through the coordinates $\cos \psi = 1$, $0 = 1/\gamma_{sw}$. The evaluation of the surface chemical properties will require a different evaluation method.

Wetting Parameters in the Detergency Equation. The detergency system has three interfaces which have three interfacial tensions at equilibrium with all three phases: γ_{sw}, γ_{fs}, and γ_{fw}. When only two phases are in equilibrium, three other surface tensions are possible: γ_{sw}^*, γ_{fs}^*, and γ_{fw}^* (however, when the fiber is an insoluble solid, $\gamma_{sw} = \gamma_{sw}^*$, and so only γ_{sw} will be referenced). Corresponding to these interfaces are Girifalco-Good interaction parameters: Φ_{sw}^*, Φ_{fs}^*, and Φ_{fw}^*. The final parameter type in a detergency system is film pressure of the two liquids on the solid fiber: π_{fs} and π_{fw}.

These ten interfacial parameters give a very complete description of the energetics of a detergency system. Further surface tension variables for a fluid-air or solid-air interface will also be used to evaluate these ten interfacial parameters. The remainder of this section will explore their evaluation.

To begin with, as many of the parameters as possible were experimentally measured. These values are given in Table 2. The contact angle ψ was determined from detergency results as described earlier.

The values of γ_s, γ_w, and γ_{sw} were measured by spinning drop interfacial tensiometry (SDIT) as described in the Experimental Section. The values of γ_f^* and γ_{fs}^* were obtained by the SDIT method using a polyethylene glycol (PEG), Jeffox PEG-300

---

‡ Note that the $\cos \psi = - \cos \theta$, hence, Equation 6 differs from References 14-16   due to the definition of the contact angle.

Table II

Hardness (ppm)	HF LAS Homo- log	MEASURED VALUES[a]			Φ_{sw}	Φ_{fw}^*
		cos ψ	γ_w	γ_{sw}		
50	10	0.741	46.7	20.1	0.756	0.978
	11	0.702	41.1	15.3	0.794	0.992
	12	0.699	30.2	7.24	0.879	1.007
	13	0.682	30.7	5.57	0.908	0.996
	14	0.677	30.2	4.04	0.932	0.985
150	10	0.897	41.7	17.6	0.764	0.984
	11	0.793	33.0	8.2	0.870	1.003
	12	0.667	26.7	2.8	0.951	0.977
	13	0.667	25.9	1.9	0.968	0.964
	14	0.660	25.8	0.4	0.995	0.911
300	10	0.929	38.3	13.3	0.811	0.997
	11	0.727	30.3	4.5	0.925	0.989
	12	0.671	26.0	1.2	0.981	0.945
	13	0.737	27.5	0.7	0.989	0.923
	14	0.764	26.0	0.4	0.994	0.911

[a]Other measured values:

$\gamma_s^* = 29.6$ dynes/cm.

$\gamma_f^* = 39.7$ dynes/cm.

$\gamma_{fs}^* = 10.5$ dynes/cm.

[b]Other calculated values:

$\Phi_{fs}^* = 0.858$.

CALCULATED VALUES[b]

γ^*_{fw}	γ_{fw}	π_{fw}	γ_{fs}	π_{fs}	b
2.19	16.1	−13.9	1.20	9.3	−23.8
0.65	12.0	−11.3	1.30	9.2	−20.8
0.19	5.95	−5.76	0.887	9.6	−15.3
0.83	4.74	−3.91	0.943	9.6	−13.52
1.71	3.65	−1.94	0.910	9.6	−11.72
1.36	16.0	−14.64	0.180	10.3	−25.4
0.08	6.96	−6.88	0.455	10.05	−16.9
2.70	2.74	−0.04	0.843	9.7	−9.94
3.79	2.17	1.62	0.901	9.6	−8.30
7.16	1.47	5.69	1.203	9.3	−3.96
0.23	12.4	−12.2	0.072	10.4	−22.7
1.41	3.94	−2.53	0.665	9.8	−12.5
4.99	1.63	3.36	0.802	9.7	−6.73
6.19	1.089	5.10	0.573	9.9	−5.24
7.17	0.782	6.39	0.477	10.0	−4.00

(Jefferson Chemical Company, Austin, Texas), for the fiber sur-face analog. This substitution was made since the fabric used in this study had a PEG fabric finish. This fabric finish differed from the Jeffox PEG-300 only in the fact that the for-mer was higher molecular weight and therefore solid, while the latter was a liquid and amenable to SDIT measurement of γ_f^* and γ_{fs}^*. With these values, the remaining eight surface parameters can be calculated.

The Girifalco-Good equation for the soil-water interfacial tension

$$\gamma_{sw} = \gamma_s^* + \gamma_w^* - 2\,\Phi_{sw}^* \sqrt{\gamma_s^*\ \gamma_w^*} \tag{12}$$

can be rearranged to calculate Φ_{sw}^* as

$$\Phi_{sw}^* = \frac{\gamma_s^* + \gamma_w^* - \gamma_{sw}^*}{2\sqrt{\gamma_s^*\ \gamma_w^*}} \tag{13}$$

where $\Phi_{sw}^* = \Phi_{sw}$, as was the case for sw interfacial tensions.

For the quantity γ_{fs}, one can combine the Young-Dupré equation

$$\gamma_{fw} = \gamma_{fs} + \gamma_{sw}\cos\psi \tag{14}$$

with the Girifalco-Good equation

$$\gamma_{sw} = \gamma_{fs} + \gamma_{fw} - 2\,\Phi_{sw}\sqrt{\gamma_{fs}\ \gamma_{fw}} \tag{15}$$

to yield the quadratic equation for γ_{fs} as

$$\gamma_{fs}^2 - \left[\frac{(1-\Phi_{sw}^2)\,\gamma_{sw}\cos\psi}{1-\Phi_{sw}^2}\right]\gamma_{fs} + \frac{\gamma_{sw}^2\,(1-\cos\psi)}{4\,(1-\Phi_{sw}^2)} = 0 \tag{16}$$

which can be solved and simplified to yield

$$\gamma_{fs} = \frac{\gamma_{sw}}{2\,(1-\Phi_{sw}^2)}\left\{\left[1-(1-\Phi_{sw}^2)\cos\psi\right]+ \right.$$

$$\left. \sqrt{\left[1-(1-\Phi_{sw}^2)\cos\psi\right]^2 - (1-\cos\psi)^2}\right\} \tag{17}$$

Using the value of γ_{fs}, the quantity γ_{fw} can be calculated from the Young-Dupré Equation 14.

The quantity γ_{fw}^* can be calculated by the simultaneous solution of two Girifalco-Good equations for γ_{fw}^*:

$$\gamma_{fw}^* = \gamma_f^* + \gamma_w^* - 2\,\Phi_{fw}^*\sqrt{\gamma_f^*\ \gamma_w^*} \tag{18}$$

and

$$\gamma_{fw}^* = \gamma_{sw} + \gamma_{fs}^* - 2 \, \Phi_{fw}^* \sqrt{\gamma_{sw} \, \gamma_{fs}^*} \tag{7}$$

to yield

$$\gamma_{fw}^* = \frac{(\gamma_{sw} + \gamma_{fs}^*) \sqrt{\gamma_f^* \, \gamma_w^*} - (\gamma_f^* + \gamma_w^*) \sqrt{\gamma_{sw} \, \gamma_{fs}^*}}{\sqrt{\gamma_f^* \, \gamma_w^*} - \sqrt{\gamma_{sw} \, \gamma_{fs}^*}} \tag{19}$$

Equation 7 can then be rearranged to allow calculation of Φ_{fw}^* as

$$\Phi_{fw}^* = \frac{\gamma_{sw} + \gamma_{fs}^* - \gamma_{fw}^*}{2 \sqrt{\gamma_{sw} \, \gamma_{fs}^*}} \tag{20}$$

The last quantity to be calculated is Φ_{fs}^*, which can be determined from the Girifalco-Good equation for γ_{fs}^* as

$$\Phi_{fs}^* = \frac{\gamma_f^* + \gamma_s^* - \gamma_{fs}^*}{2 \sqrt{\gamma_f^* \, \gamma_s^*}} \tag{21}$$

Equations 12 through 21 were used to calculate the interfacial parameters for the detergency system consisting of PEG-finished polyester/cotton cloth, mineral oil soil, and detergent baths consisting of varying homolog solutions of LAS homologs from 10- to 14-carbon-atom alkyl chain length. The calculated and experimentally determined results are listed in Table 2. These results will be the topic of the following section.

Interfacial Parameters in the Detersive System. Three phases are present in the detergency system in this study: a PEG (fiber), a hydrocarbon (soil), and an aqueous surfactant (bath) phase. These three phases meet to form three binary interfaces, whose interfacial energetic properties vary as a function of LAS homolog molecular weight.

The interfacial tension data listed in Table 2 show variations with LAS homolog molecular weight that can be ascribed to the Ferguson effect (19). This is best demonstrated at 50 ppm hardness for γ_{fw}^*, where the 12-carbon homolog gives the minimum interfacial tension. The lower and higher homologs give greater interfacial tensions (i.e., have less interfacial affinity) as a result of being too water-soluble and insoluble, respectively. Other surface and interfacial tensions, such as γ_w and γ_{sw}, show only a decrease in value with increasing LAS homolog carbon chain number. However, if tension values on LAS carbon chain homologs >14 were measured, increasing tension values with carbon chain would eventually be observed.

The film pressure values for the detergency system are also listed in Table 2. These quantities represent the difference in interfacial tension between two pure phases and the interfacial tension of the same two phases which are at saturation equilibrium with the third phase. Since the PEG fiber surface was assumed insoluble in either the bath or soil, $\pi_{sw} \equiv 0$.

The value of π_{fs} is seen to be nearly constant, indicating that the adsorption of the bath (water, surfactant, and hardness) at the fs interface is negligible. The π_{fw} value is, however, quite interesting in that it shows dramatic variations with homolog and hardness changes from negative to positive in magnitude. From the data, it can be seen that with the introduction of soil into the fw interfacial system, the value of γ_{fw} increases. While the mechanism for this behavior was not determined in this study, it would seem most likely that this behavior occurs as a result of surfactant leaving the fw interface to either emulsify or solubilize into the soil phase. Since the free energy process that corresponds to the π_{fw} value could only be spontaneous if π_{fw} was positive, another compensating process in the bath phase which had a greater free energy change, such as soil emulsification or surfactant-in-soil solubilization, must occur simultaneously.

Molecular Structure Effects and Detergency. The correlation of surfactant structure with interfacial and colloid properties is a poorly understood science. Much study in this area has been thermodynamic which has been a useful endeavor but which nevertheless fails to provide specific molecular structure/physical property correlations. The following study has also been largely thermodynamic to this point; however, since the data has been collected on pure LAS homologs, it provides an opportunity to apply some of the quasi-thermodynamic treatments that have been proffered in the literature to date.

The first of these to be discussed will be the Cohesive Energy Ratio, R concept (20). Using the concept of cohesive energy between molecules, Winsor recognized four structures. These are present in an immiscible two-phase system (O and W denoting oil and water, respectively) containing a third-surfactant component with partial solubility in both bulk phases. Each surfactant molecule has a hydrophilic (denoted by H) and a lipophilic (denoted by L) section. Conceptually then Winsor views all the possible molecular interactions in such a system in terms of their cohesive energy (denoted by C). For such a system, there are then 10 possible cohesive molecular interactions (i.e., 10 unique combinations of the letters O, W, H, and L). In the ideal case, the lipophile-oil and the hydrophile-water interaction will be the predominant interactions. The relative magnitude (R) of these two interactions then defines the location of the surfactants in the oil-water interface, and the nature of the emulsion formed, if any, as

$$R = \frac{C_{LO}}{C_{HW}} \quad \begin{cases} R > 1, \text{ W/O emulsion} \\ R = 1, \text{ planar interface} \\ R < 1, \text{ O/W emulsion} \end{cases} \qquad (22)$$

the R system corresponds to the HLB scale (20-22) where

$$\begin{aligned} R > 1 &\rightarrow HLB < 10 \\ R = 1 &\rightarrow HLB = 10 \\ R < 1 &\rightarrow HLB > 10 \end{aligned} \qquad (23)$$

The main advantage to the Winsor system is its heuristic feature of treating all cohesive interactions in a two-phase surfactant system. However, to date only the simple form of Equation 22 has been exploited quantitatively (21, 23) as

$$R = \frac{\overline{V}_L \, \delta_L}{\overline{V}_H \, \delta_H} = \frac{\overline{V}_O \, \delta_O}{\overline{V}_W \, \delta_W} \qquad (24)$$

To incorporate the surfactant structure concept, it is now convenient to introduce the group additive concept for cohesive energy densities (CED) introduced by Burrell and others (24, 25). Molecular segments are given a molar-attraction constant G. The CED is then determined for the ith compound as

$$\delta_i = \frac{\Sigma \, G}{\overline{V}_i} \qquad (25)$$

where the sum of all group molar-attraction constants in the molecule of interest is represented as $\Sigma \, G$.

To facilitate the use of CED calculations for Equation 24, a semiquantitative relationship developed by Beerbower and others (26-29) relates the surface tension of a phase to its CED as

$$\delta^2 = 16 \, \frac{\gamma}{\overline{V}^{1/3}} \qquad (26)$$

A final quasi-thermodynamic approach to molecular structure effects has been proposed by Good et al. (17-18) which relates molar volumes and the Girifalco-Good interaction parameter, Φ, as

$$\Phi_{sw} = \frac{A_{sw}}{(A_{ss} \, A_{ww})^{1/2}} \cdot \frac{4(\overline{V}_s \, \overline{V}_w)^{1/3}}{(\overline{V}_s^{1/3} + \overline{V}_w^{1/3})^2} \backsim \frac{4(\overline{V}_s \, \overline{V}_w)^{1/3}}{(\overline{V}_s^{1/3} + \overline{V}_w^{1/3})^2} \qquad (27)$$

where   A = attractive coefficient in the Lennard-Jones 6-12 po-
            tential equation.
    $\overline{V}_i$ = molar volume of ith phase.

    The values of A are not readily available for the phases in
this study, so the commonly made approximation (17) of setting
the ratio of A values to 1 will be made.
    With these concepts, expressed in Equations 24 to 26, three
types of questions can now be given semiquantitative answers:
(1) CED values can be determined from surface tension measurements, (2) the effects of particular molecular components of
surfactant molecules on surface tension and CED can be addressed,
and (3) the emulsion type and stability can be evaluated based
on either molecular structure surface tension and/or CED.
    In Table 3 are the values of surface tension for the
aqueous LAS homolog solutions. Values of molar volume used are
those for the pure LAS homolog independent of water. The justi-
fication for this comes from the Winsor R model (20, 21) and
work by Scriven and Davis (30) who showed that accurate CED
values can be obtained from a statistical mechanical treatment
of an interface using only 2 or 3 atomic or molecular layers of
that interface. For a surfactant solution, the surfactant will
predominate in the interface, hence the choice of pure LAS for
the solution molar volumes.
    Using the Beerbower Equation 26, values of the CED were
calculated and listed in Table 3. At all hardnesses, the 10 and
11 carbon LAS homologs have significantly larger CED values.
This is probably an artifact as a result of all solutions being
.024 weight/weight percent. Because the two lower homologs have
CMC values above this concentation and vice versa for the higher
homologs, the δ values for the 10 and 11 carbon homologs violate
the premise of a surfactant-saturated interface. Therefore, the
δ values for the two lowest homologs are tenuous because, among
other things, the value of $\overline{V}$ used in the Beerbower equation
would be uncertain.
    In the last columns of Table 3 are the calculated and
experimental values of Φ_{sw}. Comparison of these values shows
the calculation to do a poor job of predicting the experimental
Φ_{sw} values. This discrepancy could result from many sources.
An obvious source of error could be the approximation of setting
the A ratio to 1, thereby neglecting differences in types of
interactions between the s and w phases across the sw interface.
Another possible cause for the poor predictive power of Equation 27 could be the fact that the value of $\overline{V}_w$ used in that equation was for the LAS homolog molecule, while the phase is
actually an aqueous LAS solution. As discussed earlier, this
approximation for surfactant solutions is probably good when the
interface is saturated with surfactant molecules but should include the molar volume of water molecules as the mole fraction
of adsorbed surfactant decreases. Since the lower LAS homologs

Table III.   Quasi-Thermodynamic Analysis of Detergency Data For Molecular Structure Effects on Detergency

Hardness (ppm)	LAS Homolog	Cube Root Molar Volume $(mole/cm^3)^{1/3}$	Surface Tension for LAS at .024% Solutions γ_w (dynes/cm)	CED Constant δ_w (Hildebrands)	Molar Attraction for Sulfonate G_{sulf}	Winsor R	Experimental Detergency D_R (%)	Φ_{sw} Experimental	Φ_{sw} Calculated
50	10	6.79	46.7	10.5	1242	1.92	65	0.76	0.97
	11	6.89	41.1	9.8	1027	1.96	71	0.79	0.97
	12	6.99	30.2	8.3	524	2.22	71	0.88	0.97
	13	7.08	30.7	8.3	502	2.14	73	0.91	0.98
	14	7.17	30.2	8.2	446	2.08	74	0.93	0.98
150	10	6.79	41.7	9.9	1054	2.03	34	0.76	0.97
	11	6.89	33.0	8.8	700	2.19	56	0.87	0.97
	12	6.99	26.9	7.8	353	2.37	74	0.95	0.97
	13	7.08	25.9	7.7	289	2.30	75	0.97	0.98
	14	7.17	25.8	7.6	224	2.25	76	1.00	0.98
300	10	6.79	38.3	9.5	929	2.12	25	0.81	0.97
	11	6.89	30.7	8.4	569	2.29	67	0.93	0.97
	12	6.99	26.0	7.7	319	2.39	72	0.98	0.97
	13	7.08	27.5	7.9	360	2.25	66	0.99	0.98
	14	7.17	26.0	7.6	224	2.25	62	0.99	0.98
Mineral oil soil		9.65		7.01					
PEG cloth finish		3.42		11.1[a]					

[a] Based on tetraethylene glycol, Reference 33.

are below their CMC values, as evidenced by γ_w and γ_{sw} values, thereby having a greater mole fraction of adsorbed water, it seems reasonable that molar volumes for the lower LAS Homologs should be lower than those shown in Table 3. Directionally, this would make calculated values of Φ_{sw} lower for the lower LAS homologs, which would improve agreement between experimental and theoretical values.

Another aspect of this quasi-thermodynamic approach is to use Equation 25, along with G values from Table 4, to calculate the molar-attraction constant for the sodium sulfonate molecular group or

$$G_{sulf} = \delta_i \cdot \overline{V} - 1114 - 133 \, (n-3) \tag{28}$$

where n = number of carbons in alkyl chain of LAS molecule.
δ_i = cohesive energy density of the ith LAS homolog solution.

Using Equation 28, the values of G_{sulf} determined in this manner are listed in Table 3. The most obvious feature of the G_{sulf} values is that the 10 and 11 homologs give much higher values. However, this apparent deviation is a result of γ_w values being measured on less than saturated air/aqueous interfaces (below the CMC for that surfactant). Hence these values should be expected but are probably artifacts.

For the higher homologs (i.e., 12 to 14 LAS), the values of G_{sulf} are more nearly, but not completely, constant. They vary first due to hardness. This reflects on the fact that G_{sulf} values of γ_w at 50 ppm are higher than at 150 and 300 ppm. At the higher hardnesses, the sulfonate group repulsion for the adsorbed LAS molecules is at a minimum in the presence of excess divalent cation (i.e., hardness). This results in better LAS packing in the interface, lower surface tensions, and lower CED values which finally culminates in a lower G_{sulf}. It should be pointed out that this difference between G_{sulf} values at 50 ppm hardness and those at 150 and 300 ppm hardness is a result of the $\overline{V}$ values used in its calculation. The value of $\overline{V}$ was arrived at by dividing the specific density of pure LAS by its molecular weight. The adsorbed molecules only approach this molar volume such that if the actual $\overline{V}$ of the interfacial adsorbed density of the LAS molecules were known and could be used to calculate G_{sulf}, its value at all hardness levels should be more nearly constant.

Another trend for the G_{sulf} values is also seen within each hardness group in which this constant decreases for the 12 to 14 homologs. There is no simple explanation for this trend except to say that additivity theories like Equation 25 are not absolutely precise. Therefore, an average value of G_{sulf} for the 12 to 14 homologs at 150 and 300 ppm will be taken as the G_{sulf}

value and a 95 percent confidence interval for these six data
(given as ±) are listed in Table 4. While an uncertainty of
120 units in G_{sulf} may seem large compared to its value of 295,
for most δ calculations by the molar-attraction constant method,
the value of Σ G in Equation 25 will be > 2,000 for most molecules of interest. Therefore, the relative uncertainty will be
< 10 percent, which is about the best precision one can expect
from this method for calculating δ anyway, especially for polar
moieties.

A third molecular structure/physical property correlation
that can be made is to calculate the Winsor R for the homolog
solutions by Equation 24. The resulting values are listed in
Table 3. The R values are all clustered around the value of 2.
This would suggest the surfactant would tend to form water-in-
oil emulsions. From a detergency perspective, this data indicates that solubilization of oil in water is not the detersive
mechanism. A final observation on this data is that R does not
correlate well with detergency performance, as shown in the correlation plot of Figure 10. The quantity R is not a single
valued function of detergency as can be seen in Figure 10. The
reason for this is difficult to ascertain since the calculation
scheme for R used in this study is so approximate. Nevertheless, the value of ca. 2 is a valid observation. Therefore, as
noted above, oil-in-water emulsions, or detergency by the solubilization mechanism, is not the apparent mechanism suggested by
these data for the systems studied.

Summary

A model for detergency performance was developed based on SEM
photographs of cloth and fiber bundles within a cloth. The
detergency or relative soil removal for this model was shown to
be entirely dependent on the contact angle at the soil/fiber/
detergent bath triple-phase line. A wetting equation was then
derived so that the soil removal performance could be directly
related to the wetting characteristics of the soil, fiber, and
detergent bath interfacial properties through the contact angle.
The detergency performance equation was then applied to experimental detergency data, so that the wetting characteristics of
the pure LAS homolog solutions could be examined and molecular
structure .effects could be investigated. This segment of the
work showed that detergency performance was dependent exclusively
on the soil/bath interfacial tension so long as only the molecular weight of the LAS homolog changed. This result is consistent with other studies involving the effects of interfacial
tension on detergency performance (31, 32).

Other LAS homolog structural effects on wettability and
soil removal were found when the data were analyzed using the
cohesive energy ratio, R, the regular solution theories of the

Table IV.   Molar-Attraction Group Constants
For Cohesive Energy Density Determination

REFERENCES 23, 24		CALCULATED	
Group	G	Group	G
--CH_3	214	-SO_3N_a	295 ± 120
--CH_2-	133		
--CH	28		
-- ⬡ O --	658		

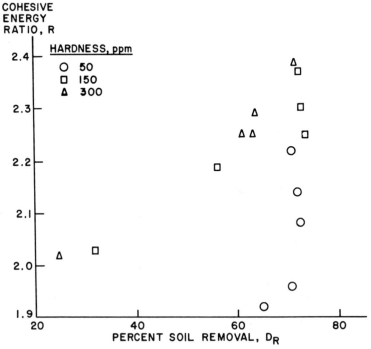

Figure 10.

Correlation Plot of Cohesive Energy Ratio for LAS Homolog
Detergency of Mineral and the Experimental Soil Removal.

Beerbower equation and the additivity concepts of molar-attraction constants for determining the cohesive energy density.
From these treatments, it was possible to calculate a molar-
attraction group constant for the sodium sulfonate group of 120
(in Hildebrand units).

Experimental

The soil used in this study was a mineral oil (see Figure 1)
manufactured by Squibb, which had a trace amount of an oil-
soluble red dye added for visual acuity. The linear alkylbenzene sulfonate sodium salts (LAS) were homologously pure
samples, each with a different linear alkyl carbon chain of 10,
11, 12, 13, or 14 carbon atoms. Each of the LAS samples was a
mixture of isomers with phenyl attachment ranging in nearly
equal amounts from 2 through 5, 6, or 7 (depending on the alkyl
chain length). These LAS samples were made with an HF catalyst.
      Inorganic ingredients were used as received from their
supplier and included; sodium tripolyphosphate or STPP (FMC
Corporation) and RU silicate or SiO_2/Na_2O (the P.Q.
Corporation).
      The detergency testing was done using a Tergotometer (U.S.
Testing Company, Inc.). Synthetic hard water was prepared by
adding varying amounts of a concentrated solution of calcium and
magnesium chlorides in a 4:1 cation mole ratio, respectively, to
the detergency bath. The Tergotometer was run for 10 minutes at
$100°F$ and 100 rpm with one liter of a 0.15 percent detergent
formulation in each pot followed by a 5-minute, 100-rpm rinse
with 1 liter of deionized water. The formulation was 16 percent
LAS, 30 percent STPP, and 10 percent SiO_2/Na_2O. The cloths were
a 35/65 blend of cotton/Dacron with a polyethylene glycol permanent press finish (Testfabrics, Inc., Cloth No. S/7406) that had
been sized. The cloths were soiled in a Benz soiling unit
(Ernst Benz AG) type KLFM "K" and KTF/M. The reflectance readings on clean, soiled, and washed cloths were performed on a
Gardner Colorimeter (Gardner Laboratory, Inc.) using the Y axis
of the YXZ scale. The cloth readings were made on individual
cloths using a backing plate then corrected graphically for the
backing plate effect.
      Surface and interfacial tension measurements were made at
$40°C$ with a Model 300 Spinning Drop Tensiometer obtained from
the University of Texas at Austin, Chemistry Department.

Acknowledgments

The author would like to thank Ms. M. Debra Davis for making the
surface and interface tension measurements and Dr. K. Lee
Matheson for the detergency data. In addition, the author

wishes to thank Conoco Inc. for permission to publish this study.

Literature Cited

1. W. G. Cutler and R. C. Davis, "Detergency," Parts I, II, and III, Marcel Dekker, Inc., New York, 1972 through 1981.
2. E. Kissa, Textile Res. J., 41, 760 (1971).
3. E. Kissa, Textile Res. J., 45, 736 (1975); Pure and Appl. Chem., 53(11), 2255-68 (1981).
4. A. F-C Chan, "Solubilization Kinetics in Detergency," Diss. Abstr. Int. B, 38(4), 1,808 (1977).
5. O. C. Bacon and J. E. Smith, Ind. Eng. Chem., 40, 2,361-2,370 (1948).
6. W. J. Popiel, "Introduction to Colloid Science," Exposition Press, Hicksville, New York, pp. 37-46 (1978).
7. J. Chem., J. Chem. Ed., 56(9), 610-611 (1979).
8. A. M. Schwartz, "Surf. and Colloid Sci.," Vol. 11, Plenum Press, New York, p. 307 (1979).
9. A. W. Adamson, "Physical Chem. of Surfaces," 3d ed., Wiley, New York, pp. 474-484 (1976).
10. H. Renmuth, "Haus der Technik Vor. Essen No. 68," pp. 7-24 (1965, published 1966).
11. J. A. Wingrave, presented at the 57th National Colloid and Interface Science Symposium, University of Toronto, Toronto,Canada, June 12-15, 1983.
12. P. Kubelka and F. Munk, Z. Tech. Physik, 12, 593 (1931).
13. Cutler and Davis, ibid., pp. 386-92.
14. A. W. Adamson, ibid., pp. 339-42.
15. T. Young, "Miscellaneous Works," Vol. I, G. Peacock, ed., Murray, London, 1855, p. 418.
16. A. Dupre´, "Theorie Mecanique de la Chaleur," Paris, 1869, p. 368.
17. L. A. Girifalco and R. J. Good, J. Phys. Chem., 61, 904-9 (1957).
18. R. J. Good and E. Elbing, Ind. Eng. Chem., 62(3), 54-78 (1970).
19. J. Ferguson, Proc. Roy. Soc. (London), 127B, 387 (1939).
20. P. Winsor, "Solvent Properties of Amphiphilic Compounds," Butterworth Sci. Publ., London, 1954.
21. A. Beerbower and M. W. Hill, Detergents and Emulsifiers, 223-36 (1971).
22. W. C. Griffin, J. Soc. Cosmet. Chem., 5, 249 (1954).
23. A. Beerbower and J. Nixon, Div. Petr. Chem. Preprints, ACS, 14(1), 62-71 (1969).
24. H. Burrell, Official Digest of the Federation of Societies for Paint Technology, 27, 726 (1955); 29, 1069 (1957); 29, 1159 (1957).
25. P. A. Small, J. Appl. Chem., 3, 71 (1953).

26. A. Beerbower, J. Colloid Interface Sci., 35(1), 126-32 (1971).
27. J. H. Hildebrand and R. L. Scott, "Solubility of Nonelectrolytes," 3d ed., Reinhold, N.Y. (1950).
28. A.F.M. Barton, Chem. Rev., 75(6), 731-53 (1975).
29. C. Hansen and A. Beerbower in "Kirk-Othmer Encyclopedia of Chemical Technology," Supl. Vol., 2nd ed., A. Standen (ed.), p. 889 (1971).
30. L. E. Scriven and H. T. Davis, J. Phys. Chem., 80(25), 2805-6 (1976).
31. M. P. Aronson, M. L. Gum, and E. D. Goddard, J. Am. Oil Chem. Soc., 60(7), 1333-9 (1983).
32. K. W. Dillan, E. D. Goddard, and D. A. McKenzie, J. Am. Oil Chem. Soc., 57, 230 (1980).
33. "The Handbook of Chemistry and Physics; 49th Ed.", R. C. Weast, Ed., CRC Co., Cleveland, Ohio, p. C-562 (1968).

Appendix
Pore Volume Between Fibers

The pore volume, shown as a cross section of two parallel fibers of length, L, and radius, R, has a half-surface dimension, r_p, and meets the fiber surface with an angle, ψ (see Figure 5). The volume of such a pore, V_D, will be

$$V_D = 2L\left[\,hr_p - \int_{R-r_p}^{R} Y\,dX\right] \tag{29}$$

Then since

$$Y = \sqrt{R^2 - X^2} \tag{30}$$

Equations 29 and 30 combine to yield

$$V_D = 2L\left[hr_p - \int_{R-r_p}^{R} \sqrt{R^2 - X^2}\,dX\right] \tag{31}$$

which can be integrated to yield

$$V_D = 2Lhr_p + 2L\left(\frac{X}{2}\sqrt{R^2 - X^2} + \frac{R^2}{2}\sin^{-1}\frac{X}{R}\right)\Bigg|_{R}^{R-r_p} \tag{32}$$

or

$$V_D = 2Lhr_p + L\left[(R - r_p)\sqrt{R^2 - (R - r_p)^2} + \right.$$

(33)

$$\left. R^2 \sin^{-1}\left(\frac{R - r_p}{R}\right) - R^2 \frac{\pi}{2}\right]$$

Then noting

$$h = R \cos \psi \qquad (34)$$

$$R - r_p = R \sin \psi \qquad (35)$$

$$r_p = R(1 - \sin \psi) \qquad (36)$$

Equations 33 through 36 can be combined and simplified to yield the desired expression of V_D as

$$V_D = LR^2 \cos \psi \; (2 - \sin \psi) - LR^2 \; \frac{\pi}{2} - \psi \qquad (37)$$

The total pore volume, V_T, can be determined from Equation 37 when $\psi = 0$ as,

$$V_T = LR^2\left(2 - \frac{\pi}{2}\right) \qquad (38)$$

If a cloth is soil-saturated, then the relative soil removal of the duplex cloth-held soil, V_R, will be,

$$V_R = \frac{V_D}{V_T} \qquad (39)$$

or combining Equations 37 and 39 yields,

$$V_R = \frac{2 \cos \psi \; (2 - \sin \psi) - (\pi - 2\psi)}{4 - \pi} \qquad (40)$$

RECEIVED January 20, 1984

Relationship Between Surfactant Structure and Adsorption

P. SOMASUNDARAN, R. MIDDLETON, and K. V. VISWANATHAN

School of Engineering and Applied Science, Columbia University, New York, NY 10027

Adsorption of a surfactant on solids is dependent, among other things, on the structure of both the hydrophobic and hydrophilic portions of it. There are a number of mechanisms proposed for surfactant adsorption and an understanding of the effects of the structure of the surfactant can help in elucidating the role of these mechanisms. In this study, the effect on adsorption on alumina of some structure variations of sulfonates (chain length and the branching and the presence of ethyoxyl, phenyl, disulfonate and dialkyl groups) is examined above and below CMC as a function of surfactant concentration, pH and salinity. Co-operative action between an ionic alkylsulfonate and a nonionic ethoxylated alcohol is also studied.

Surfactant adsorption on solids from aqueous solutions plays a major role in a number of interfacial processes such as enhanced oil recovery, flotation and detergency. The adsorption mechanism in these cases is dependent upon the properties of the solid, solvent as well as the surfactant. While considerable information is available on the effect of solid properties such as surface charge and solubility, solvent properties such as pH and ionic strength (1,2,3), the role of possible structural variations of the surfactant in determining adsorption is not yet fully understood.

Adsorption is governed by a number of forces: covalent bond formation or electrostatic attraction or hydrogen bond formation between the adsorbate and the adsorbent, electrostatic repulsion among the adsorbate species, lateral associative interaction among adsorbed species, solvation of adsorbate or adsorbent surface species. Structural modifications can affect one or more of the above interactions that might be predominant in different concentration regions, and it is the cumulative effect of all of these modifications on all interactions in various concentration regimes that will determine the overall adsorption behavior of a surfactant (4,5). Thus while in practice chain length or branching will affect only the lateral interactions in the hemi-micellar region,

0097–6156/84/0253–0269$06.50/0

presence of multifunctional groups (such as disulfonates with an
ether linkage) can affect both electrostatic and lateral association interactions. Coadsorption between different surfactant
species also can be expected to be influenced significantly by such
structural variations.

In this paper the adsorption characteristics of a series of
structurally modified surfactants will be analyzed. Figure 1
summarizes this series showing the following structural variations:
aryl addition, chain length variation, branching, xylene alkyl
addition, ether linkage and etheylene oxide addition, with alumina
as the adsorbent. By understanding the effect of structural
variations upon the adsorption mechanism a guideline may be established by which a surfactant may be tailored with specific
structural modifications for certain situations.

Experimental

Surfactants. n-sodium dodecylsulfonate specified to be 99.4%
pure was purchased from Aldrich Chemicals.

n-sodium octyl, decyl, dodecyl and tetradecylbenzene sulfonates
were synthesized and purified in our laboratory. Characterization
of these chemicals using p-NMR, C-13 NMR, mass spectrometry and
ALC showed these compounds to be isomerically pure. Branched
hexadecyl benzene sulfonate was obtained from Conoco and used
as received after characterization.

Alkyl aryl orthoxylene sulfonates were also investigated.
The first being a linear nonyl orthoxylene sulfonate and the second
being a branched dodecyl one. Both were supplied by the Exxon
Corporation and contain known amounts of unsulfonated hydrocarbons
(14% and 25.2% respectively).

The disulfonate, Dowfax 3B2, used was a didecylphenoxy-
disulfonate containing 10% monosulfonated impurities. HPLC
analysis showed this surfactant to be a mixture of several compounds.

Triton X-200 was used to study the effects of ethoxylation on
the sulfonates. Nonionic ethoxylated surfactants were investigated
using Synfac 8216 obtained from the Milliken Corporation. This
was stated to be 100% active with a molecular weight of 1100-1200.
HPLC showed this surfactant to be a mixture of several components.

Mineral. Alumina used in this study was a high purity α- Linde
sample purchased from the Union Carbide Corporation. BET surface
area was determined to be 15.0 m^2/g.

Procedure. A gram of the mineral was preconditioned for 90 minutes
with 5cc of a 0.2 $kmol/m^3$ sodium chloride solution at 75°C on a
wrist action shaker. Then a 5cc solution of known surfactant
concentration is added and allowed to shake for four hours. Four
hours mixing was found to be sufficient to reach equilibrium
from adsorption test conducted as a function of mixing time. The

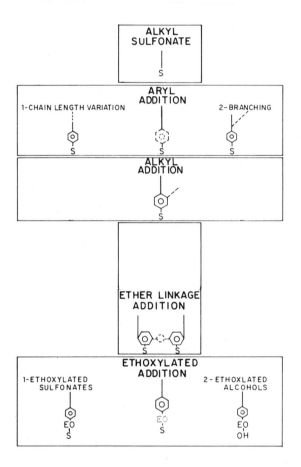

Figure 1.   General overview.

supernatant was obtained using a thermostated centrifuge at 4500 rpm. Analyzation of surfactant concentration was completed using a two-phase titration technique and U.V. spectroscopy.

Results and Discussion

The adsorption isotherm obtained for dodecylsulfonate (DDS) on alumina is given in Figure 2. This isotherm is similar to that obtained in the past for sulfonate/alumina systems (4). This isotherm behaves in an s-shaped manner (6) revealing its four characteristic regions of adsorption: 1) Electrostatic interaction 2) Lateral association (hemimicellization), 3) Electrostatic hindrance and 4) micellization.

In Region 1, under low surfactant concentration conditions, adsorption occurs mainly due to the electrostatic attraction between the surfactant ion and the charged sites on the solid surface (3). The beginning of Region 2 is characterized by the onset of lateral association among the adsorbed surfactant species (4). This process is analogous to micellization but takes place at a lower surfactant concentration assisted by the electrostatic attractive forces between the hemi-micelle and the charged solid (7, 4). While sufficient surfactant ions have been adsorbed to neutralize the surface charge of the mineral, further adsorption progresses owing to continued hemimicellization. Charge reversal occurs and adsorption in this concentration region is hindered by the increasing electrostatic repulsion between the now similarly charged solid and surfactant species (Region 3) (4). Region 3 ends and a new region begins with the onset of micellization (8). In the micellar region, adsorption is nearly constant due to the absence of a marked increase in activity of the surfactant species in this region with an increase in its concentration (8). In this region apparent "adsorption" can, however, undergo drastic changes owing, among other things, to precipitation or micellar redissolution of the surface precipitate (5, 9), resulting in adsorption maxima and minima. While adsorption in the hemimicellar region, like that in the micellar region, is dependent on the hydrophobic properties of the surfactant, adsorption in the electrostatic region and alterations in adsorption owing to precipitation, redissolution and reprecipitation will depend upon solution properties such as pH, salinity and hardness. Thus factors to be considered while examining the effect of surfactant structure will include changes in the electrostatic attraction and repulsion of the adsorbate and adsorbent, solvent power of the medium for the surfactant and its precipitates, and lateral association as well as electrostatic repulsive interactions among the adsorbed species.

Aryl Addition. Surfactants having an aryl group added between the sulfonate and alkyl chain are studied using the standard adsorption procedure, and compared with an alkyl sulfonate. Figure

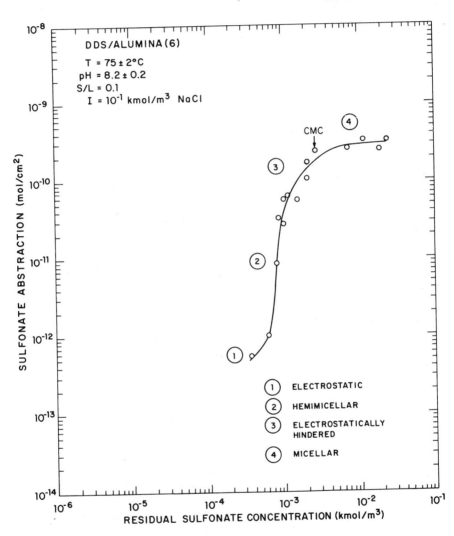

Figure 2. Alkyl sulfonate adsorption.

3 shows the adsorption isotherms of DDS, OBS and DDBS. Under the
assumption that one benzene ring equals 3.5 alkyl units (10), the
equivalent alkyl chain length of OBS and DDBS can be calculated:
11.5 and 13.5 respectively. DDS and OBS, both having similar
equivalent alkyl chain lengths (.5 alkyl difference), do be-
have with similar adsorption characteristics. Comparison of
DDS and DDBS shows the large increase in adsorption associated
with the aryl addition.

Chain Length Variation.   Adsorption isotherms obtained for
alkylbenzene sulfonates of varying chain length are given in
Figure 4. Adsorption clearly increases as the chain length of the
alkyl group is increased from 8 to 14. This is found to be the
case even in Region 1 where adsorption has been proposed to take
place owing to electrostatic attraction only. Since in all cases,
the charge of the sulfonate species is 1, adsorption in the electrostatic region should have been invariant with chain length.
The fact that it is, to the contrary, dependent on the chain
length suggests the possible influence of the reduced dielectric
constant in the interfacial region(11). Transfer of the monomer chain
into a less dielectric region should result in a lowering of
free energy, with the reduction being larger for the longer chains.
Adsorption of the long chain sulfonates in Region 1 should therefore be considered to be the result of electrostatic attraction as
well as increased solvent power of the interfacial water.
Increase in adsorption seen in the lower three regions (Figure
4) can be attributed to the increase in hydrophobicity
with chain length resulting in stronger hemi-micellization
as well as monomer transfer. The shift in the micellar
regions of the AAS isotherms is in effect a measure of the
increased hemimicellization versus micellization as the chain
length is increased (12). When the hemi-micellization is
essentially complete at the end of Region 3, the end methylene
group of the peripheral chains in each hemimicelle is still
exposed to bulk water; however, the fraction of exposed groups
per chain will decrease with increase in chain length. Any
such change in the case of micellization could be expected to be
of a lower magnitude. Thus the higher adsorption of longer
chains achieved at the beginning of the Region 4 can conceivably
be the result of the greater energy of hemimicellization energy in
relation to that of micellization.
Effect of Branching.   The effect of branching was investigated
by comparing the adsorption of various hexadecylbenzene sulfonates
(Figure 5).   The first had the benzene group occupying the
number 2 position upon the alkyl chain and the second had it
at the 8 position. Adsorption of the 2 ϕ HDBS is significantly
higher than that of the 8 ϕ HDBS. In the low concentration region
the former appears to adsorb almost an order of magnitude more than
the latter; it is however to be noted that adsorption in this region
borders upon the experimental limitations; even in the micellar
region, adsorption of it is higher. 2ϕ HDBS which is least branched

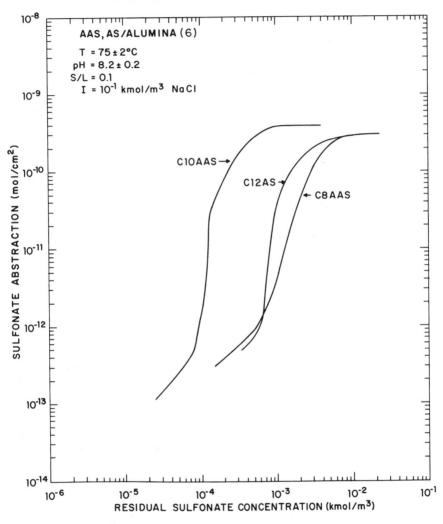

Figure 3.   Structural comparison: AAS vs. AS.

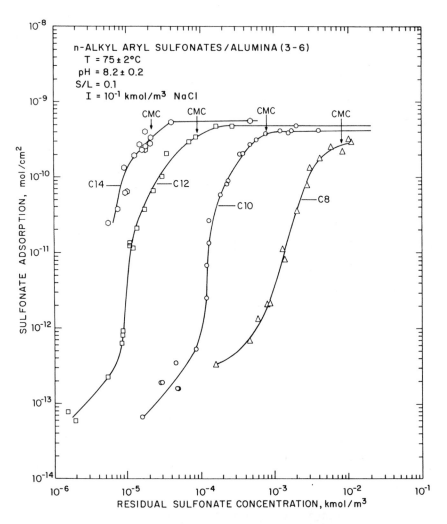

Figure 4.   Effect of chain length.

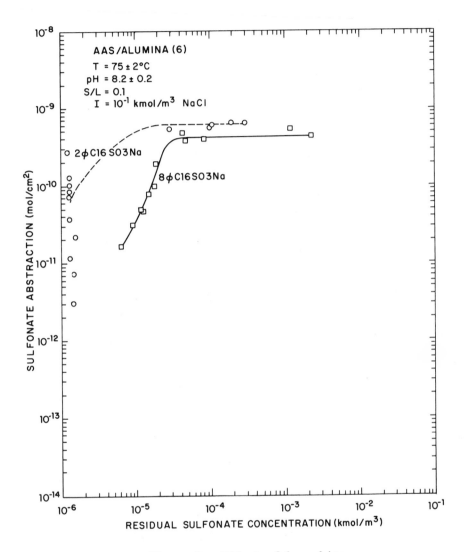

Figure 5. Effect of branching.

is considered to have a larger effective alkyl chain than the 8 ϕ
HDBS (13), since the CMC of the former is considerably lower than
that of the latter. It must, however, be pointed out that while
formation of a spherical micelle might be more easily accomplished
with the 2 ϕ HDBS than with the 8 ϕ HDBS, it is not fully evident
as to why a planar hemi-micelle should be more easily achieved
with the 2ϕ HDBS.

Comparison of these adsorption isotherms with those obtained
for the linear alkyl aryl sulfonates (Figure 6) reveals the behavior of the 2 ϕ HDBS to be close to that which would be expected
for a 1 ϕ HDBS and that of the 8 ϕ HDBS to be equivalent to that
of a tridecyl benzene sulfonate. Development of a quantitative
model that can account for the effect of the position of the benzene
group on the chain warrants additional data for a variety of
surfactants with branched chains.

Alkyl Addition on the Aromatic Ring. The two surfactants chosen
to be representative of this section were alkylaryl sulfonates
having an additional alkyl groups substituted in the aromatic
ring. These compounds were commercially named alkylarylorthoxylene sulfonates and referred to as such. The adsorption
behavior of the following xylene sulfonates was examined: linear
nonylorthoxylene sulfonate and branched dodecylorthoxylene sulfonate. Such orthoxylene sulfonates have been reported to
be very effective in reducing interfacial tensions; however,
adsorption characteristics of these compounds are not clearly
known. Results obtained for the two alkylorthoxylene sulfonates
are given in Figure 7. Since the position of the aryl group
in the dodecyl chain was unknown for the branched orthoxylene
sulfonate, an estimate of the relative hydrophobicity of the two
compounds was obtained using HPLC. Two peaks were analyzed
for both of the compounds giving retention times of 2.4 and
3.0 minutes for NXS and 2.8 and 3.74 minutes for DDXS. Therefore,
DDXS, even though branched can be considered to possess an
effective chain length longer than that of NXS; the larger
adsorption of the former is in accord with the results of the
above HPLC analysis. The above samples of the commercial xylene
sulfonates were, however, reported to contain significant
amounts of unsulfonated hydrocarbons (oil), NXS containing 14%
and DDXS containing 25.2%. As oil is known to produce marked
effects on adsorption, it becomes necessary to determine the adsorption behavior of the deoiled samples of the above sulfonates
in order to more precisely identify the properties of xylene
sulfonates. The adsorption isotherms of the deoiled xylene
sulfonates are given in Figure 8 along with those of OBS and DBS
for comparative purposes.

Firstly, deoiling is seen to produce lower adsorption; secondly, the isotherms for these deoiled sulfonates also show lower adsorption levels when compared with the straight chain alkylaryl
sulfonates. In Figure 8, the nonylxylene sulfonate acts

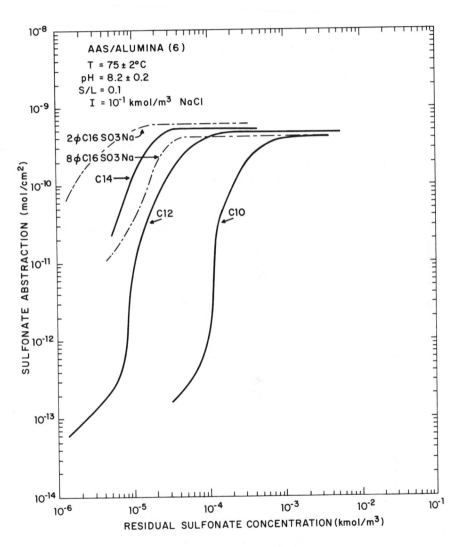

Figure 6.   Structural comparison: branched vs. linear AAS.

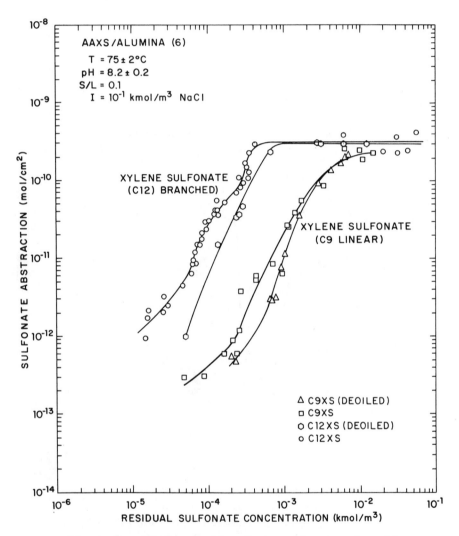

Figure 7.   Alkylarylorthoxylene sulfonate adsorption.

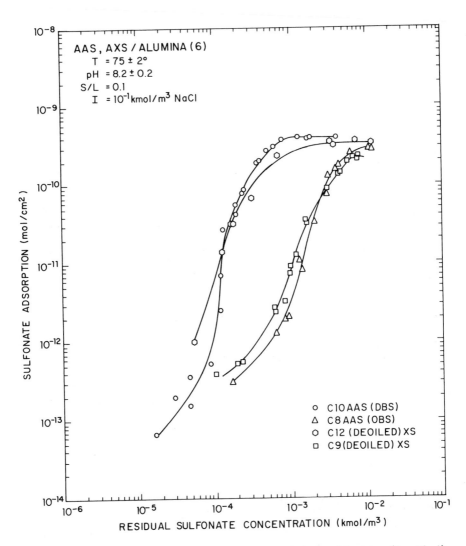

Figure 8. Structural comparison: xylene sulfonate (deoiled) vs. AAS.

equivalently with the OBS and the branched dodecylxylene sulfonate
approximates the behavior of DBS. Thirdly the slope of the hemi-
micellar region did decrease without any measurable effect in the
micellar region itself. Evidently the presence of the alkyl
group on the benzene ring creates some steric hindrance in the
two-dimensional packing of the surfactant species into the hemi-
micelles. The factors responsible for these different effects of
the xylene sulfonates in different regions could similarly yield
different interfacial effects (interfacial tension vs. adsorption)
and a full understanding of the mechanisms responsible for it
should prove useful.

Ether Linkage (Disulfonate). The disulfonate used here is essentially two decylbenzene sulfonates connected through an oxygen
(14). The isotherm obtained with the dodecylphenoxy sulfonate
is compared in Figure 9 with that for the decylbenzene sulfonate.
The disulfonates isotherm is characterized by the absence of
multiple regions obtained in all other cases. Adsorption in the
"Electrostatic region" is comparable, but at higher concentrations,
adsorption of the disulfonate is markedly lower than that of the
DBS. Importantly, there is an absence of the sharp rise in ad-
sorption(attributed to hemi-micellization). Evidently, the
disulfonate with the oxygen linkage in between the two AAS prevents
the alkyl chains from packing tightly to form the two-dimensional
aggregate. Also, a molecular model of the compound suggests the
possibility of coiling of the two alkyl chains, even in the bulk,
minimizing the driving force for aggregation on the surface (Figure
10). Adsorption under these conditions should be considered to be
the result of electrostatic attraction only. The fact that the
disulfonates adsorption was greater than the monosulfonates in the
electrostatic attraction region helps prove these assumptions.
Also, a value of 1.1 was obtained for the slope of the isotherm
(3×10^{-6} – 2×10^{-3} $kmol/m^3$ range) further supports this consideration.

Ethylene Oxide Addition. Anionic and nonionic alkylaryl compounds
containing amound of thylene oxide were used in this study. Addition
of ethylene oxide groups is known to impart salt tolerance to the
surfactant and therefore these compounds are of particular
interest for micellar flooding purposes.
     The anionic alkylarylpolyether sulfonate used in this
study was Triton X-200. The chemical composition of this compound
is such that the sulfonate is connected to a small alkyl group
which in turn is connected to the ethylene oxide group. Then
comes the aryl and alkyl groups. Due to the alkyl sulfonate
link comparison with an alkyl sulfonate might prove more helpful
than comparisons made with alkylaryl sulfonates. The isotherm
obtained for Triton X-200 is similar in shape to that of the
alkyl sulfonate, of equivalent chain length (Figure 11), with
the electrostatic, hemi-micellar, electrostatic hindered and micellar regions all being comparable. When a 10^{-2} $kmol/m^3$ solution of

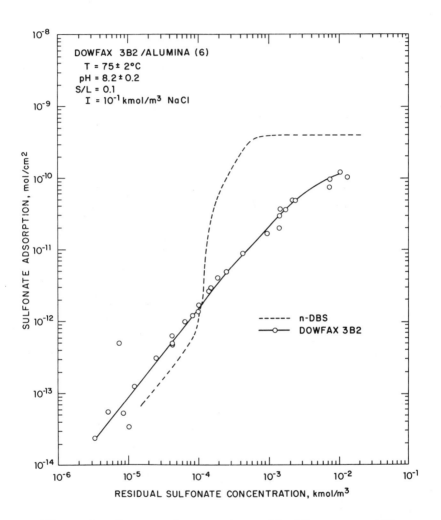

Figure 9. Structural comparison: disulfonate vs. AAS.

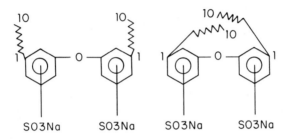

Figure 10.   Disulfonate adsorption model.

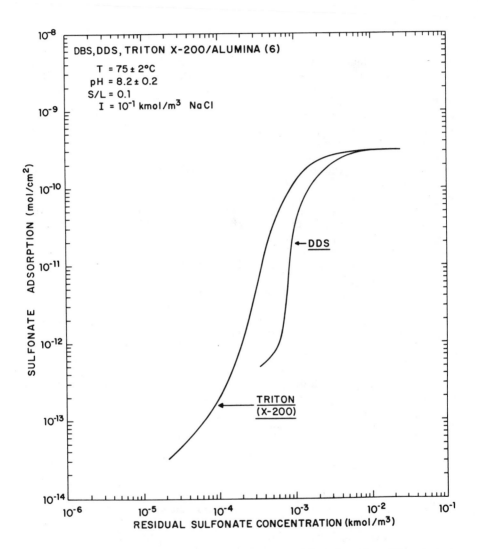

Figure 11. Structural comparison: ethoxylated sulfonate vs. AS.

calcium chloride was added to each of these systems the alkyl
sulfonate precipitated while the Triton X-200 adsorption values
remained relatively unchanged proving the salt tolerance of the
ethelyne oxide groups.

The nonionic surfactant used is Synfac 8216, an alkylaryl-
ethoxylated alcohol. This compound did not adsorb by itself on
alumina at a concentration of 4.3×10^{-4} kmol/m^3 even after a salt
level of 30% at 27°C. At 75°C measurements were conducted in up
to 5% NaCl and no significant adsorption was obtained. Synfac
did however, undergo significant adsorption in the presence of
dodecylsulfonate with maximum adsorption under the tested conditions at 1.5×10^{-3} kmol/m^3 DDS (see Figure 12). The adsorption
of sulfonate was also enhanced by the addition of Synfac except in
the micellar region (Figure 13). Synergetic interaction between
Synfac and dodecylsulfonate suggests chain-chain interaction between the chains of the two reagents. Co-adsorption of nonionics
can enhance hemi-micellization of ionic surfactants on account
of the reduced lateral electrostatic repulsion between the ionic
heads and thereby also increase the overall adsorption of both
the compounds. Indeed, if most or all of the adsorption sites
are occupied, the two species will have to share the sites, and
under these conditions the individual adsorption of both the compounds will undergo a decrease unless multilayer adsorption can
take place. It is clear from the results given in Figures 12 and
13 that interactions between various surfactants and co-surfactants
can play a major role in determining the adsorption in the system
depending particularly on their chemical structure.

Conclusions

Structural variations of sulfonate surfactants in terms of
chain length and branching, incorporation of phenyl, ethoxyl and
multiple functional groups are found to produce specific effects
in various adsorption regimes (electrostatic, hemimicellar,
electrostatic hindered and micellar regions):

a) Incorporation of the phenyl group between the α- CH$_2$ and
the sulfonate of the octyl and decyl sulfonate increased the effective chain length by 3 to 4 CH$_2$ groups with respect to its adsorption on alumina;

b) Increase in the chain length of the alkyl groups increased
the adsorption in all the regions except the micellar region.
These effects suggest the role of hemi-micellization as well as
the enhanced solvent power of the interfacial region for the
longer chains;

c) The position of the branching of the sulfonate has a
measurable effect on adsorption. The nature of the effect of
the positioning on adsorption, particularly in hemi-micellar
region, is suggestive of the type of packing of the surfactant
species in the two-dimensional aggregates;

d) Presence of alkyl substitutions on the aromatic ring

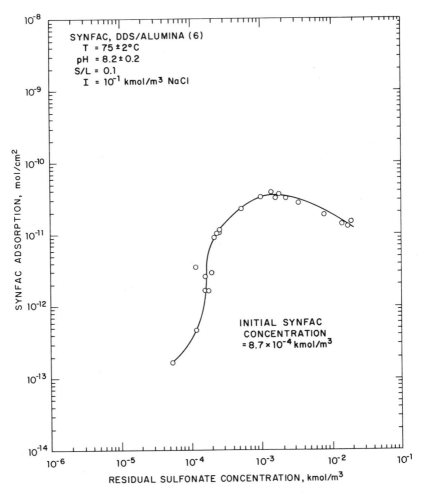

Figure 12. Ethoxylated alcohol coadsorption.

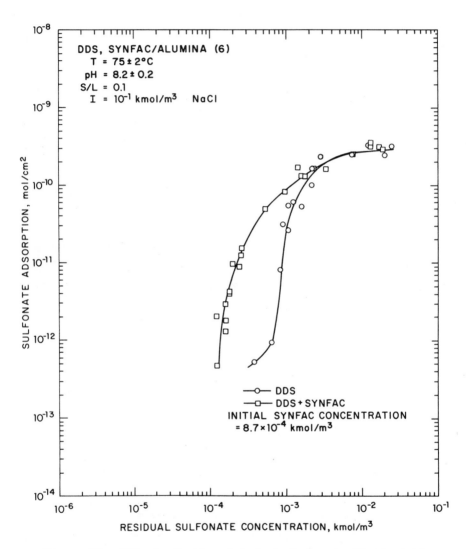

Figure 13.   Effect of ethoxylated alcohol upon AS adsorption.

of the alkylaryl sulfonate decreased the adsorption.   Furthermore, the hemi-micellization effect was also reduced apparently
due to the steric hindrance produced by the xylene group.

e) Adsorption of the disulfonate, while comparable in the
electrostatic region, was markedly lower at higher concentrations
than that of the monosulfonate. Non-appearance of the sharp
transition observed normally between Regions 1 and 2 suggests
absence of hemi-micellization in the present case due to the head
group bulkiness introduced by the ether linkage connecting the two
sulfonates;

f) Adsorption characteristics of an anionic and a nonionic
surfactant containing ethylene oxide groups were also studied.
Even though salt (Ca) tolerance of the alkylarylpolyether
sulfonate was markedly higher than that of the alkyl sulfonate
of similar chain length, their adsorptions were not significantly
different. The nonionic polyether alcohol, on the other hand,
did not adsorb on alumina under the conditions studied. However,
its adsorption was significant when the alkyl sulfonate was
present in the system. This co-adsorption was also found to
enhance the adsorption of the sulfonate.

The important role of the structure of the surfactants in
determining adsorption is evident. Some of the surfactants
discussed above can produce low interfacial tension and some
others have excellent salt tolerance. A knowledge of the
structure of such surfactants in adsorption can be helpful in
developing surfactants that will meet different requirements
simultaneously for special applications such as in enhanced oil
recovery.

Acknowledgments

Support of the Department of Energy, National Science
Foundation (CPE-82-01216), Amoco Production, Chevron Oil Field
Research, Exxon Research and Engineering, Gulf Research and
Development, Marathon Oil, Shell Development, Standard Oil of
Ohio, Texaco, and Union Oil, is gratefully acknowledged.

Literature Cited

1.   P. Somasundaran and K.P. Ananthapadmanabhan, "Physico Chemical
     Aspects of Adsorption on Surface Active Agents on Minerals",
     Croatica Chemica Acta, 52, 1979, p. 67-86.
2.   E.D. Goddard and P. Somasundaran, "Adsorption of Surfactants
     on Solids", Croatica Chemica Acta, 48, 1976, p. 451-61.
3.   H.S. Hanna and P. Somasundaran, "Physico-Chemical Aspects
     of Adsorption at Solid/Liquid Interfaces, Part II. Berea
     Sandstond/Mahogony Sulfonate System", in Improved Oil
     Recovery by Surfactants and Polymer Flooding, D.O. Shah and
     R.S. Schecter, eds., Academic Press, 1977, p. 253-274.

4. P. Somasundaran and D.W. Fuerstenau, "Mechanism of Alkyl Sulfonate Adsorption at the Alumina-Water Interface", J. Phys. Chem., 70, 1966, p. 90-96.
5. P. Somasundaran, M. Celik, and A. Goyal, "Precipitation and Redissolution of Sulfates and Their Role in Adsorption on Minerals", in Surface Phenomena in Enhanced Oil Recovery, D.O. Shah, ed., Plenum, 1981, p. 641.
6. M. Nakagaki, T. Hamda, and Shimabapashi, in J. Colloid and Interface Science, 1973, Vol. 43, p. 521-9.
7. C. Giles, T. MacEwan, H. Nakhwa, and D. Smith, J. ACS, 3973, 1969.
8. B. Tamamushi, Colloidal Surfactants, Academic Press, New York, 1963, p. 125.
9. M. Celik, A. Goyal, E. Manev, and P. Somasundaran, "The Role of Surfactant Precipitation and Redissolution in the Adsorption of Sulfonates on Minerals", S. of Petroleum Eng., SPE 8263, 1979.
10. K. Shinoda, Colloidal Surfactants, Academic Press, New York, 1963.
11. P. Somasundaran, "The Relationship Between Adsorption at Different Interfaces and Flotation Behavior", Trans. AIME, 241, 1968, p. 105-108.
12. G. Parks, "Adsorption in the Marine Environment" in Chemical Oceanograpny, 2nd Ed., Riley and Shirrow, Academic Press, 1974.
13. S. Dick, D.W. Fuerstenau, and T. Healy, J. Colloid and Interface Science, 37, 1971.
14. R. Marriot, C. Kao, and F. Kristal, SPE J., 1982, p. 993-997.

RECEIVED March 6, 1984

The Effect of Preadsorbed Polymers on Adsorption of Sodium Dodecylsulfonate on Hematite

J. E. GEBHARDT[1] and D. W. FUERSTENAU

Department of Materials Science and Mineral Engineering, University of California, Berkeley, CA 94720

The presence of pre-adsorbed polyacrylic acid significantly reduces the adsorption of sodium dodecylsulfonate on hematite from dilute acidic solutions. Nonionic polyacrylamide was found to have a much lesser effect on the adsorption of sulfonate. The isotherm for sulfonate adsorption in absence of polymer on positively charged hematite exhibits the typical three regions characteristic of physical adsorption in aqueous surfactant systems. Adsorption behavior of the sulfonate and polymer is related to electrokinetic potentials in this system. Contact angle measurements on a hematite disk in sulfonate solutions revealed that pre-adsorption of polymer resulted in reduced surface hydrophobicity.

Technological problems caused by fine particulates occur in many industrial situations including mineral processing circuits. The processing of fine particle suspensions has been the topic of several technical conferences in recent years (1-2). One method exhibiting significant potential for recovering valuable minerals from fine-particle suspensions is selective flocculation. This involves the use of a polymer or long-chained organic molecule to selectively aggregate particles of one of the minerals prior to a separation stage. Depending on the process and mineral composition, either the valuable or gangue mineral may be flocculated. To obtain a suitable concentrate, the flocculated particles must be separated from the suspension. The usual method is sedimentation of the flocs combined with elutriation of the dispersed particles. Flotation of the flocculated particles is a possible method to achieve that separation. The effect of polymers used as flocculants on the flotation of a few minerals has received

[1]Current address: U.S. Bureau of Mines, Avondale Metallurgy Research Center, 4900 LaSalle Road, Avondale, MD 20782

0097-6156/84/0253-0291$06.00/0

limited investigation (3-5), and details on the mechanism of
polymer and flotation collector coadsorption and relation to
flotation are lacking. This paper reports the results of an
investigation undertaken to delineate the role of polymer and
collector interactions with fine particles and their relation to
resultant surface wettability. In particular, the adsorption of
an anionic surfactant on a surface with previously adsorbed nonionic or anionic polymer is examined.

Experimental Methods and Materials

Materials. Synthetic hematite was obtained from J. T. Baker
Chemical Company, Phillipsburg, NJ. Particle size analysis using
a HIAC instrument (Montclair, CA) indicated the particles to be
80 percent (number) finer than 2 microns. Using nitrogen as the
adsorbate, the B.E.T. specific surface area was found to be 9
square meters per gram. The point of zero charge, as obtained
from electrophoretic measurements in the presence of indifferent
electrolytes, occurred at pH 8.3.

Polyacrylic acid (PAA) was obtained from Scientific Polymers,
Inc., Ontario, NY, as a secondary standard with a mass-averaged
molecular weight of two million. The polyacrylamide (PAM) used
was Separan MGL obtained from Dow Chemical Company, Midland, MI.
Its reported molecular weight was in the range of 500,000 to
5,000,000. The monomer structures of PAA and PAM are illustrated
in Figure 1.

Sodium dodecylsulfonate (SDS) was prepared from dodecyl alcohol by Ben Den Chemical Company, Naperville, IL. The material
was recrystallized by dissolution in hot ethanol solution, fil-
tering and cooling to crystallize the SDS. The precipitate was
filtered, washed with cold ethanol, and dried in a dessicator
under vacuum. Analytical-grade HCl and NaOH were used for pH
adjustment and NaCl for controlling ionic strength.

Adsorption Methods. Five grams of hematite were first conditioned in 0.001 M NaCl at pH 4.1. After the SDS had been added
to the slurry and the pH adjusted as required, the samples were
conditioned on a rotating shaker for two hours. The solutions
were then centrifuged, and the supernatant liquid analyzed for
its SDS content. The amount of SDS adsorbed was calculated as
the difference between the initial amount added and the residual
amount measured. Experimental results showed that two hours was
sufficient time for equilibrium to be reached. Somasundaran (6)
observed similar equilibrium adsorption times for sulfonate
adsorption on aluminum oxide.

For adsorption on flocculated particles, the polymer was
added in a drop-by-drop-wise manner from a burette containing a
50 cc solution to a 50 cc solution containing the solids. Flocculation was performed in an unbaffled vessel, 58 mm in diameter.
Agitation was achieved with a 3-bladed propeller, 35 mm in

POLYACRYLIC ACID (PAA)

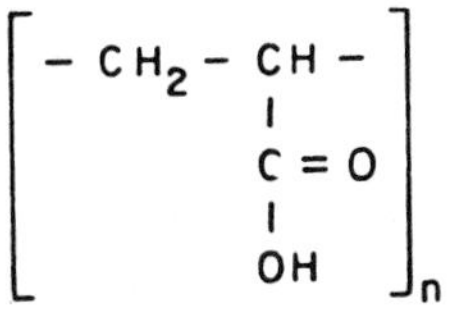

$$\left[-CH_2 - \underset{\underset{\underset{OH}{|}}{\underset{C=O}{|}}}{CH} - \right]_n$$

POLYACRYLAMIDE (PAM)

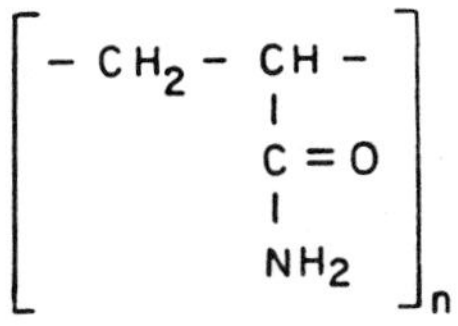

$$\left[-CH_2 - \underset{\underset{\underset{NH_2}{|}}{\underset{C=O}{|}}}{CH} - \right]_n$$

SODIUM DODECYLSULFONATE (SDS)

$$CH_3(CH_2)_{10} \, CH_2SO_3Na$$

Figure 1. Chemical structure of polyacrylic acid and polya-crylamide monomers and sodium dodecylsulfonate.

diameter, which was located approximately 10 mm from the bottom
of the vessel. Each blade was 13 mm across having a pitch of
about 30 degrees. After agitating the slurry for 10 minutes at
600 rpm, the solution was decanted and the residual polymer
determined. Polymer adsorption was rapid, and the method of
polymer addition to these well-mixed suspensions was found to
yield reproducible results. The flocs were transferred to 70-cc
glass sample bottles to which SDS was added. The pH was adjusted
as required, and the samples were conditioned on a rotating
shaker (approximately 60 rpm) for two hours. Two hours was
observed to be adequate for constant adsorption values to be
reached in the case of particles that had been previously flocculated with polymer.

Electrokinetic Measurements. Electrophoretic mobilities were
measured with a flat-cell apparatus manufactured by Rank
Brothers, Cambridge, England. In addition, several mobility
values were checked for accuracy with a Zeta Meter, New York.
Mobilities were determined with a small volume of the suspension
(approximately 25 cc) that had been prepared for the adsorption
experiments. The pH of the solution was measured prior to determining the electrophoretic mobilities, which involved measuring
the velocities of five to ten particles in each direction. An
average value of the mobilities was recorded. Samples containing
the flocculated particles were dipped into an ultrasonic bath for
approximately one second prior to making the pH and mobility
measurements.

Contact Angle Measurements. Contact angles were measured on a
mounted disk of synthetic hematite. The disk was prepared by
cold pressing hematite powder in a steel die and sintering it at
1,000 degrees Celsius for 8 hours in air. X-ray diffraction
patterns indicated that the sample was still α-hematite. The
angle formed by a bubble attached to the disk was measured using
a goniometer. To form a bubble, a J-shaped glass capillary was
placed under the disk, which had been positioned surface-side
down in the solution. Advancing and receding angles were
obtained by manipulating air in and out of the capillary tube.
"Equilibrium" angles were determined by detaching the bubble from
the capillary. The disk was cleaned between each measurement by
several turns on a polishing wheel.

Surface Tension Measurements. The surface tension of surfactant
solution was measured by the capillary rise method with a glass
capillary tube, 18.66 cm in length and 0.0531 cm internal diameter. This was obtained by filling the capillary with a column
of mercury and weighing the capillary with and without the mercury. The density of mercury was taken to be 13.5939 grams per
cubic centimeter at 20 degrees Centigrade (7). The height of
capillary rise was measured with a cathetometer manufactured by

Gaertner Scientific Corp., Chicago, IL, in a temperature-
controlled room at 21 degrees Centigrade.

<u>Analytical Methods</u>. It was found that the concentration of both
polymers could be analyzed by determining the total carbon in
solution with a carbon dioxide coulometer, Coulometrics, Inc.,
Wheat Ridge, CO. The accuracy of this method is not good in the
low polymer concentration range, that is less than 10-15 ppm.
For higher accuracy in the low polymer concentration range, two
different methods were used. In the case of PAA, potentiometric
titrations of solutions of PAA were performed with 0.01 N NaOH
using a Brinkman model, Westbury, NY, automated titrator. Blank
tests indicated no interfering species. Known amounts of PAA
were used to prepare a calibration curve immediately after
titration of the samples containing unknown amounts of polymer.
The starting point of the titration was pH 4.0, and the end point
was reached near pH 8. Total volumes of 75 or 100 cc were used
for the titrations, and the ionic strength was controlled at 0.01
M NaCl.

For the determination of PAM, UV adsorbance at 189 nm was
measured for various PAM concentrations. The nephelometric
technique of Attia and Rubio (<u>8</u>) as modified by Pradip (<u>9</u>) was
used to check the calibration curve obtained by the UV method.

The amount of SDS in solution was determined by the method of
Jones (<u>10</u>) modified in a manner similar to that of Somasundaran
(<u>6</u>). Methylene blue chloride was added to a solution containing
S<u>D</u>S. The resulting blue complex was extracted into an organic
phase, and the absorbance in the visible range was measured.
One-half cc of a one percent (by weight) methylene blue chloride
solution was added to a 4.5 cc sample of solution containing 5-20
micrograms of SDS. Five cc of chloroform was then added and the
solutions shaken by hand for 30 seconds. For the flocculated
system, the polymer content was determined first and sufficient
amounts of polymer were added prior to SDS determination to maintain the polymer level constant at 200 ppm. A calibration curve
was made using absorbance at 652 nm as a function of SDS
concentration.

<u>Results and Discussion</u>

<u>Polymer/Surfactant Interactions</u>. Interaction between polymers
and surfactants was recently reviewed by Robb (<u>11</u>) and surfactant
association with proteins by Steinhardt and Reynolds (<u>12</u>).
Polymer/surfactant interactions are highly dependent on the chemical nature of the polymer and the surfactant. In general, surfactants tend to associate with uncharged polymers in aggregates
rather than individual surfactant molecules interacting with the
macro-molecule. The ability of surfactants to form micelles is
thought to be an important factor in the role of surfactant behavior in interactions with polymers. Individual surfactant

molecules can interact directly with polyelectrolytes or polymers
capable of some degree of ionization, such as proteins, while
association between polymers and surfactant aggregates follows at
higher surfactant concentrations. Electrostatic forces are primarily responsible for the different interaction behavior of
surfactants with the charged and uncharged polymers.

The measurement of the surface tension of SDS solutions at
constant polymer additions was performed to investigate any possible interactions between SDS and the polymers used in these
experiments. The results, shown in Figure 2, indicate no interaction between SDS and either PAA or PAM. Interactions between
similarly charged surfactant and polyelectrolyte are not common
as electrical effects frequently dominate to prevent any hydrophobic or hydrogen bonding interaction. The hydrophilic nature
of the amide dipole of polyacrylamides has been suggested (11) as
a possible factor in preventing interaction with sodium
dodecylsulfate.

Adsorption of SDS on Hematite. Adsorption of SDS on hematite at
pH 4.1 is illustrated in Figure 3 as the zero polymer (top line-
filled circles) addition. The results indicate three distinct
regions of adsorption and in this respect agree well with previous investigations (13, 14) of systems involving adsorption of
an anionic surfactant from an aqueous solution of constant ionic
strength onto a positively charged oxide surface. Fuerstenau and
Raghavan (15) have summarized the proposed mechanism of adsorption by describing the three regions of differing slope. In
brief, at low SDS equilibrium concentration the adsorption process is described as being an electrostatic exchange of surfactant
ions for counter ions in the double layer. The increased slope
of the next region is attributed to the onset of hemi-micelle
formation. At high SDS equilibrium concentration, the zeta
potential is of similar sign as the surfactant and electrostatic
interactions oppose specific interactions resulting in a reduced
slope of the adsorption isotherm.

Adsorption on Hematite With Pre-Adsorbed PAA. The adsorption of
SDS on hematite particles which were previously flocculated by
the addition of PAA is illustrated in Figure 3. The adsorption
of SDS decreases as increasing amounts of PAA occupy surface
sites. The adsorption of PAA on hematite, which had been investigated previously (16), is given in Figure 4 as a function of
equilibrium PAA concentration for various pH values. These
results show that the adsorption of PAA is characterized by high
adsorption densities at low equilibrium concentrations. The
plateau or surface saturation value for pH 4.1 would be approximately 0.41-0.42 mg PAA per square meter. This corresponds
approximately to the highest PAA addition shown in Figure 3, 0.44
mg PAA per square meter, where it was determined that 0.41 mg PAA

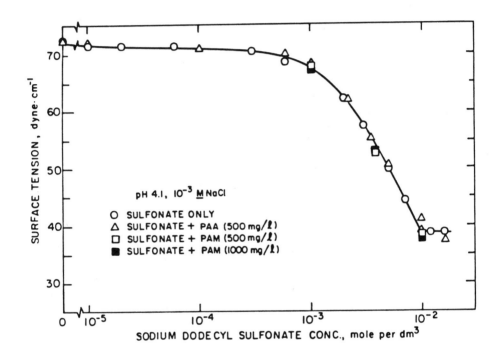

Figure 2.   Surface tensions of sodium dodecylsulfonate solutions with and without polymer addition as measured by the capillary rise method.

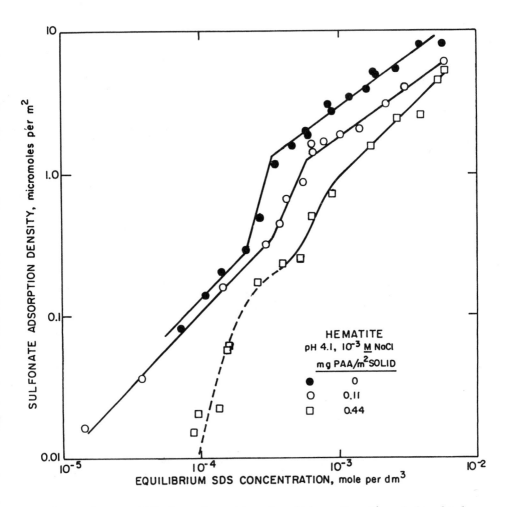

Figure 3.   Equilibrium adsorption densities of sodium dodecylsulfonate on hematite at pH 4.1 and 0.001 M NaCl in the presence and
absence of pre-adsorbed polyacrylic acid.

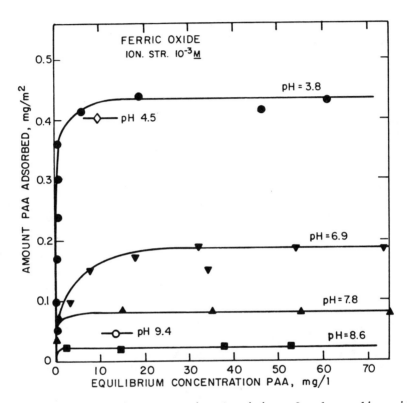

Figure 4. Equilibrium adsorption densities of polyacrylic acid on hematite at various pH values. Reproduced with permission from Ref. 22. Copyright 1983, Colloids and Surfaces.

per square meter had been adsorbed. No residual polymer was
measured in solution for the 0.11 mg PAA addition.

In general, the adsorption of a surfactant on particles with
previously adsorbed polymer can be influenced by (i) a reduction
of surface area available for adsorption as a result of the
presence of adsorbed polymer, (ii) possible interactions between
polymer and surfactant in the bulk solution or in the interfacial
region (that is, surfactant with loops, tails or trains of adsorbed polymer molecules), (iii) the steric effect of adsorbed
polymer, preventing approach of surfactant molecules for adsorption at the surface, or (iv) possible electrostatic effects if
polymer and/or surfactant are charged species.

Since there was no evidence of surfactant/polymer interactions, the two most important factors affecting SDS adsorption
on PAA-flocculated hematite particles are believed to be electrostatic effects and a reduction in the number of surface adsorption sites due to adsorbed polymer segments. Polyelectrolyte
adsorption on oppositely charged surfaces can involve strong
attractive forces (as evidenced by the high affinity isotherm)
and is expected to result in a somewhat flat surface configuration (17). Using the surface saturation value of 0.41 mg PAA
per square meter and assuming a 25 square angstrom PAA monomer
segment size, 87 pct of the total surface area would be occupied
by polymer segments if the segments were free to make contact
individually. In actuality, polymer segments are not free to
make individual contact with the surface because of the nature of
the molecular chain structure. However, the adsorbed amounts
shown in the adsorption isotherms of SDS on PAA-flocculated hematite indicate that the polymer chains may have a significant
number of segments in contact with the surface, thereby reducing
the number of sites available for SDS adsorption.

The effect of polymer charge on the zeta potential is given
in Figure 5 where the electrophoretic mobility of hematite particles flocculated with PAA and conditioned in SDS is indicated as
a function of SDS adsorption density. The electrophoretic mobilities of hematite particles without polymer in the presence of
SDS are also shown in Figure 5 (filled circles). In the presence
of the anionic surfactant, the electrophoretic mobility decreases
with increasing SDS adsorption density, reverses in sign and
becomes negative in value at high SDS adsorption densities. This
behavior is typical for anionic surfactant interaction with an
oppositely charged surface and has been observed by other
researchers (18). At pH 4.1 in the absence of surfactant (and
polymer), hematite particles are positively charged and have a
mobility of 3.9 electrophoretic mobility units.

For the addition of 0.11 mg PAA per square meter hematite,
the mobility decreased at SDS adsorption densities less than 0.3
micromoles per square meter but remained positive in value.
Above this adsorption density, negative mobilities were recorded.
At PAA additions of 0.22 and 0.44 mg per square meter hematite,

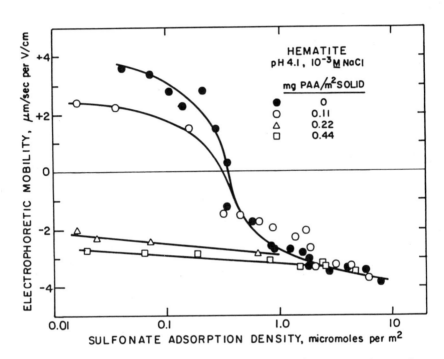

Figure 5.   Electrophoretic mobility of hematite at pH 4.1 and
0.001 M NaCl as a function of sulfonate adsorption density in the
absence and presence of pre-adsorbed polyacrylic acid.

mobility values were negative for all SDS concentrations and
became slightly more negative with increasing surfactant adsorption. The presence of PAA has a dramatic effect on the mobility
values, although surfactant adsorption also contributed to
changes in the mobility, especially at PAA additions less than
0.22 mg per square meter. Above SDS adsorption densities of 1-2
micromoles per square meter, the mobilities observed in all the
experiments were the same in the absence or presence of pre-
adsorbed polymer.

The surface charge of the hematite used here was measured by
titration (16) and found to be approximately 18 microcoulombs per
square meter, equivalent to a charge of 5.40×10^8 esu per square
meter of hematite. Assuming one equivalent per monomer segment,
the total charge of PAA would be 4.02×10^9 esu per mg PAA for
the totally ionized polymer. Approximately 0.11 mg PAA (if
totally ionized) would be almost sufficient to neutralize the
charge of one square meter of hematite. The transition of positive mobility to negative mobility occurs between 0.11 and 0.22
mg PAA per square meter hematite, in accordance with the amount
of polymer charge required to neutralize the solid surface
charge.

Adsorption on Hematite With Pre-Adsorbed PAM. Adsorption of SDS
on hematite flocculated with PAM was investigated with 0.44 mg
PAM per square meter hematite. SDS adsorption density as a
function of equilibrium SDS concentration is shown in Figure 6
for the same conditions as used in the PAA/SDS system. In the
low SDS concentration range (that is, less than 0.0003 M SDS),
adsorption of SDS was unaffected by the presence of PAM. At
equilibrium concentrations greater than 0.0003 M SDS, adsorption
of SDS on the flocculated particles was reduced from that on the
unflocculated particles. Adsorption of PAM on hematite as a
function of PAM solution concentration is shown in Figure 7.
Adsorption of PAM does not exhibit the strong affinity observed
for PAA, although the plateau or saturation value is similar and
slightly higher for PAM. These results suggest that PAM adsorption may occur in a more loosely packed surface configuration.

Some polymer molecules can be regarded to maintain their
approximate solution conformation upon adsorption (19). Adsorption of a nonionic polymer would lead to a coiled adsorbed
polymer configuration with a small number of polymer segments in
actual contact with the surface. The number of surface sites
available for surfactant adsorption would remain quite large.

At equilibrium surfactant concentrations of less than 0.0003
M SDS where the hematite surface is still positively charged,
adsorption of surfactant follows its normal pattern due to the
electrostatic forces which provide the driving force for adsorption. Sufficient effective surface area must be available for
this level of SDS adsorption density. As surfactant adsorption

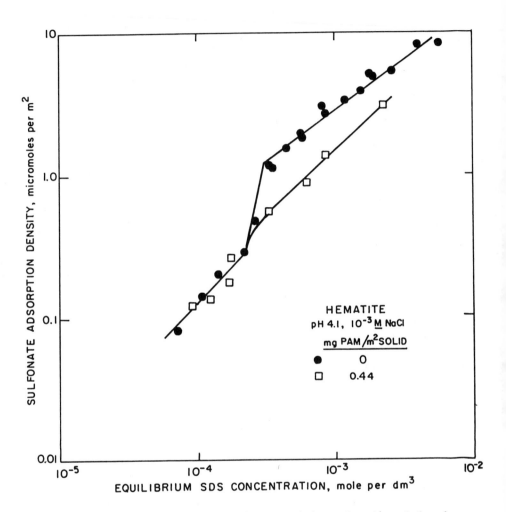

Figure 6. Equilibruium adsorption densities of sodium dodecyl-sulfonate on hematite at pH 4.1 and 0.001 M NaCl in the presence and absence of pre-adsorbed polyacrylamide.

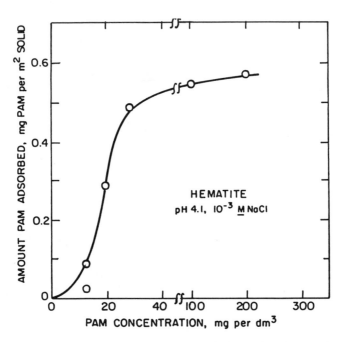

Figure 7. Equilibrium adsorption densities of polyacrylamide on
hematite at pH 4.1 and 0.001 M NaCl.

increases with increasing SDS concentration, the surface charge
is reversed (see Figure 8), and additional surfactant adsorption
is reduced due to electrical forces and also to a reduction of
available surface area due to adsorbed PAM segments. No significant change in slope is observed, and it is possible that the
presence of adsorbed polymer prevents or reduces the effect of
hemi-micelle formation.

The number of polymer segments in contact with the surface
can be estimated from Figure 6 at high SDS concentrations. It is
assumed that reduction in sulfonate adsorption, for the nonionic
polymer case, is due only to the presence of adsorbed polymer
segments. The amount of reduction in SDS adsorption at high SDS
concentrations is taken (from Figure 6) to be an average of 50
percent, that is the effective area available for SDS adsorption
is reduced by 50 percent. Using a monomer size of 25 square
angstroms, a total of 10.7 mg of PAM would be required to occupy
the area reduced which in this case is 22.5 square meters (total
sample surface area was 45 square meters). Since the uptake of
PAM was an average of 18 mg PAM, it is estimated that, at the
most, approximately 50 to 60 percent of the PAM segments are in
contact with the surface.

Surface Wettability. Contact angles on hematite at pH 4.1 were
measured as a function of PAA and PAM concentration at constant
surfactant concentration of 0.001 M SDS. The results are given
in Figures 9 and 10 for PAA and PAM, respectively. The hematite
disk was conditioned for 15 minutes in a solution containing the
polymer and then transferred to the SDS solution for contact
angle measurement. The angles were measured after 15 minutes
conditioning with SDS and were not observed to vary with longer
conditioning times. With no polymer present, the bubble contact
angle was high, indicative of a hydrophobic surface. It is
evident that adsorbed polymer caused a dramatic decrease in the
contact angle. No bubble contact, for the equilibrium angle,
could be achieved above polymer solution concentrations of 2 ppm
for either PAA or PAM. Both polymers, being hydrophilic in character, prevent bubble attachment when a sufficient amount of
polymer has been adsorbed. These results parallel findings by
Somasundaran and coworkers (20, 21) that adsorption of a large
hydrophilic molecule masks any effect that adsorbed surfactant
molecules may have on the surface wettability.

Additional contact angles were measured at 0.001 M SDS concentration. The hematite disk was conditioned in the SDS solution before adding polymer to the solution. The contact angle
was measured after adding sufficient quantities of polymer for
concentrations of 1, 3, and 5 ppm. For both PAA and PAM, the
contact angle remained constant and identical to the value
obtained in the absence of polymer. No polymer was able to
adsorb on the surfactant-coated disk. It can be concluded that
whichever species, surfactant or polymer, adsorbed first was not

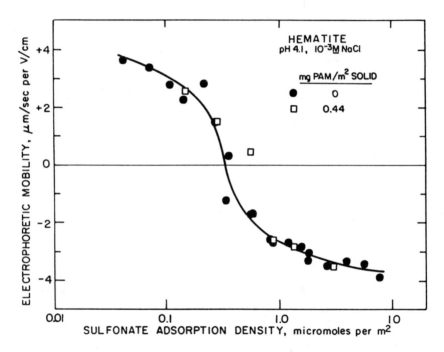

Figure 8.   Electrophoretic mobility of hematite at pH 4.1 and 0.001 M NaCl as a function of sulfonate adsorption density in the absence and presence of pre-adsorbed polyacrylamide.

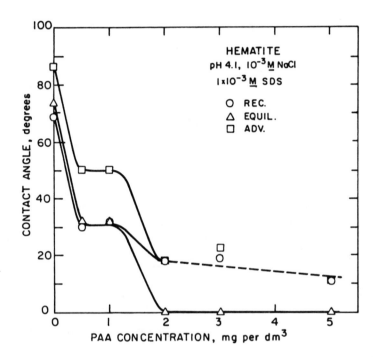

Figure 9. Contact angles on a hematite disk at 0.001 M SDS, pH 4.1, and 0.001 M NaCl as a function of the polyacrylic acid concentration in which the hematite disk had been preconditioned.

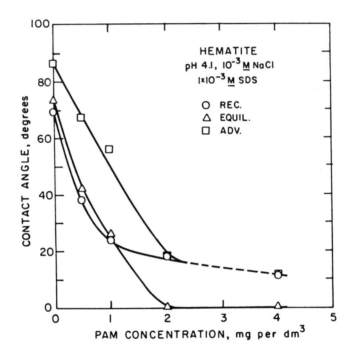

Figure 10.   Contact angles on a hematite disk at 0.001 SDS pH
4.1, and 0.001 M NaCl as a function of the polyacrylamide concentration in which the hematite disk had been preconditioned.

replaceable by the other, at least under these particular
conditions.

Summary

The adsorption behavor of surfactant onto particles in the
absence and presence of pre-adsorbed polymer was determined.
Electrokinetic studies of the system were made. Contact angle
measurements yielded information on the level of hydrophobicity
achieved at various additions of polymer and collector.
    The results are summarized as follows:
1. Surface tension measurements indicated no bulk interaction
    between the anionic surfactant and the anionic or nonionic
    polymer.
2. The adsorption of an anionic surfactant on a positively
    charged oxide surface is significantly reduced by the
    presence of a pre-adsorbed anionic polymer.
3. The presence of a pre-adsorbed nonionic polymer has almost
    negligible effects on surfactant adsorption except at high
    surfactant concentrations where surfactant adsorption is
    reduced.
4. Electrokinetic studies revealed that the mobilities of parti-
    cles with pre-adsorbed anionic polymer in the presence of
    surfactant were controlled by the charge associated with the
    polymer, while the mobilities were unaffected by the presence
    of pre-adsorbed nonionic polymer.
5. Contact angle measurements indicated that conditioning with
    increasing amount of polymer before conditioning with surfac-
    tant resulted in reduced surface hydrophobicity. Pre-
    conditioning with surfactant resulted in a hydrophobic
    surface which was not affected by subsequent polymer
    additions.

Acknowledgments

The authors wish to acknowledge support of this research by the
National Science Foundation and the Department of the Interior
for a Grant to the California MMRRI, University of California.

Literature Cited

1. "Fine Particles Processing"; Somasundaran, P., Ed.; Society
    of Mining Eng. of AIME: New York, 1980; 1865 pp.
2. "Benefication of Minerals Fines"; Somasundaran, P.; Arbiter,
    N., Eds.; American Institute of Mining & Metallurgical
    Engineers: New York, 1979; 406 pp.
3. Usoni, L.; Rinelli, G.; Ghigi, G. Proc. 8th International
    Mineral Processing Congress 1968, 14 pp.
4. Osborne, D. G. Trans. Instn. Min. Metal. 1978, 87, C189-
    C193.

5. Balajee, S. R.; Iwasaki, I., Trans. A.I.M.E., 244, 401-406; 407-411 (1969).
6. Somasundaran, P., Ph.D. Thesis, University of California, Berkeley, 1961.
7. "CRC Handbook of Chemistry and Physics"; 60th Edition, 1981.
8. Attia, Y. A.; Rubio, J. Br. Polymer J. 1975, 7, 135-138.
9. Pradip, M. S. Thesis, University of California, Berkeley, 1977.
10. Jones, J. H. J. Assoc. Official Agricultural Chem. 1945, 28, 398-409.
11. Robb, I. D. in "Anionic Surfactants: Physical Chemistry of Surfactant Action"; Luscassen-Reynders, E. H., Ed.; Marcel Dekker, Inc.: New York, 1981; Chap. 3, pp. 109-142.
12. Steinhardt, J.; Reynolds, J. A. "Multiple Equilibria in Proteins" Academic Press: New York, 1969; 234 pp.
13. Wakamatsu, T.; Fuerstenau, D. W., in ADVANCES IN CHEMISTRY SERIES Vol. 79, Gould, R. F., Ed.; 1968, pp. 161-172.
14. Somasundaran, P.; Fuerstenau, D. W. J. Phys. Chem. 1966, 70, 90-96.
15. Fuerstenau, D. W.; Raghavan, S. "Flotation: A.M. Gaudin Memorial Volume"; Fuerstenau, M. C., Ed.; American Institute of Mining, Metallurgical, and Petroleum Engineers, Inc.: New York, 1976: Vol. 1, pp. 21-65.
16. Gebhardt, J. E., M.S. Thesis, University of California, Berkeley, 1979.
17. Eirich, F. R. J. Colloid Interface Sci. 1977, 58, 423-435.
18. Hunter, R. J. "Zeta Potential in Colloid Science"; Academic Press: New York, 1981.
19. Rowland, F.; Bulas, R.; Rothstein, E.; Eirich, F. R. in "Chemistry and Physics of Interfaces"; Ross, S., Ed.; American Chemical Society: Washington, D.C., 1965.
20. Somasundaran, P. J. Colloid Interface Sci. 1969, 31, 557-565.
21. Somasundaran, P.; Lee, L. T. Separation Sci. & Tech. 1981, 16, 1475-1490.
22. Gebhardt, J. E.; Fuerstenau, D. W. Colloids and Surfaces 1983, 7, 221-231.

RECEIVED February 3, 1984

19

Adsorption and Electrokinetic Effects of Amino Acids on Rutile and Hydroxyapatite

D. W. FUERSTENAU, S. CHANDER[1], J. LIN[2], and G. D. PARFITT[3]

Department of Materials Science and Mineral Engineering, University of California, Berkeley, CA 94720

The mechanism of interaction of amino acids at solid/
aqueous solution interfaces has been investigated
through adsorption and electrokinetic measurements.
Isotherms for the adsorption of glutamic acid, proline
and lysine from aqueous solutions at the surface of
rutile are quite different from those on hydroxyapatite.
To delineate the role of the electrical double layer in
adsorption behavior, electrophoretic mobilities were
measured as a function of pH and amino acid concentrations. Mechanisms for interaction of these surfactants with rutile and hydroxyapatite are proposed,
taking into consideration the structure of the amino
acid ions, solution chemistry and the electrical
aspects of adsorption.

Interest in the nature of interactions between shortchain organic
surfactants and large molecular weight macromolecules and ions
with hydroxyapatite extends to several fields. In the area of
caries prevention and control, surfactant adsorption plays an important role in the initial states of plaque formation (1–5) and
in the adhesion of tooth restorative materials (6). Interaction
of hydroxyapatite with polypeptides in human urine is important
in human biology as hydroxyapatite has been found as a major or
minor component in a majority of kidney stones (7). Hydroxyapatite is used in column chromatography as a material for separating proteins (8–9). The flotation separation of apatite from

[1]Current address: Mineral Processing Section, Department of Mineral Engineering, The
Pennsylvania State University, University Park, PA 16802
[2]Current address: Research Laboratory, IBM, San Jose, CA 95113
[3]Current address: Department of Chemical Engineering, Carnegie–Mellon University,
Pittsburgh, PA 15213

gangue minerals as an important industrial operation in which surfactants are used to effect the separation (10). Dewatering of
the colloidal phosphatic slimes that are generated in large quantities in the processing of phosphate rock is a major industrial
problem which has been studied by a number of researchers in recent
years (11).

Only a few systematic studies have been carried out on the
mechanism of interaction of organic surfactants and macromolecules.
Mishra et al. (12) studied the effect of sulfonates (dodecyl),
carboxylic acids (oleic and tridecanoic), and amines (dodecyl and
dodecyltrimethyl) on the electrophoretic mobility of hydroxyapatite. Vogel et al. (13) studied the release of phosphate and
calcium ions during the adsorption of benzene polycarboxylic acids
onto apatite. Juriaanse et al.(14) also observed a similar release of calcium and phosphate ions during the adsorption of polypeptides on dental enamel. Adsorption of polyphosphonate on
hydroxyapatite and the associated release of phosphate ions was
investigated by Rawls et al. (15). They found that phosphate ions
were released into solution in amounts exceeding the quantity of
phosphonate adsorbed.

The present investigation was undertaken to study the mechanism of adsorption of selected amino acids in order to understand
their interfacial behavior at the hydroxyapatite surface. The
presence of two or more functional groups in amino acids give rise
to surface properties which are quite different from the interfacial properties for the adsorption of simple surfactants, that
is, those containing only one charged group. The adsorption behavior of hydroxyapatite is further complicated because of its
partial solubility. Accordingly, the interfacial properties were
also determined for TiO_2 and compared with those of hydroxyapatite.

Properties of Amino Acids

Amino acids are characterized by the presence of adjacent carboxylic ($-CO_2H$) and amino ($-NH$) functional groups. The equilibrium
constant for protonation or dissociation of these groups is a
function of their position in the amino acid molecule. Therefore,
widely differing acid-base properties of amino acids occur, depending upon the number of functional groups and their relative
position in the molecule. The dissociation constants for various
amino acids used in this investigation are given in Table I.

The first dissociation constant for the $-CO_2H$ group is the
more acidic group with a pK of 1.8 to 2.4. This group in amino
acids is substantially more acidic than acetic acid, which has a
pK_a = 4.76 due to the large inductive effect of the adjacent $-NH_3^+$
group. The pK for the protonation of the amino group has a value
of 9.0 to 10.0 which is slightly lower than that for the conjugate
acid of methylamine with a pK of 10.4. Thus, the amino acids have
a positively charged $-NH_3^+$ group acidic pH's (pH < 2) a positive
$-NH_3^+$ and a negative $-CO_2^-$ group at neutral pH's (3 < pH < 9) and

Table I.   Stability Constants of Amino Acids

Amino Acid	Formula	pK_1	pK_2	pK_3
Glutamic acid	$\overset{\overset{\displaystyle NH_2}{\displaystyle |}}{HOOC(CH_2)_2CHCOOH}$	2.13	4.32	9.95
Lysine	$\overset{\overset{\displaystyle NH_2}{\displaystyle |}}{H_2N(CH_2)_4CHCOOH}$	2.16	9.20	10.80
Proline	$\begin{array}{c} H_2C \text{———} CH_2 \\ | \qquad\qquad | \\ H_2C \qquad CHCOOH \\ \diagdown N \diagup \\ H \end{array}$	1.95	10.64	

a negative $-CO_2^-$ group in alkaline pH's (pH > 10). In addition, one must consider dissociation of other functional groups present as part of the amino acid. For example, glutamic acid contains another $-CO_2$ which has a pK (dissociation constant) of 4.32, and lysine has another amine group with the pK of its conjugate acid being 10.80. Proline, because of its nitrogen being part of the cyclic ring, has pK's of 1.95 and 10.64. The charge on various functional groups as a function of pH is schematically illustrated in Table II. It is evident that an amino acid molecule may have strongly charged positive and negative sites even though the molecule has an overall charge neutrality. The present study shows that this zwitterionic character of amino acids plays an important role in the adsorption process.

Experimental Methods and Materials

Synthetic hydroxyapatite prepared by mixing stoichiometric amounts of aqueous solutions of calcium nitrate and ammonium phosphate was used in this study. The pH of the boiling suspension was maintained at about 10 by flowing a mixture of NH_3 and N_2 throughout the precipitation process. The precipitate was repeatedly washed until the conductivity of the supernatant liquid was observed to be constant. The washed sample was freeze-dried and analyzed. An elemental analysis of the batch preparation showed the Ca/P molar to be ratio 1.64, with the predominant impurity being Si (0.12% SiO_2). The sample had a BET surface area of 19.6 m^2/g and a density of 2.96 g/cm^3. The x-ray diffraction pattern was sharp and characteristic of synthetic hydroxyapatite. "Analytical grade" reagents and CO_2-free double-distilled water were used throughout the investigation. The hydroxyapatite suspensions for the adsorption studies were prepared by adding 0.25 g hydroxyapatite to 25 ml of 10^{-2} M KNO_3 solution (unless otherwise indicated) and the requisite amount of the amino acid. After equilibration for one hour, the solids were separated in a centrifuge and the liquid was analyzed for its equilibrium amino acid concentration by the minhydrin method. The suspension was maintained at 37 ± 0.5°C throughout the adsorption equilibration period. Electrophoretic mobilities were measured with a Zeta-Meter electrophoresis apparatus. Lower concentrations of KNO_3 (generally 2×10^{-3} M) were used in these tests to avoid excessive compression of the electrical double layer and to maintain thermal stability of the suspension in an electrical field.

Results and Discussions

Adsorption and Electrokinetic Behavior of Rutile. Isotherms for the adsorption of lysine, proline and glutamic acid on rutile (TiO_2) are given in Figure 1. There is no simple relationship between the adsorption density and the equilibrium concentration. The adsorption does not obey the Langmiur, Freundlich or Stern-Grahame relationships. The leveling-off of the adsorption

Table II. Schematic Illustration of Charges on Functional Groups in Amino Acids

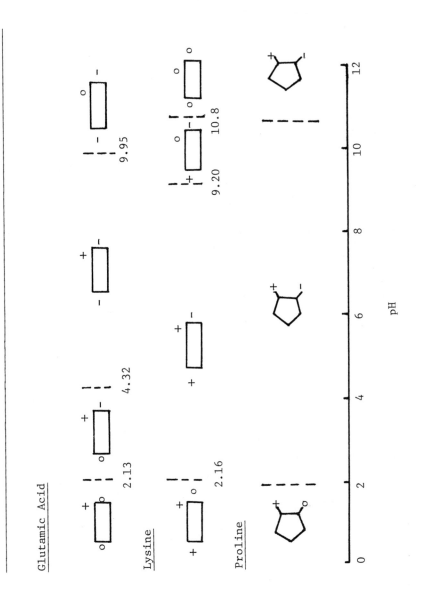

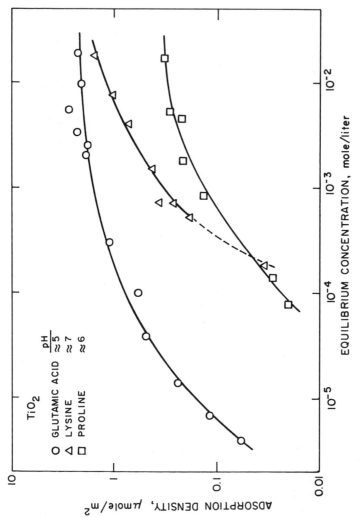

Figure 1.   The isotherms for adsorption of glutamic acid, lysine and proline on rutile.

isotherms at high concentrations shows an apparent decrease in the free energy of adsorption with increase in concentration. This decrease in overall adsorption energy may be attributed to repulsion between the adsorbed molecules. Figure 2 further illustrates the electrostatic nature of interactions between the rutile surface and the amino acids. At pH's below the PZC of rutile (pH 6.7) the solid is positively charged and it is negatively charged at higher pH's. Glutamic acid which has two negative charges between pH 4.3 and 10.0 adsorbs mainly at the positive surface. Adsorption through the amine group on the negative surface is much weaker, perhaps because of the repulsion between the two negatively charged carboxylate groups and the negative surface. Lysine adsorbs only through electrostatic interactions also, and there is no indication of adsorption through carboxylic groups. Proline exhibits a very complex behavior, however. Although the uptake of proline can be considered to involve specific adsorption, the results can be interpreted by an electrostatic model if reorientation of the surfactant molecules is taken into consideration. As compared to glutamic acid and lysine, proline molecules contain only two functional groups. At a positive surface the adsorption could occur through the carboxylate group, whereas at a negative surface the molecule adsorbs through the amine group. Stronger repulsion between the identically charged groups in the adsorbing molecule and the surface is perhaps the reason for the relatively lower adsorption density of proline compared to glutamic acid.

The electrophoretic behavior of TiO_2 in glutamic acid, lysine and proline solutions are given in Figures 3, 4 and 5, respectively. Below the PZC of TiO_2 (pH < 6.7) adsorption of glutamic acid makes the electrophoretic mobility more negative as anticipated. At pH's greater than the PZC (pH > 6.7) a weak adsorption through amine groups makes the electrophoretic mobility even more negative because for each of the glutamic acid molecule adsorbed, two carboxylate groups are attached to the surface. At pH's near 10 or higher, the adsorption of amine groups ceases because they hydrolyze to the neutral form. Therefore all electrophoretic mobility curves merge at pH 10.

Lysine does not exhibit any significant influence on the electrophoretic mobility at pH values below the PZC of TiO_2. At higher pH's, the electrophoretic mobility slightly increases because two amine groups are adsorbed for each adsorbed molecule. Proline makes the electrophoretic mobility slightly more negative in the intermediate pH range. Apparently, carboxylate ions in the adsorbing molecules are oriented in an outword direction such that they make the electrophoretic mobility more negative even though the proline molecule is overall neutral.

Adsorption and Electrokinetic Behavior of Hydroxyapatite. The adsorption densities of glutamic acid and lysine on hydroxyapatite are shown in Figures 6 and 7. The change in slope of the adsorption isotherm at 10^{-3} M glutamic acid is considered to be due to a

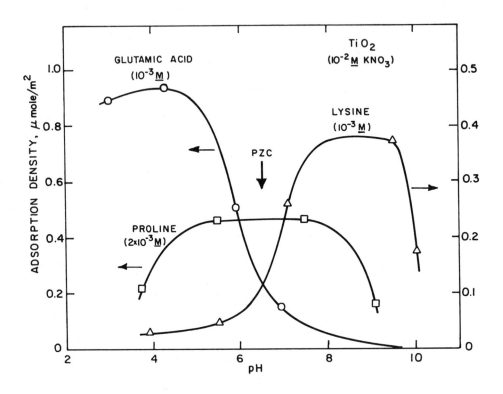

Figure 2. The effect of pH on the adsorption of glutamic acid, lysine and proline on rutile.

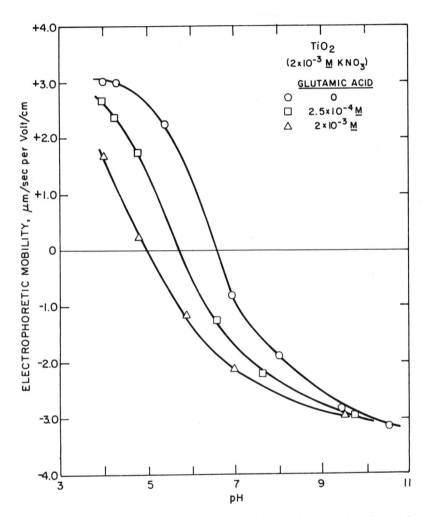

Figure 3. The effect of glutamic acid on the electrophoretic mobility of rutile.

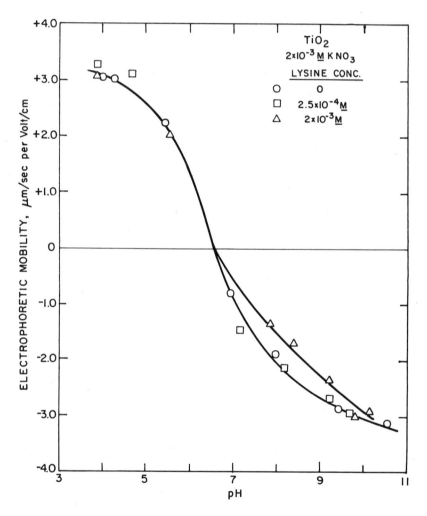

Figure 4. The effect of lysine on the electrophoretic mobility of rutile.

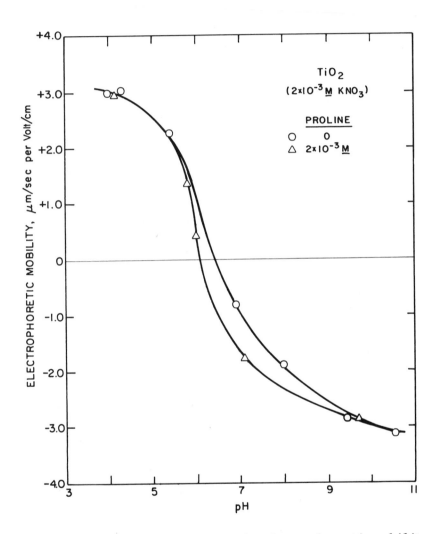

Figure 5. The effect of proline on the electrophorectic mobility of rutile.

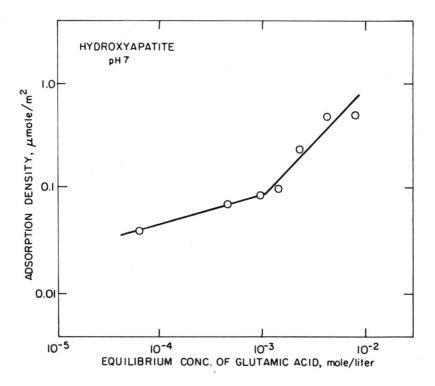

Figure 6.   The isotherm for adsorption of glutamic acid on
hydroxyapatite.

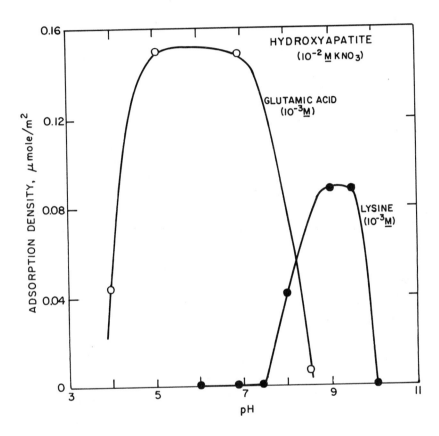

Figure 7. The effect of pH on the adsorption of glutamic acid and lysine on hydroxyapatite.

change in the adsorption mechanism. At low concentrations, we propose that the glutamic acid adsorbs whereas at high concentration the system undergoes chemical reaction involving the calcium ions at the surface. This hypothesis may explain the complex electrokinetic behavior of hydroxyapatite shown in Figure 8. At low concentration of glutamic acid, the electrophoretic mobility becomes more negative as would be expected if glutamic acid adsorbs. At higher concentrations, a chemical reaction occurs with the surface becoming positively charged, apparently because of a high concentration of coadsorbed calcium ions. Further studies are required to completely understand this highly complex system.

The adsorption of both glutamic acid and lysine is almost an order of magnitude smaller on hydroxyapatite than it is on TiO_2 (compare Figures 2 and 7). Glutamic acid adsorbs even at pH 7, suggesting that the adsorption can occur in different orientations of the glutamic acid ion. Lysine adsorbs only at a negative surface through the amine groups and adsorption ceases at pH 10 because the amine groups hydrolyze. The influence of lysine on the electrophoretic mobility of hydroxyapatite is presented in Figure 9. The mobility becomes positive because of the presence of two positively charged amine groups for each adsorbing ion.

Concluding Discussion

The adsorption of amino acids on rutile and hydroxyapatite exhibits some characteristics of specific adsorption. The results can be interpreted in terms of electrostatic models of adsorption, however, if reorientation of adsorbed molecules is taken into consideration. The electrokinetic behavior of hydroxyapatite in glutamic acid is complicated because of a chemical reaction, possibly involving calcium ions. The study shows that it is necessary to take into consideration the orientation of adsorbed molecules, particularly for zwitterionic surfactants.

Acknowledgments

The authors wish to acknowledge the National Science Foundation and the National Institute of Health, Grant NIH R01 DE 03708-06 for the support of this research.

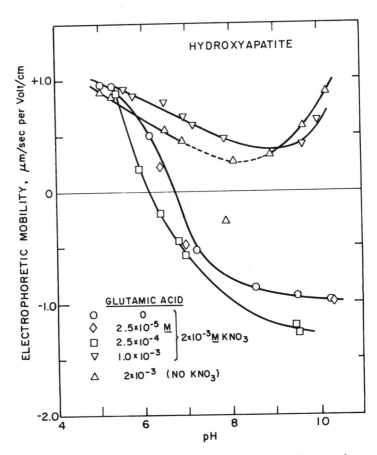

Figure 8. The effect of glutamic acid on the electrophoretic
mobility of hydroxyapatite.

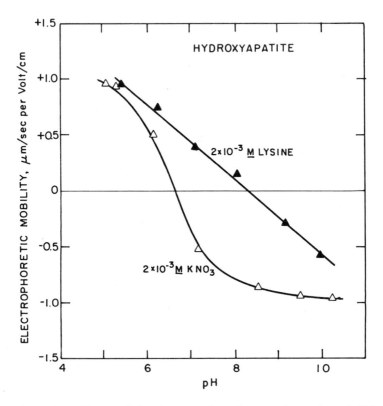

Figure 9. The effect of lysine on the electrophoretic mobility of hydroxyapatite.

Literature Cited

1. Hay, D. I. Arch. Oral Biol. 1967, 12, 937.
2. Francis, M. D. Calif. Tissue Res. 1969, 3, 151.
3. Anbar, M.; St. John, G. A.; Elward, T. E. J. Dental Res. 1974, 53, 1240.
4. Quintana, R. P. in "Applied Chemistry at Protein Interface"; Baier, R. E., Ed.; ACS SYMPOSIUM SERIES No. 145, American Chemical Society: Washington, D. C., 1975; p. 290.
5. Bartels, T.; Schudhof, J.; Arends, J. J. Dentistry 1979, 7, 221.
6. Farley, E. P.; Jones, R. L.; Anbar, M. J. Dental Res. 1977, 56, 943.
7. Malek, R. S.; Boyce, W. H. J. Urol. 1977, 117, 336.
8. Bernardi, G.; Kawasaki, T. Biochem. Biophys. Acta 1968, 160, 301.
9. Bernardi, G. in "Methods in Enzymology"; Hirs, Ch. W.; Timasheff, S. N., Eds.; Academic: New York, 1972; Vol. 27, p. 471.
10. Smith, P. R., Jr., in "Flotation, A. M. Gaudin Memorial Volume"; Fuerstenau, M. C., Ed.; AIME: New York, 1976; Vol. 2, p. 1265.
11. Nagraj, D. R.; McAllister, L.; Somasundaran, P. Int. J. Mineral Processing 1977, 4, 111.
12. Mishra, R. K.; Chander, S.; Fuerstenau, D. W. Colloids and Surfaces 1980, 1, 105.
13. Vogel, J. C.; Frank, R. M. J. Colloid Interface Sci. 1981, 83, 26.
14. Juriaanse, A. C.; Arends, J.; Tenbasch, J. J. J. Colloid Interface Sci. 1980, 76, 212.
15. Rawls, H. R.; Bartels, T.; Arends, J. J. Colloid Interface Sci. 1982, 87, 339.

RECEIVED January 20, 1984

Interfacial Tension of Aqueous Surfactant Solutions by the Pendant Drop Method

K. S. BIRDI and E. STENBY

Fysisk–Kemisk Institut, Technical University of Denmark, DK–2800 Lyngby, Denmark

The low interfacial tensions between two liquids have been measured for different systems by using the pendant drop method. In the case of the quaternary system: $C_{12}H_{25}SO_4Na+H_2O+n-Butanol+Toluene$, the interfacial data as measured by pendant drop method are compared with reported literature data, using other methods (with varying NaCl concentration). In order to understand the role of co-surfactant, ternary systems were also investigated. The pendant drop method was also used for measuring the interfacial tension between surfactant-H_2O/n-alcohol (with number of carbon atoms in alcohol varying from 4-10). The interfacial tension variation was dependent on both the surfactant and alcohol.

In the current literature one finds that the knowledge of interfacial tension, γ_{ij}, at liquid$_i$-liquid$_j$ is of much importance in many different systems, e.g. emulsions, microemulsions, ehnanced oil recovery etc.

If one considers a system consisting of: water (with or without added electrolyte) + oil + surfactant (with or without a co-surfactant) at equilibrium, there will most likely be present more than two phases (due to the formation of emulsion or microemulsion). The determination of the interfacial tension, γ_{ij}, between the two liquid phases is, therefore, of much importance, in order to understand the forces which stabilize these emulsions or microemulsions. The interfacial tension can be measured by using a variety of methods, as described in detail in surface chemistry text-books (1-3). If the magnitude of γ_{ij} is of the order of few mN/m (=dyne/cm), then the methods generally used are: Wilhelmy plate method or the drop volume (or weight) method (1-4). However, in certain systems ultra-low (or low) interfacial tensions have been reported. Since these low values are reported to be essential in order to mo-

0097–6156/84/0253–0329$06.00/0

bilize the residual oil, these studies are of much importance in
the enhanced oil recovery processes (5,6,7). In the current literature, almost all the low or ultra-low γ_{ij} values have been measured by using the "spinning drop" method (8-10). The purpose of
this study is to report comparative data of a system (11,12), i.e.
NaDDS $(C_{12}H_{25}SO_4Na)$+H_2O+n-Butanol+Toluene, by using the pendant
drop method. Further, the pendant drop method is used to determine
the γ_{ij} of various ternary systems. The general application feasi-
bility of the pendant drop method is described. As regards these
various methods used for measuring the magnitudes of γ_{ij} at two liquid interfaces, we must mention that the only low γ_{ij} value of a
known system is for the interface, n-Butanol/H_2O (γ_{ij}=1.8 mN/m at
20 °C) (1,13). In other words, whether one uses the spinning drop
method or the pendant drop method (or any other), one will have to
extrapolate from 1.8 mN/m to very low values of γ_{ij} (about 10^{-6}
mN/m as reported from spinning drop method). This general procedure of extrapolation by few decades prompted us to reinvestigate
some ternary and quaternary systems by using the pendant drop method.

Theory

The detailed theoretical derivations for the determination of the
interfacial tension by the pendant drop are given in different literature reports (1-3,14-19). The profile of any liquid drop, Figure 1, of a given volume hanging from the smooth horizontal circular syringe is an unique function of the tube radius, interfacial tension (surface tension), density difference between the liquid in the syringe and the surrounding (air or another liquid)
and the gravitational acceleration. As is well known, when one
creates such a drop, it increases to a certain height (volume or
weight) until it detaches itself. It is known that all such drops
are in stable equilibrium, so that the drop falls when its theoretical maximum weight (volume) is attained.

   If we consider the pendant drop as shown in Figure 1, we can
write from the Laplace equation (20) for the point P in terms of
the pressure differences across the interface at a reference point
B:

$$(P_{B,i}-g\rho_i Z) - (P_{B,j}-g\rho_j Z) = \gamma_{ij}(1/R_{i,P}+1/R_{j,P}) \qquad (1)$$

where $P_{B,i}$ and $P_{B,j}$ are the pressures on the concave and the convex
side of the surface at point B, respectively; γ_{ij} is the interfacial tension, $R_{i,P}$ and $R_{j,P}$ are the two principal radii of curvature
of the surface at any reference point P in the coordinate system, where point B is chosen as the origin.

   From geometry we find that at point B: $R_{B,i} = R_{B,j} = R_o$, and
$(P_{B,i}-P_{B,j}) = 2\gamma_{ij}/R_o$.
From Equation 1 and these relations we get:

$$\gamma_{ij}(1/R_{1,P}+1/R_{2,P}) = 2\gamma_{ij}/R_o + (gZ\Delta\rho) \qquad (2)$$

where $\Delta\rho = \rho_j-\rho_i$.

This equation is exact, and therefore the determination of the principal radii of curvature at two points on the interface enables one to estimate γ_{ij} and R_o. However, it is also clear that the photographic image if analyzed, as such, is not accurate enough to give a satisfactory accuracy.

Another procedure is to rewrite the relation in Equation 2 in terms of quantities which can be accurately measured from a photographic image (1,2,20). At point P we can let the radius of curvature be $R_{1,P}$ of curve V_1, Figure 1. The curve V_2, which is perpendicular to V_1, and passes through P, will be such that OP is normal to both curves at P. Further, since OP is on the axis of revolution, P remains on curve V_1 when OP rotates about the axis BO. This gives the relation : OP = $X/\sin\phi$, which is the other radius of curvature of the interface at point P = R_{2P}. We can now rewrite Equation 2 as:

$$\gamma_{ij}(1/R_{1,P}+\sin\phi/X) = 2-\beta Z/R_o \qquad (3)$$

where $\beta = -\Delta\rho gR_o^2/\gamma_{ij}$.

The standard relation which gives the radius of curvature of a curve in the X-Z plane (1) is as follows:

$$R_{1,P} = (1+(dZ/dX)^2)^{3/2}/(d^2Z/dX^2) \qquad (4)$$

Since $\tan\phi = dZ/dX$, and $\sin\phi = \tan\phi/(1+\tan^2\phi)^{1/2}$, we can obtain:

$$d^2Z/dX^2 + (dZ/dX)/X(1+(dZ/dX/^2) = (2/R_o-\beta Z/R_o^2)(1+(dZ/dX)^2)^{3/2} \qquad (5)$$

These derivations have been described in detail (15). However, the relationship in Equation 5 was found to be very unsuitable for the determination of γ_{ij}, since the curvatures are not easily evaluated from the photographic images. Especially, the older studies were unsatisfactory, arising from the inadequate optical and photographic techniques. In a later analysis an empirical procedure was described (21) which defined a function, S, which determines the drop shape as:

$$S = d_e/d_s \qquad (6)$$

where d_e is the maximum (equatorial) diameter of the pendant drop, and d_s is the diameter of the drop at a selected plane at a distance d_e from the apex of the drop, Figure 2. As depicted in Figure 1, the distance O-E and OZ are different, which arises from the gravity forces. This observation (21) was described by another quantity in terms of d_e and R_o:

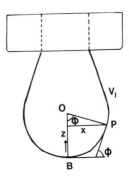

Figure 1: The Profile of a Pendant
Drop.

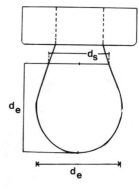

Figure 2: The selected planes of
pendant drop.

$$H = -\beta(d_e/R_o)^2 \qquad (7)$$

Further, rewriting Equation 7 by eliminating β we get:

$$\gamma_{ij} = \Delta\rho \cdot g \cdot (d_e^2/H) \qquad (8)$$

This requires the knowledge of d_e, d_s, $\Delta\rho$, and $(1/H)$. In literature, tables are found which give $(1/H)$ as a function of S (19). However, for general accuracy (less than 10 %), the following relationship is found to be satisfactory (22):

$$(1/H) = 0.31270 \; (S)^{-2.6444} \qquad (9)$$

Experimental and Methods

All chemicals used were of very high purity grade (>99 %). The quaternary systems were mixed and allowed to reach equilibrium over a week at constant temperature, with occasional stirring. The oil phase (top phase) was placed in the measuring cell of the pendant drop apparatus. The bottom (aqueous) phase was filled in the syringe for the measurement.

In the ternary systems, the aqueous phase was filled in the syringe and the drop was formed, with over 10 minutes interval, in the oil (toluene or n-heptane or the n-alcohols) phase.

$C_{12}H_{25}SO_4Na$ (NaDDS) from B.D.H. or Serva (Germany); Na-deoxycholate (NaDOC) from Sigma; Cetyl trimethylammonium bromide (CTAB) from Schuchardt, were used as supplied by the respective manufacturer (over 99 % purity). Distilled water was used after treatment with the Millipore Q filter.

Apparatus

The apparatus consists of a light source which is placed next to a cell with thermostat. The oil phase is contained in the cell while the aqueous phase is filled in a syringe (with thermostat jacket) with diameters as:

   outside, 0.45 or 0.2 mm;   inside, 0.2 or 0.10 mm, respectively.

The drop formed is photographed by a suitable camera (Linhof, W.Germany) with a magnification of about 20 times. This magnification is sufficient for the range of γ_{ij} measured, since the diameters were measured by using a suitable microscope (with an accuracy of ±0.01 mm). The whole setup was mounted on a vibration-free optical bench.

Results

System A: NaDDS + H_2O + n-Butanol + Toluene

The phase diagram of the system NaDDS+H_2O+n-Butanol+Toluene has

STRUCTURE/PERFORMANCE RELATIONSHIPS IN SURFACTANTS

334

been extensively described in the literature (11,12). Further, the interfacial tensions of this system when it separates into two or three phases has been reported, depending on the concentration of the added electrolyt, NaCl, Tables I and II. (12)

Table I. Composition of the system (12)

	weight	mol
Salt water	48 %	2.7 (H$_2$0)
Toluene	46 %	0.5
NaDDS	2 %	0.042
n-Butanol	4 %	0.051
Total	100 g	3.29 mol/100 g mixture

Table II. Number and Composition of the Phases in Equilibrium as a function of salinity (11,12, this study) at 20 ^{O}C

2-Phases	3-Phases	2-Phases
Oil rich phase	Oil rich phase	Water in oil microemulsion
Oil in water	Microemulsion (inversion zone)	
Microemulsion	Aqueous Phase	Aqueous phase
0	6	8	10
% NaCl			

In this system, in the aqueous phase, the micellar system, NaDDS, on addition of butanol would change in free energy due to mixed micelle formation (i.e. NaDDS+n-Butanol), as we showed in an earlier study (23). The cahnge in free energy is also observed from the fact that both the critical micelle concentration (c.m.c.) and the Krafft point of NaDDS solution change on addition of n-Butanol (23, 24). The addition of electrolyte, NaCl, to this system (11,12, present study) gives rise to the formation of three-phases, when the salinity is between 5.8 % to 7.8 %. On the other hand, the system consists of two-phases from 0 - 5.8 % and 7.8 to 10 % NaCl, at 20 ^{O}C (experiments at 25 ^{O}C gave the same results; unpublished) Table II.

Further, it is well known that the addition of electrolytes to ionic surfactant aqueous solutions increase the Krafft point (24, 25). This indicates that as one increases the NaCl concentration, the Krafft point is most likely higher than the experimental tem-

perature, 20 $^\circ$C, in this case. However, the meaning of Krafft point
is ambiguous at lower NaCl concentrations where both butanol and
some toluene (solubilized in NaDDS micelles) are present in the
aqueous phase.

From these observations we will argue that the transition from
the three-phase to the two-phase region with increasing NaCl takes
place where the Krafft point is most likely higher than the experimental temperature (24).

The interfacial tensions, γ_{om} and γ_{mw}, where oil (top), middle
and bottom (water) phases are designated as o, m, and w, respectively, are given in Figure 3. It is seen that the magnitude of γ_{ow}
decreases on addition of NaCl all the way down to where the system
separates into three-phases, about 5.8 % NaCl (about 1.0 M/L in
aqueous phase).

The pendent drop interfacial tensions, γ_{ow}, are given in Figure 3 together with the literature data. It is seen that the two
sets of data agree with each other for γ_{ow} to as low as 0.02 mN/m.
In the three-phase region, where ultra-low γ_{ij} values have been
reported by the spinning drop and other methods, the pendant drop
could not provide any useful data when using a syringe of the diameter of 0.2 mm. However, investigations are in progress where much
thinner syringes will be used. Theoretically one should not expect
any reason why the pendant drop theory should not be valid for
very small drops (19). This is the first time in the current literature that pendant drop has been used to investigate the phase behaviour of multicomponent systems. The magnitude of γ_{ow} in the absence of NaCl is 1.3 mN/m, which compares with the value of 1.8
mN/m for n-butanol-H_2O interface. In order to understand the effect of NaCl, we further investigated the interfacial tensions of
other systems where n-butanol was absent.

System B: Aqueous Phase: (Surfactant + H_2O with or without n-Buta-
                                        nol)
         Oil Phase   : Toluene or n-Heptane

The purpose of this series was to determine the variation of γ_{ow}
at the aqueous phase (with the addition of NaCl to the surfactant
(with or without butanol)), and the oil phase (n-heptane or toluene) interface. The data in Figure 4 (a,b) show clearly that NaDOC
gives a higher γ_{ow} in both systems, in comparison with NaDDS. The
general effect, i.e. decrease in magnitude of γ_{ow}, on addition of
NaCl is the same in all the systems.

The addition of 8 % butanol reduces the interfacial tensions
by about 2 mN/m in all the systems. The systems show divergent
curves as regards the effect of added NaCl in these systems.

In comparison with System A, we thus find that different surfactants would give divergent phase behaviour, due to the dependence of γ_{ow} on surfactant characteristics. Further, the addition
of n-butanol gives rise to a lowering of γ_{ow} by about 2 mN/m (Fi-

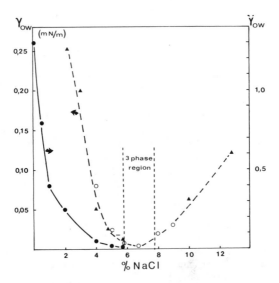

Figure 3: Variation of γ_{ow} for the system:
NaDDS + H_2O + n-Butanol + Toluene by Pendant Drop (,)
and other (O; Ref. 12) methods.
γ_{ow} is the interfacial tension between oil (top) and
bottom (water) phase.

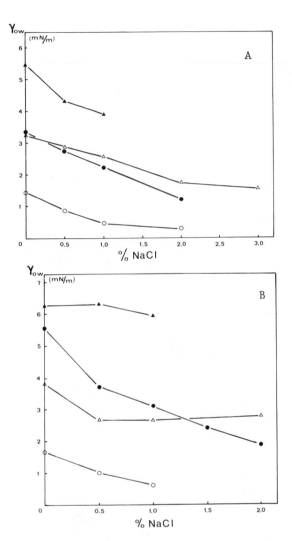

Figure 4. Variation of γ_{ow} of the systems. A, aqueous
phase: surfactant + H_2O; oil phase: toluene; with pendant drop method (24 °C). Concentration of detergent =
20 g/L. NaDOC (▲); NaDOC + 8% n-butanol (△); NaDDS (●);
NaDDS + 8% n-butanol (○). B, aqueous phase: surfactant
+ H_2O; oil phase: n-heptane; by pendant drop method
25 °C). Key same as for 4A.

gure 4 (a,b)). This is important difference, since the role of co-surfactant, i.e. n-butanol, is obviously related to this observation.

System C: Aqueous Phase: Surfactant + H₂0

Oil Phase   : n-Alcohol

The purpose of this series of measurements was to study the interfacial tension, γ_{wa} between phases:
Aqueous phase: Surfactant + H₂0:Oil phase: n-alcohol (after the phases were allowed to reach equilibrium), as a function of surfactant and number of carbon atoms, n_C, in the alcohol. The different surfactants used were: NaDDS, NaDOC (Na-deoxycholate) and CTAB (cetyl trimetyl ammonium bromide). The data in Figure 5 show that the magnitude of γ_{wa} for CTAB is larger than that of the system with NaDDS or NaDOC. Further, all systems show a "liquid-crystalline" structure formation when n-decanol is the second phase (n-dodecanol gives the same results (Birdi, unpublished)). These results show tha the "liquid-crystal" formation is not dependent on the surfactant, but mostly on the characteristics of alcohol. Similar structure formations have been reported by other investigators (25) from spinning drop studies of quaternary systems. This observation is very significant, since in the development of equations under theoretical section, the two-phases were assumed to be liquids. However, if one of the phases is transformed into a liquid-crystalline state, the equations used would need modifications. The same will be true for the application of spinning drop analyses. However, no such modifications have been reported in literature, and this will be pursued as more data becomes available in the authors laboratory.

Discussion

These studies, carried out by measuring interfacial tensions, γ_{ow}, between aqueous and oil phases, by using the pendant drop method, show that this method is very useful for ternary and quaternary systems. In one system (A), e.g. NaDDS + H₂0 + n-butanol + Toluene the γ_{ow} data as measured by pendant drop method agreed with the literature data (where another method was used (12)).
The pendant drop method was satisfactory for low γ_{ow} values, i.e. 0.02 mN/m. Typical data are given in Table III. In the systems where ultra-low γ_{ij} values have been reported by other methods (like spinning drop), the pendant drop needs further investigations before it can be applied, since this would require syringes with much smaller diameters (i.e. 10^{-3} mm). As regards the theoretical analyses, we cannot find any concern why pendant drop should have any limitations for such studies. (Same is valid for spinning drop method).
The presence of liquid-crystalline structures (25-28) at aque-

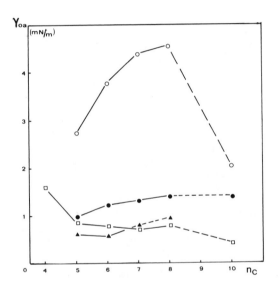

Figure 5: Variation of γ_{oa} of the systems:
Aqueous Phase: Surfactant; Oil Phase: n-alcohol;
CTAB, 20 g/L (O); NaDOC, 20 g/L (●); NaDDS, 10 g/L (□);
20 g/L (▲).

ous-oil (n-alcohol) was observed between (surfactant + H_2O) and n-decanol interfaces. The liquid-crystalline structures were found to be dependent on the magnitude of n_C in n-alcohol, with no dependence on the characteristics of the surfactant. In such systems the interfacial tensions measured are not valid, since the pendant drop theory is derived for liquid$_i$-liquid$_j$. The presence of liquid-crystalline phases at the interface would thus require modifications in the applications of pendant drop theory. This we will argue, is also the same if one uses other methods, such as the spinning drop.

In conclusion we will argue that if liquid-crystalline phases are indeed present at oil-aqueous phases, then the magnitudes of γ_{ij} reported by any method must be accepted with caution. This observation needs further analyses, which is the subject of research in the authors laboratory.

Table III. Interfacial Tension (γ_{ij}) Values by Pendant Drop Method: d_e, d_s, $(1/H)$ and $\Delta\rho$ values for different interfaces (this study

	tip diameter (mm)	factor of magnification	$\Delta\rho$ (kg/m³)	d_e (mm)	d_s (mm)	$1/H$	γ_{ij} (mN/m)
NaDDS+H_2O+n-Butanol +Toluene	0.20	21.05	153.0	0.369	0.244	1.666	0.238
(2.33 % NaCl)	0.45	20.18	153.0	0.555	0.458	0.5198	0.240
NaDDS+H_2O+n-Butanol +Toluene (4.76 % NaCl)	0.20	21.25	146.7	0.218	0.207	0.3583	0.0246
10 g/ℓ SDS vs. n-Heptane	0.20	20.00	182.6	0.819	0.600	0.7120	0.854`
	0.45	21.25	182.6	0.675	0.425	1.063	0.868

Acknowledgments

Erling Stenby likes to thank the Danish Council for Scientific & Industrial Research (STVF) for a research grant. (Licentiatstip.).

Literature Cited

1. Adamson, A. W. "Physical Chemistry of Surfaces"; Interscience Publishers: New York, 3rd ed., 1976.
2. Jaycock, M. J.; Parfitt, G. D. "Chemistry of Interfaces"; John Wiley & Sons; New York, 1981.
3. Chattoraj, D. K.; Birdi, K. S. to be published.
4. Birdi, K. S.; Nikolov, A. J. Phys. Chem. 1979, 83, 365.
5. Taber, J. J. Soc. Pet. Eng. J. March 1979, 3; Trans., AIME, Vol. 246.

6. Foster, W. R. J. Pet. Tech. Feb. 1973, 205; Trans., AIME, Vol. 255.
7. Melrose, J. C.; Brandner, C. F. J. Cdn. Pet. Tech. 1974. 54.
8. Vayias, K. S.; Schechter, R. S.; Wade, W. H., in "adsorption at Interfaces"; ACS SYMPOSIUM SERIES No. 8, American Chemical Society: Washington, D.C., 1975, 234.
9. Manning, C. D. MS Thesis, University of Minnesota: Minneapolis, August 1976.
10. Puig, J. E.; Franses, E. I.; Davies, H. T.; Miller, W. G.; Scriven, L. E. Soc. Petr. Eng. J. April 1979, 71.
11. Bellocq, A. M.; Biais, J.; Clin, B.; Lalanne, P.; Lemanceau, B. J. Colloid Interface Sci. 1979, 70, 524.
12. Pouchelon, A.; Meunier, J.; Langevin, D.; Cazabat, A. M. J. Phys. Letts. 1980, 41, 239.
13. Ross, J. "Chemistry and Physics of Interfaces, II"; ACS Publications, Washington, D.C., 1971.
14. Fordham, S. Proc. Roy. Soc. (London) A, 1948, 194.
15. Bashforth, F.; Adams, J. C. "An Attempt to Test the Theories of Capillary Action"; University Press, Cambridge, England, 1983.
16. Stauffer, C. E. J. Phys. Chem. (1965) 1933, 69.
17. Arundel, P. A.; Bagnall, R. D. J. Phys. Chem. 1977, 81, 2077.
18. Bagnall, R. D.; Arundel, P. A. J. Phys. Chem. 1978, 82, 898.
19. Ambwani, D.; Fort, T. Surface Colloid Sci. 1979, 11, 93.
20. Laplace, P. S. "Mecanique Celeste", Supplement to the 10th Book, Duprat, Paris, 1806.
21. Andreas, J. M.; Hansen, E. A.; Tucker, W. B. J. Phys. Chem. 1938, 42, 1001.
22. Stegemeier, G. I.. Ph. D. Thesis, University of Texas, Austin, Texas, 1959.
23. Backlund, S.; Rundt, K.; Birdi, K. S.; Dalsager, S. J. Colloid Interface Sci. 1981, 79, 578.
24. Birdi, K. S. unpublished.
25. Frances, E. I.; Puig, J. E.; Talmon, Y.; Miller, W. G.; Scriven, L. E.; Davis, H. T. J. Phys. Chem. 1980, 84, 1547.
26. Ekwall, P.; Mandell, L.; Foutell, K. Mol. Cryst. Liquid Cryst. 1969, 8, 157.
27. Ekwall, P.; Danielsson, I.; Stenius, P. in "Surface Chemistry and Colloids"; Kerker, M., Ed.; MTP Intern. Rev. Sci., Phys. Chem. Ser. 1. Vol. 7, p.97, Butterworths, London, 1972.
28. Friberg, S., Ed. "Food Emulsions", Marcel Dekker Inc., New York, 1976.

RECEIVED January 10, 1984

Author Index

Subject Index

Production by Frances Reed
Indexing by Susan Robinson
Jacket design by Anne G. Bigler

Elements typeset by Hot Type Ltd., Washington, D.C.
Printed and bound by Maple Press Co., York, Pa.

RECENT ACS BOOKS

"Chemistry and Characterization of Coal Macerals"
Edited by Randall E. Winans and John C. Crelling
ACS SYMPOSIUM SERIES 252; 192 pp.; ISBN 0-8412-0838-7

"Conformationally Directed Drug Design:
Peptides and Nucleic Acids as Templates or Targets"
Edited by Julius A. Vida and Maxwell Gordon
ACS SYMPOSIUM SERIES 251; 288 pp.; ISBN 0-8412-0836-0

"Ultrahigh Resolution Chromatography"
Edited by S. Ahuja
ACS SYMPOSIUM SERIES 250; 240 pp.; ISBN 0-8412-0835-2

"Chemistry of Combustion Processes"
Edited by Thompson M. Sloane
ACS SYMPOSIUM SERIES 249; 286 pp.; ISBN 0-8412-0834-4

"Geochemical Behavior of Disposed Radioactive Waste"
Edited by G. Scott Barney, James D. Navratil, and W. W. Schulz
ACS SYMPOSIUM SERIES 248; 470 pp.; ISBN 0-8412-0831-X

"NMR and Macromolecules:
Sequence, Dynamic, and Domain Structure"
Edited by James C. Randall
ACS SYMPOSIUM SERIES 247; 282 pp.; ISBN 0-8412-0829-8

"Geochemical Behavior of Disposed Radioactive Waste"
Edited by G. Scott Barney, James D. Navratil, and W. W. Schulz
ACS SYMPOSIUM SERIES 246; 413 pp.; ISBN 0-8412-0827-1

"Size Exclusion Chromatography: Methodology and
Characterization of Polymers and Related Materials"
Edited by Theodore Provder
ACS SYMPOSIUM SERIES 245; 392 pp.; ISBN 0-8412-0826-3

"Industrial-Academic Interfacing"
Edited by Dennis J. Runser
ACS SYMPOSIUM SERIES 244; 176 pp.; ISBN 0-8412-0825-5

"Characterization of Highly Cross-linked Polymers"
Edited by S. S. Labana and Ray A. Dickie
ACS SYMPOSIUM SERIES 243; 324 pp.; ISBN 0-8412-0824-9

"Archaeological Chemistry--III"
Edited by Joseph B. Lambert
ADVANCES IN CHEMISTRY SERIES 205; 324 pp.; ISBN 0-8412-0767-4

"Molecular-Based Study of Fluids"
Edited by J. M. Haile and G. A. Mansoori
ADVANCES IN CHEMISTRY SERIES 204; 524 pp.; ISBN 0-8412-0720-8

David L. Andrews

Lasers in Chemistry

Springer-Verlag

David L. Andrews

Lasers in Chemistry

With 115 Figures

Springer-Verlag Berlin Heidelberg New York
London Paris Tokyo   (1986)

Dr. David L. Andrews

University of East Anglia, School of Chemical Sciences, Norwich NR4 7TJ/England

ISBN-3-540-16161-9 Springer-Verlag Berlin Heidelberg New York
ISBN-0-387-16161-9 Springer-Verlag New York Heidelberg Berlin

Library of Congress Cataloging-in-Publication Data.
Andrews, David L. Lasers in chemistry.
Bibliography: p. Includes index.
1. Lasers in chemistry. I. Title.
QD715.A53 1986 542 86-10042
ISBN 0-387-16161-9 (U.S.)

Typesetting and Offsetprinting: Interdruck, Leipzig; Bookbinding: Lüderitz & Bauer, Berlin
2152/3020-543210

To my wife Karen
whose love and encouragement are inexpressibly precious

Preface

Let us try as much as we can, we shall still unavoidably fail in many things
'The Imitation of Christ', Thomas a Kempis

Since the invention of the laser in 1960, a steadily increasing number of applications has been found for this remarkable device. At first it appeared strangely difficult to find any obvious applications, and for several years the laser was often referred to as 'a solution in search of a problem'. The unusual properties of laser light were all too obvious, and yet it was not clear how they could be put to practical use. More and more applications were discovered as the years passed, however, and this attitude slowly changed until by the end of the 1970's there was scarcely an area of science and technology in which lasers had not been found application for one purpose or another. Today, lasers are utilised for such diverse purposes as aiming missiles and for eye surgery; for monitoring pollution and for checking out goods at supermarkets; for welding and for light-show entertainment. Even within the field of specifically chemical applications, the range extends from the detection of atoms at one end of the scale to the synthesis of vitamin D at the other.

In this book, we shall be looking at the impact which the laser has made in the field of chemistry. In this subject alone, the number of laser applications has now grown to the extent that a book of this size can provide no more than a brief introduction. Indeed, a number of the topics which are here consigned to a brief sub-section are themselves the subjects of entire volumes by other authors. However, this itself reflects one of the principal motives underlying the writing of this book; much of the subject matter is elsewhere only available in the form of highly sophisticated research-level treatments which assume that the reader is thoroughly well acquainted with the field. It is the intention in this volume to provide a more concise overview of the subject accessible to a more general readership, and whilst the emphasis is placed on chemical topics, most of the sciences are represented to some extent. The level of background knowledge assumed here corresponds roughly to the material covered in a first-year undergraduate course in chemistry. There has been a deliberate effort to avoid heavy mathematics, and even the treatment of the fundamental laser theory in Chap. 1 has been simplified as much as possible.

The link between lasers and chemistry is essentially a three-fold one. Firstly, several important chemical principles are involved in the operation of most lasers. This theme is explored in the introductory Chap. 1 and 2; Chap. 1 deals with the chemical and physical principles of laser operation in general terms, and Chap. 2 provides a more detailed description of specific laser sources. Secondly, a large number of techniques based on laser instrumentation are used to probe systems of chemical interest. The systems may be either chemically stable or in the process of chemical reaction, but in each case the laser can be used as a highly sensitive analytical device. There are numerous ways in which laser instrumentation is now put to use in chemical laboratories, and some of the general principles are discussed in Chap. 3. However, by far the most widely used methods are specifically spectroscopic in nature, and Chap. 4 is therefore entirely given over to a discussion of some of the enormously varied chemical applications of laser spectroscopy.

The third link between lasers and chemistry concerns the inducing of chemical change in a system through its irradiation with laser light. In such applications, the chemistry in the presence of the laser radiation is often quite different from that observed under other conditions; thus, far from acting as an analytical probe, the radiation is actively involved in the reaction dynamics, and acts as a stimulus for chemical change. This area, discussed in Chap. 5, is one of the most exciting and rapidly developing areas of chemistry, and one in which it appears likely that the laser may ultimately make more of an impact than any other.

An attempt has been made throughout this book to provide examples illustrating the diversity of laser applications in chemistry across the breadth of the scientific spectrum from fundamental research to routine chemical analysis. Nonetheless the emphasis is mostly placed on applications which have relevance to chemical industry. The enormous wealth of material from which these illustrations have been drawn means that this author's choice is inevitably idiosyncratic, although each example is intended to provide further insight into the underlying principles involved. Since this is an introductory textbook, references to the original literature have been kept to a minimum in order to avoid swamping the reader; needless to say, this means that a great many of the pioneers of the subject are not represented at all. I can only record my immense debt to them and all who have contributed to the development of the subject to the state in which I have reported it.

I should like to express my thanks to the editorial staff of Springer-Verlag, and especially Almut Heinrich who has been most sympathetic and helpful at all times. Thanks are also due to the staff of Technical Graphics for their work on the original Figures. I am indebted to all who have contributed spectra and consented to the reproduction of diagrams, and I am very grateful to Peter Belton, Godfrey Beddard, Nick Blake, John Boulton, Colin Creaser, Allan Dye, U. Jayasooriya and Ron Self for comments on various parts of the manuscript. Lastly I should like to express my thanks to my good friend and severest critic John Sodeau, whose comments have done more than any other to help remould my first draft into its completed from.

Norwich, June 1986   David L. Andrews

Table of Contents

1 Principles of Laser Operation

And the atoms that compose this radiance do not travel as isolated
individuals but linked and massed together
'De Rerum Naturae', Lucretius

1.1 The Nature of Stimulated Emission

The term *laser*, an acronym for *l*ight *a*mplification by the *s*timulated *e*mission of *r*adiation, first appeared in 1960 and is generally held to have been coined by Gordon Gould, one of the early pioneers of laser development. Since the device was based on the same principles as the *maser*, a microwave source which had been developed in the 1950's, the term 'optical maser' was also in usage for a time, but was rapidly replaced by the simpler term. In order to appreciate the concepts of laser action, we need to develop an understanding of the important term 'stimulated emission'. First, however, it will be helpful to recap on the basic quantum mechanical principles associated with the absorption and emission of light. Although these principles apply equally to individual ions, atoms or molecules, it will save unnecessary repetition in the following discussion if we simply refer to molecules.

According to quantum theory, molecules possess sets of discrete energy levels, and the energy which any individual molecule can possess is limited to one of these values. Broadly speaking, the majority of molecules at any moment in time exist in the state of lowest energy, called the ground state, and it is often by the absorption of light that transitions to states of higher energy can take place. Light itself consists of discrete quanta known as photons, and the absorption process thus occurs as individual photons are intercepted by individual molecules; in each instance the photon is annihilated and its energy transferred to the molecule, which is thereby promoted to an excited state. For this process to occur the photon energy E, which is proportional to its frequency v, (E = hv, where h is Planck's constant,) must match the gap in energy between the initial and final state of the molecule. Since only discrete energy levels exist, there results a certain selectivity over the frequencies of light which can be absorbed by any particular compound; this is the principle underlying most of spectroscopy.

What we now need to consider in more detail is the reverse process, namely the emission of light. In this case we start off with a system of molecules many of which must already exist in some excited state. Although the means for producing the excitation is not important, it is worth noting that it need not necessarily be the absorption of light; for example in a candle it is the chemical energy of combustion which provides the excitation. Molecules in excited states generally have very short lifetimes (often between 10^{-7}s and 10^{-11}s) and by releasing energy they rapidly undergo relaxation processes. In this way, the molecules undergo transitions to more stable states of lower energy, frequently the ground state. There are many different mechanisms for the release of energy, some of which are radiative, in the sense that light is emitted, and some of which are non-radiative. However, although chemical distinctions can be made between different types of radiative decay, such as fluorescence and phosphorescence, the essential physics is precisely the same — photons are emitted which precisely match the energy difference between the excited state and the final state involved in the transition. Because this kind of photon emission can occur without any external stimulus, the process is referred to as *spontaneous emission*.

We are now in a position to come to terms with the very different nature of *stimulated emission*. Suppose, once again, we have a system of molecules, some of which are in an excited state. This time, however, a beam of light is directed into the system, with a frequency such that the photon energy exactly matches the gap between the excited state and some state of lower energy. In this case, each molecule can relax by emitting another photon of the same frequency as the supplied radiation. However, it turns out that the probability of emission is enhanced if other similar photons are already present. Moreover, emission occurs preferentially in the direction of the applied beam, which is thereby *amplified* in intensity. This behaviour contrasts markedly with the completely random directions over which spontaneous emission occurs when no beam is present. This type of emission is therefore known as 'stimulated emission': it is emission which is stimulated by other photons of the appropriate frequency. The two types of emission are illustrated in Fig. 1.1.

The concept of stimulated emission was originally developed by Einstein in a paper dealing with the radiation from a heated black body, a subject for which it had already been found necessary to invoke certain quantum ideas (Planck's hypothesis) in order to obtain a theory which fitted experimental data. Despite publication of this work in 1917, and subsequent confirmation of the result by the more comprehensive quantum mechanics in the late 1920's, little attention was paid to the process by experimental physicists for many years. It was only in the early 1950's that the first experimental results appeared, and the practical applications began to be considered by Townes, Basov and Prochorov, and others. The key issue in the question of applications was the possi-

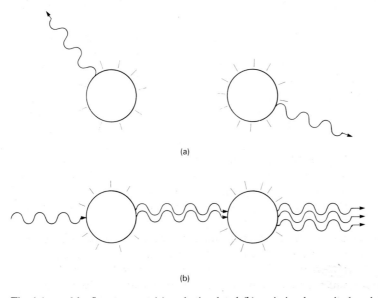

(a)

(b)

Fig. 1.1a and b. Spontaneous (a) and stimulated (b) emission by excited molecules

bility of amplification implicit in the nature of the stimulated emission process. These considerations eventually led to development of the maser, the forerunner of the laser which produced microwave, rather than optical frequency radiation. It was undoubtedly the success in producing an operational device based on stimulated microwave emission which spurred others on towards the goal of a visible light analogue.

1.2 Resonators and Pumping Processes

We can now consider some of the fundamental practical issues involved in the construction of a laser. For the present, we shall consider only those features which directly follow from the nature of the stimulated emission process on which the device is based; other considerations will become evident later in this chapter, and a full account of particular laser systems is given in the following chapter. At this stage, however, it is clear that the first requirement is a suitable substance in which stimulated emission can take place, in other words an *active medium*. We also require an external stimulus in order to promote atoms or molecules of this medium to an appropriate excited state from which emission can occur.

The active medium can take many forms, gas, liquid or solid, and the substance used is determined by the type of output required. Each substance has its own unique set of permitted energy levels, so that the frequency of light emitted depends on which levels are excited, and how far these are separated in energy from states of lower energy. The first laser, made by Maiman in 1960, had as its active medium a rod of ruby, and produced a deep red beam of light. More common today are the gas lasers, in which gases such as argon or carbon dioxide form the active medium; the former system emits various frequencies of visible light, the latter infra-red radiation.

There is an equal diversity in the means by which the initial excitation may be created in the active medium. In the case of the ruby laser, a broadband source of light such as a flashlamp is used; in gas lasers an electrical discharge provides the stimulus. Later, we shall see that even chemical reactions can provide the necessary input of energy in certain types of laser. Two general points should be made, however, concerning the external supply of energy. First, if electromagnetic radiation is used, then the frequency or range of frequencies supplied must be such that the photons which excite the laser medium have an energy equal to or larger than that of the laser output. Secondly, because of heat and other losses no laser has 100 % efficiency, and the energy output is always less than the energy input in the same way as in any electronic amplifier. These rules are simply the results of energy conservation. One other point is worth mentioning at this juncture. So far, we have implicitly assumed that the excitation precedes laser emission; indeed this is true for many kinds of laser system. However, for a continuous output, we are faced with a requirement to sustain a population of the generally short-lived excited molecules. We shall see shortly that this imposes other requirements on the nature of the active medium.

As it stands, we have described a set-up in which a laser medium can become excited, and relax by the emission of photons. We do not have to supply another beam of light for stimulated emission to occur; once a few molecules have emitted light by spontaneous emission, these emitted photons can stimulate emission from other excited molecules. However, a single passage of photons through the medium is not gen-

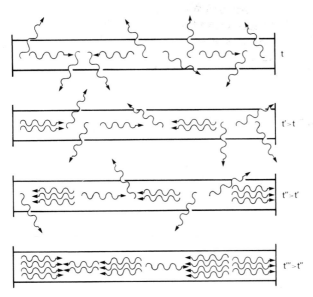

Fig. 1.2. Diagram illustrating the development of laser action from the onset of photon emission at time t through three later instants of time

erally sufficient for stimulated emission to play a very significant role. This is because the rate of stimulated emission is proportional to the initial intensity, which as we have pictured the process so far cannot be very great.

For this reason, it is generally necessary to arrange for the multiple passage of light back and forth through the active medium in a *resonator*, so that with each traversal the intensity can be increased by further stimulated emission. In practice this can be arranged by placing parallel mirrors at either end of the laser medium, so that light emitted along the axis perpendicular to these mirrors is essentially trapped, and bounces backwards and forwards indefinitely, growing in intensity all the time. By contrast, photons which are spontaneously emitted in other directions pass out of the active medium and no longer contribute to stimulated emission. Figure 1.2 illustrates how the intensity of light travelling between the mirrors builds up at the expense of spontaneous emission over a short period of time; since emission along the laser axis is self-enhancing, we very rapidly reach the point at which nearly all photons are emitted in this direction.

We have seen in this Section that the resonator plays an important role in the operation of any laser. To conclude, we briefly mention two other practical considerations. Although we have talked in terms of two parallel plane end-mirrors, the difficulty of ensuring sufficiently precise planarity of the mirrors, coupled with difficulty of correct alignment, has led to several other configurations being adopted. One of the most common is to have two confocal concave mirrors, which allow slightly off-axis beams to propagate back and forth. The other crucial consideration is, of course, the means of obtaining usable output of laser light from the resonator beam; this is generally accomplished by having one of the end-mirrors made partially transmissive. Figure 1.3 schematically illustrates the basic components of the laser as discussed so far.

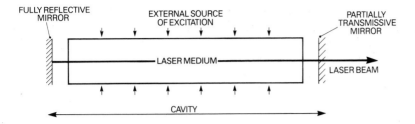

Fig. 1.3. Essential components of a laser

1.3 Coherent Radiation, Standing Waves and Modes

Stimulated emission, as we have seen, results in the creation of photons with the same frequency and direction as those which induce the process. In fact, the stimulated photons are identical to the incident photons in every respect, which means that they also have the same phase and polarisation; this much is implied in Fig. 1.2. It is the fact that the inducing and emitted photons have a phase relationship which lies behind use of the term *coherent* to describe laser radiation. In the quantum theory of light, it is usual to refer to *modes* of radiation, which simply represent the possible combinations of frequency, polarisation and direction which characterise photons. In the laser, only certain modes are permitted, and these are the modes which can result in the formation of standing waves between the end-mirrors. This means that these mirrors, together with everything inbetween them including the active medium, must form a *resonant cavity* for the laser radiation; it is for this reason that the term 'cavity' is often used for the inside of a laser.

The standing wave condition means that nodes, in other words points of zero amplitude oscillation, must be formed at the ends of the laser cavity. This condition arises from the coherent nature of the stimulated emission process; since there is a definite phase for the radiation, then it is necessary to ensure that only constructive interference, and not destructive interference occurs as light travels back and forth inside the cavity. The only way to satisfy the standing wave condition is if it is possible to fit an integer number of half-wavelengths into the cavity, as illustrated in Fig. 1.4. Hence the wavelengths λ which can resonate inside the laser cavity are determined by the relation

$$m\lambda/2 = L, \tag{1.1}$$

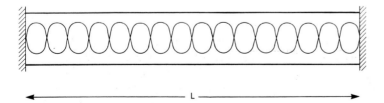

Fig. 1.4. A standing wave of electromagnetic radiation between two mirrors

where m is an integer, and L is the cavity length. Standing waves, of course, crop up in many different areas of physics; the vibrations of a violin string, for example, are subject to the same kind of condition. However in the laser, we are typically dealing with several million half-wavelengths in the cavity, and the range of wavelengths emitted is constrained by the molecular transitions of the active medium, and not just the standing wave condition; we shall examine this more fully in Sect. 1.6.4.

Although we have been discussing a standing wave based on propagation of radiation back and forth exactly parallel to the laser axis, as shown in Fig. 1.5a, lasers can also sustain resonant oscillations involving propagation slightly off-axis, as shown in Fig. 1.5b. Such modes are distinguished by the label TEM_{pq}, in which the initials stand for 'transverse electromagnetic mode', and the subscripts p and q take integer values determined by the number of intensity minima across the laser beam in two perpendicular directions. Figure 1.5c illustrates the intensity patterns for three simple cases. Although it can be possible to increase the intensity of laser output by allowing *multimode* operation, it is generally more desirable to suppress all but the axial TEM_{oo} mode, known as the *uniphase* mode, which has better coherence properties. Uniphase operation is generally ensured by adopting a narrow laser cavity.

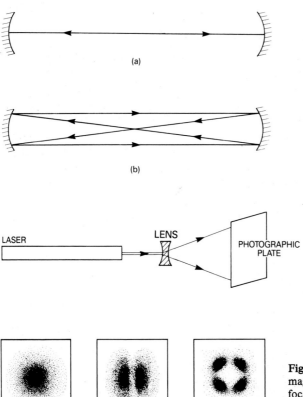

Fig. 1.5a–c. Transverse electromagnetic modes formed with confocal concave mirrors; (a) TEM_{00}, (b) TEM_{10}. In (c), the distribution of intensity in the laser beam cross-section is illustrated for three different modes

1.4 The Kinetics of Laser Emission

1.4.1 Rate Equations

We are now in a position to take a look at some crucial theoretical constraints on laser operation. The simplest approach to a more detailed understanding of laser dynamics is to take a model system in which there are only two energy levels E_1 and E_2 involved; to start with, let us also assume that both are non-degenerate. There are, as we discussed earlier, three distinct radiative processes that can accompany transitions between the two levels; absorption, spontaneous emission, and stimulated emission, illustrated in Fig. 1.6. For simplicity, we shall for the present ignore non-radiative processes. The transition rates corresponding to the three radiative transitions, due to Einstein, may be written as follows;

$$R_{absorption} = N_1 \varrho_v B_{12} ; \tag{1.2}$$

$$R_{spont.emission} = N_2 A_{21} ; \tag{1.3}$$

$$R_{stim.emission} = N_2 \varrho_v B_{21} . \tag{1.4}$$

Here, N_1 and N_2 are the numbers of molecules in each of the two energy levels E_1 and E_2 respectively, ϱ_v is the energy density of radiation with frequency v, and A_{21}, B_{21} and B_{12} are known as the Einstein coefficients. (Note that originally the Einstein coefficients were defined as the constants of proportionality in rate equations for polychromatic radiation, and were thus expressed in different units in terms of the energy density *per unit frequency interval*. However, the definition used here is gaining acceptance since it is more appropriate for application to essentially monochromatic laser radiation.) For a non-degenerate pair of energy levels, it is readily shown that the two B coefficients are equal, and we can therefore drop the subscripts.

Now we need to consider the intensity of light travelling along the laser axis. For a parallel beam, this is best expressed in terms of the *irradiance* I, defined as the energy

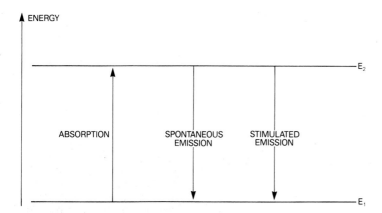

Fig. 1.6. The three radiative processes in a two-level system

crossing unit cross-sectional area per unit time. This can be directly related to the instantaneous photon density, as follows. Consider a small cube of space of side length l, and volume V, through which the beam passes. If at any instant this cube contains N photons, the photon density is N/V, and the cube contains an energy Nhv. Now if the laser medium has refractive index n, then since each photon takes a time l/c′ to traverse the cube, where c′ = c/n is the velocity of light within the medium, it clearly takes the same time l/c′ for the energy Nhv to pass through the cube. Hence the irradiance is given by

$$I = \frac{\text{energy per unit time}}{\text{area}} = \frac{Nhv/(l/c')}{l^2}, \tag{1.5}$$

and since $l^3 = V$, we obtain the following relation for the instantaneous photon density:

$$N/V = I/hc'v. \tag{1.6}$$

We can also relate this to the energy density ϱ_v, since this is simply hv times the above result;

$$\varrho_v = I/c'. \tag{1.7}$$

Returning to the rate Eqs. (1.2) to (1.4), we note that since spontaneous emission occurs in random directions, it cannot appreciably contribute to an increase in the intensity of light propagating along the laser axis, since very few photons will be spontaneously emitted in this direction. Therefore if we write down an equation for the rate of increase of intensity along the laser axis, we have only two major terms, namely a positive term due to stimulated emission, and a negative term due to absorption. Using the results (1.6) and (1.7), we thus have;

$$(hc'v)^{-1}dI/dt \approx N_2 B(I/c') - N_1 B(I/c'), \tag{1.8}$$

and hence

$$dI/dt \approx (N_2 - N_1) BIhv. \tag{1.9}$$

Strictly speaking, there are other terms, such as the intensity loss due to imperfect reflectivity of the end-mirrors, which ought to be included in Eq. (1.9), but for the moment we shall ignore them.

What is now apparent is that the rate of increase of intensity is proportional to the difference in the excited state and ground state populations, $(N_2 - N_1)$. Clearly for the amplification we require, we need to have a positive value for dI/dt, which leads to the important condition for laser action;

$$dI/dt > 0 \Rightarrow N_2 > N_1. \tag{1.10}$$

Physically, this means that unless we have more molecules in the excited state than the ground state, absorption will win over stimulated emission. If we extend our theory to include the effects of degeneracy in the two energy levels, the counterparts to eqs. (1.9) and (1.10) are;

$$dI/dt \approx (N_2 - (g_2/g_1)N_1) \, BIh\nu, \qquad (1.11)$$

$$dI/dt > 0 \Rightarrow N_2 > N_1 \, (g_2/g_1), \qquad (1.12)$$

where g_1 is the degeneracy of level 1, and g_2 that of level 2. The interesting thing about the latter inequality is that it can never be satisfied under equilibrium conditions. For a system in thermal equilibrium at absolute temperature T, the ratio of populations is given by the Boltzmann relation;

$$N_2/N_1 = (g_2/g_1) \exp \left((E_1 - E_2)/kT \right), \qquad (1.13)$$

where k is Boltzmann's constant. Since $E_2 > E_1$, it follows that (1.13) can only hold for the physically meaningless case of a negative absolute temperature! The result illustrates the fact that laser operation cannot occur under the equilibrium conditions for which the concept of temperature alone has meaning. Indeed, at thermal equilibrium, the converse of Eq. (1.12) always holds true — in other words the population of the state with lower energy is always largest. The situation we require in a laser is therefore often referred to as a *population inversion*, and energy must be supplied to the system to sustain it in this state. The means by which energy is provided to the system is generally known as *pumping*.

The solution to Eq. (1.11) may be written in the form

$$I = I_0 \exp (at), \qquad (1.14)$$

where

$$a = (N_2 - (g_2/g_1)N_1) \, Bh\nu. \qquad (1.15)$$

Alternatively, the intensity may be expressed in terms of the distance x travelled by the beam, since $t = x/c'$. Hence we may write

$$I = I_0 \exp (kx), \qquad (1.16)$$

where k is known as the small signal gain coefficient, and is defined by

$$k = (N_2 - (g_2/g_1)N_1) \, Bh\nu/c' \qquad (1.17)$$

1.4.2 Threshold Conditions

We can now make our theory a little more realistic by considering the question of losses. Quite apart from the imperfect reflectivity of the end-mirrors, there must always be some loss of intensity within the laser cavity due to absorption, non-radiative decay, scattering and diffraction processes. In practice, then, even with perfectly reflective mirrors the distance-dependence of the beam should more correctly be represented by the formula

$$I = I_0 \exp((k - \gamma)x),$$ (1.18)

where γ represents the net effect of all such loss mechanisms. However, since one end-mirror must be partially transmissive in order to release the radiation, and the other will not in practice be perfectly reflective, then in the course of a complete round-trip back and forth along the cavity (distance 2L, where L is the cavity length) the intensity is modified by a factor

$$G = I/I_0 = R_1 R_2 \exp(2L(k - \gamma)),$$ (1.19)

where R_1 and R_2 are the reflectivities of the two mirrors. The parameter G is known as the *gain*, and clearly for genuine laser amplification we require $G > 1$. For this reason, there is always a certain *threshold* for laser action; if $G < 1$, the laser acts much more like a conventional light source, with spontaneous emission responsible for most of the output. The threshold value of the small signal gain coefficient k is found by equating the right-hand side of Eq. (1.19) to unity, giving the result;

$$k_{\text{threshold}} = \gamma + (2L)^{-1} \ln(1/R_1 R_2)$$ (1.20)

For a laser with known loss and mirror reflectivity characteristics, the right-hand side of Eq. (1.20) thus represents the figure which must be exceeded by the gain coefficient, as defined by Eq. (1.17), if laser action is to occur. This in turn places very stringent conditions on the extent of population inversion required.

1.4.3 Pulsed Versus Continuous Emission

Thus far, the conditions for laser operation have been discussed without considering the possibility of pulsed emission. However, we shall discover that whilst some lasers operate on a *continuous-wave* (CW) basis, others are inherently pulsed. In order to understand the reason for this difference in behaviour, we have to further consider the kinetics of the pumping and emission processes, and the physical characteristics of the cavity. To start with, we make the obvious remark that if the rate of pumping is exceeded by that of decay from the upper laser level, then a population inversion cannot be sustained, and pulsed operation must ensue, with a pulse duration governed by the decay kinetics. For continuous operation, we therefore require a pumping mechanism which is tailored to the laser medium.

Next, we note that the round-trip time for photons travelling back and forth between two parallel end-mirrors is given by

$$\tau = 2L/c'.$$ (1.21)

If the two end-mirrors were perfectly reflective, the number of round trips would be infinite, if we ignore absorption and off-axis losses. However, with end-mirror reflectivities R_1 and R_2, the mean number of round-trips is given by

$$z = (1 - R_1 R_2)^{-1}. \tag{1.22}$$

Hence the mean time spent by a photon before leaving the cavity is

$$t = \frac{2L}{c'(1 - R_1 R_2)} . \tag{1.23}$$

Continuous-wave operation is therefore possible provided the mean interval between the emission of successive photons by any single atom or molecule of the active medium is comparable to t.

In the next section, we shall leave the simple two-level model and consider more realistic energy level schemes on which lasers can be based. When there are more than two levels involved in laser action, there are also many more transitions participating, and the cycle which starts and ends with the active species in its ground state generally includes a number of excitation, emission and decay processes. Under such circumstances, the criterion for cw operation is that the timescale for a complete cycle of laser transitions must be similar to the mean cavity occupation time. We also note that whilst the converse is not feasible, it is possible to modify continuous-wave lasers so as to produce pulsed output. This is accomplished by the inclusion of electro-optical devices within the cavity; the principles are discussed in detail in Section 3.3.

1.5 Transitions, Lifetimes and Linewidths

We have seen that the competition between absorption and stimulated emission provides a problem in a two-level system. Indeed if the initial excitation is optically induced it is a problem which cannot be solved, for absorption will always win over stimulated emission. For a more workable proposition we therefore need to look at systems where more than two levels can be involved in laser action. In this section we begin by considering three- and four-level systems, both of which are widely used in lasers, before making some further general remarks concerning emission linewidths. Because of the greatly increased complexity of the rate equations when more than two levels are involved, a less mathematical approach is adopted for this Section.

1.5.1 Three-level Laser

In a three-level laser, two levels (E_1 and E_2) must of course be coupled by the downward transition which results in emission of laser light. Suppose then, that a third, and higher energy level E_3 is present, and that it is this level which is first populated by a suitable pumping mechanism. (As noted earlier, the energy of the emitted photon cannot exceed the energy supplied to excite a molecule of the laser medium—hence it has to be the state of highest energy which is initially populated in this case.) What we require for laser action is that a population inversion is established between levels E_1 and E_2, and this can be accomplished as follows.

With three different energy levels, there are three pairs of levels between which transitions can occur. It is the possibility of one of the downward transitions being a non-radiative process which makes all the difference; such radiationless processes can

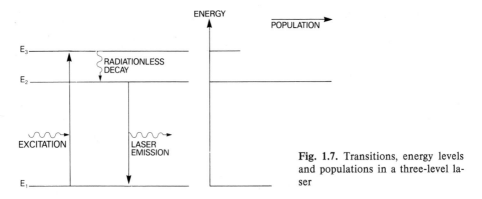

Fig. 1.7. Transitions, energy levels and populations in a three-level laser

result from a number of different mechanisms such as bimolecular collisions or wall collisions in gases, or lattice interactions in the solid state. Thus if level E_3 rapidly undergoes radiationless decay down to level E_2, then provided the laser emission is slower, the population of this level can increase while that of the ground state is diminishing because of pump transitions to E_3. Obviously, for the laser emission to be slow we require state E_2 to be metastable, with a comparatively long lifetime (say 10^{-3} s). Hence we can obtain the required population inversion, and laser action results.

Figure 1.7 illustrates the energy levels and populations for this arrangement. One difficulty is that lasing automatically repopulates the ground state, making it somewhat difficult to sustain the population inversion. In such a laser, once a population inversion is established all the atoms or molecules of the active medium may cascade down to the ground state together, resulting in the emission of a short sharp burst of light. The earliest ruby laser (see Sect. 2.1.1) was of this type, emitting pulses of about one millisecond duration, and requiring something like a minute between pulses to build up the population inversion again; this provides us with a good example of a laser which inherently operates in a pulsed mode.

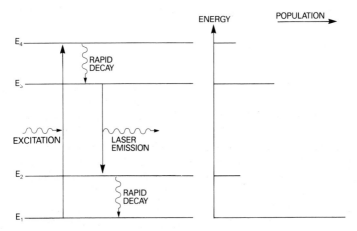

Fig. 1.8. Transitions, energy levels and populations in a four-level laser

1.5.2 Four-level Laser

The problems associated with ground state repopulation may be overcome in a four-level laser. Here the energy level scheme is quite similar to the three-level case we have just considered; the only difference is that the lower of the two laser levels involved in the radiative transition lies above the ground state, as shown in Fig. 1.8. If this state rapidly decays to the ground state below, then it can never build up an appreciable population, and so population inversion *between the levels involved in lasing*, i.e. E_2 and E_3, is retained. Thus even though the ground state population may exceed that of any other level, laser action can still occur. In practice, this kind of scheme is quite common; a good example is the neodymium-YAG laser which we shall look at in more detail in the next Chapter (Sect. 2.1.2). Since a population inversion can be sustained in this kind of laser, it can be made to operate in either a continuous or a pulsed mode.

1.5.3 Emission Linewidths

Ideally, each transition responsible for laser emission should occur at a single well-defined frequency determined by the spacing between the energy levels it connects. However there are several *line-broadening* processes which result in statistical deviations from the ideal frequency, so that in practice the emission has a frequency distribution like that shown in Fig. 1.9. The quantity normally referred to as the *linewidth*, or *half-width* is a measure of the breadth of this distribution, and is more accurately defined as the *full width at half maximum*, or FWHM for short. The various line-broadening mechanisms responsible for the frequency distribution can be divided into two classes. Those which apply equally to all the atoms or molecules in the active medium responsible for emission are known as *homogeneous* line-broadening processes and generally produce a Lorentzian frequency profile, and those based on the statistical differences between these atoms or molecules are known as *inhomogeneous* line-broadening processes, and generally result in a Gaussian profile. Usually a number of these mechanisms operate simultaneously, each contributing to an increase in the emission linewidth.

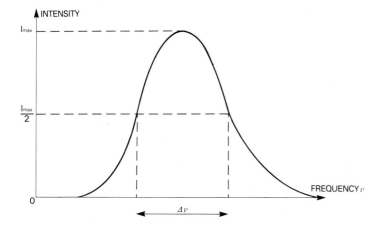

Fig. 1.9. Lineshape of laser emission, with FWHM linewidth Δv

The particulars of these processes vary according to the nature of the active medium. The one homogeneous mechanism which applies universally is that which arises from a frequency-time uncertainty relation. If laser photons are emitted from an excited state associated with a lifetime Δt, then there is a minimum uncertainty $\Delta \nu \geqslant (1/2\,\pi\Delta t)$ in the frequency of the photons emitted, resulting in emission line-broadening. This process is known as *natural line-broadening*. The lifetime of any excited state is, of course, influenced by the various decay pathways open to it, including radiative decay (both spontaneous and stimulated emission), radiationless decay and collisional energy transfer. The same frequency-time uncertainty relation also has an important bearing on the linewidth in lasers producing pulses of very short duration, as we shall see in Sect. 3.3.3.

The other major homogeneous line-broadening mechanisms are briefly as follows. In crystalline media, lattice vibrations cause a time-dependent variation in the positions and hence the electrostatic environments experienced by each atom. Similar perturbations occur in liquids, although on a larger scale due to the translational, rotational and vibrational motions of the molecules. In gases the atoms or molecules are perturbed by collisions with other atoms or molecules, with the walls of the container, and with electrons if an ionising current is used. Whilst the last two of these may be controlled or eliminated, the rate of collisions at a given temperature between atoms or molecules of the gas itself is subject only to the pressure; the line-broadening it gives rise to is thus called either *collision-* or *pressure-broadening*.

Inhomogeneous line-broadening is principally caused in the solid state by the presence of impurity atoms in various crystallographic sites; a similar effect occurs in the case of liquid solutions. There is, however, a quite different inhomogeneous mechanism in gases. This results from the fact that the frequencies of the photons emitted are shifted by the well-known Doppler effect. If a molecule travelling with velocity v emits a photon of frequency ν in a direction k, as shown in Fig. 1.10, then the apparent frequency is given by

$$\begin{aligned} \nu' &= \nu\,(1 + v_k/c)^{1/2}\,(1 - v_k/c)^{-1/2} \\ &\approx \nu\,(1 + v_k/c), \end{aligned} \qquad (1.24)$$

where v_k is the component of velocity in direction k, which is generally small compared to the speed of light c. Since there is a Maxwellian distribution of molecular velocities, there is a corresponding range of Doppler-shifted frequencies given by Eq. (1.24). This effect is known as *Doppler broadening*.

As an illustration of the magnitudes of the various linewidths, a gas laser operating on an electronic transition at a pressure of 1 torr would have a natural linewidth typi-

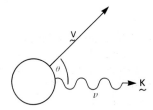

Fig. 1.10. Emission by a translating molecule. The velocity component v_k in the direction of emission is $v \cos \theta$

cally of the order of 10 MHz, a collisional linewidth 100 MHz − 1 GHz, and a Doppler linewidth in the GHz range. For a similar laser operating on a vibrational transition, the natural linewidth would be a factor of about 10^5 smaller; hence the natural linewidth is seldom the dominant line-broadening mechanism. Other line-broadening mechanisms are discussed at the beginning of Chap. 4, in connection with laser spectroscopy.

1.6 Properties of Laser Light, and Their Applications

It has been recognized since the earliest days of laser technology that laser light has characteristic properties which distinguish it from that produced by other sources. In this Section we shall look at how these properties arise from the nature of the lasing process, and briefly consider examples of how they are utilized for particular laser applications. Such a discussion obviously has to be highly selective, and those examples which are presented here have been chosen more to illustrate the diversity of laser applications than for any other reason.

1.6.1 Beamwidth

Since stimulated emission produces photons with almost precisely the same direction of propagation, the end-mirror configuration results in selective amplification of an axial beam which is typically only 1 mm. in diameter. The laser thus emits a narrow, and essentially parallel beam of light from its output mirror, usually with a Gaussian distribution of intensities across the beam. A typical figure for the angle of beam divergence is 1 mrad, which would illuminate an area only one metre across at a distance of one kilometre; exciplex lasers are now available with beam divergences of less than 200 microradians. Although the extent of beam divergence is initially determined by the diffraction limit of the output aperture, the little divergence that there is can to a large extent be corrected by suitable optics. A striking illustration of how well collimated a laser beam can be is provided by the fact that it has been possible to observe from Earth the reflection of laser light from reflectors placed on the surface of the Moon by astronauts during the Apollo space programme. One highly significant application of the collimation and narrow beamwidth of lasers is *optical alignment* in the construction industry, for example in tunnel boring. Pulsed lasers can also be used for *tracking and ranging*, and for *atmospheric pollution monitoring*, about which we shall have more to say in Sect. 3.6.5. In the latter case it is the narrow beamwidth that makes it possible to monitor from ground level, by analysis of the scattered light, gases escaping from high factory chimneys.

1.6.2 Intensity

The property which is most commonly associated with laser light is a high intensity, and indeed lasers do produce the highest intensities known on Earth. Since a laser emits an essentially parallel beam of light in a well-defined direction, rather than in all directions, the most appropriate measure of intensity is the irradiance, as defined in Sect. 1.4.1. Since energy per unit time equals power, we thus have;

Irradiance I = Power/Area.   (1.25)

In using this equation, however, it must be stressed that 'power' refers to the output power, and not the input power of the laser. To put things into perspective when we look at typical laser irradiances, we can note that the mean intensity of sunlight on the Earth's surface is of the order of one kilowatt per square metre, i. e. 10^3 W m^{-2}.

Let us consider first a moderately powerful argon laser which can emit something like 10 W power at a wavelength of 488 nm. Assuming a cross-sectional area for the beam of 1 mm^2, this produces an irradiance of $(10 \text{ W})/(10^{-3} \text{ m})^2 = 10^7$ W m^{-2}. In fact we can increase this irradiance by focussing the beam until we approach a diffraction limit imposed by the focussing optics. In this respect, too, laser light displays characteristically unusual properties, in that by focussing it is possible to produce intensities that exceed that of the source itself; this is not generally possible with conventional light sources. As a rough guide, the minimum radius of the focussed beam is comparable with the wavelength, so that in our example a cross-sectional area of 10^{-12} m^2 would be realistic, and give rise to a focussed intensity of 10^{13} W m^{-2}.

Not surprisingly, however, it is in lasers which first accumulate energy as a population inversion is built up, and then release it through emission of a pulse of light, that we find the highest output intensities, though we have to remember that the peak intensity is obtained only for a very short time. A good Q-switched ruby laser, for example, which emits 25 ns pulses (1 ns = 10^{-9} s) at a wavelength of 694 nm, can give a peak output of 1 GW = 10^9 W in each pulse, though typically in a somewhat broad beam of about 500 mm^2 cross-sectional area. The mean irradiance of each pulse is thus approximately 2×10^{12} W m^{-2}, which can easily be increased by at least a factor of 10^6 by appropriate focussing. It should be noted that in all these rough calculations, it has been implicitly assumed that the intensity remains constant throughout the duration of each pulse, whereas in fact there is a definite rise at the beginning and decay at the end; in other words, there is a smooth *temporal profile*. Because the peak intensity from a pulsed laser is inversely proportional to its pulse duration, there are various methods of reducing pulse length so as to increase the intensity, and we shall examine these in Sect. 3.3.

Let us briefly look at a few applications of lasers which hinge on the high intensities available. A fairly obvious example from industry is *laser cutting and welding*. For such purposes, the high-power carbon dioxide and Nd-YAG lasers, which produce infra-red radiation, are particularly appropriate. Such lasers can cut through almost any material, though it is sometimes necessary to supply a jet of inert gas to prevent charring, for example with wood or paper; on the other hand, an oxygen jet facilitates cutting through steel. A focussed laser in the 10^{10} W m^{-2} range can cut through 3 mm steel at approximately 1 cm s^{-1}, or 3 mm leather at 10 cm s^{-1}, for example. Applications of this kind can be found across a wide range of industries, from aerospace to textiles, and there are already several thousand laser systems being used for this purpose in the U.S.A. alone.

One of the most promising areas of medical applications is in *eye surgery*, for which several clinical procedures have already become quite well established. A detached retina, for example, which results in local blindness, can be 'spot welded' back onto its support (the choroid) by treatment with high intensity pulses of light from an argon laser. There are a great many advantages in the use of lasers for such surgery; the laser technique is non-invasive, and does not require the administration of anaesthetics, nor the need for prolonged fixation by the eye during treatment, in view of the short duration of the pulses. When we return to a fuller discussion of pulsed laser systems in the next Chapter, we shall see that there are far more exotic applications of high laser intensities than these.

1.6.3 Coherence

Coherence is the property which most clearly distinguishes laser light from other kinds of light, and it is again a property which results from the nature of the stimulated emission process. Light produced by more conventional thermal sources which operate by spontaneous emission is often referred to as *chaotic*; there is generally no correlation between the phases of different photons, and appreciable intensity fluctuations result from the essentially random interference which ensues. By contrast in the laser, photons emitted by the excited laser medium are emitted *in phase* with those already present in the cavity. The timescale over which phase correlation is retained is known as the *coherence time*, t_c, and is given by

$$t_c = 1/\Delta v, \tag{1.26}$$

where Δv is the emission linewidth. Directly related to this is the *coherence length*,

$$l_c = ct_c. \tag{1.27}$$

Thus two points along a laser beam separated by less than the coherence length should have related phase. The coherence length of a single mode output from a gas laser may be 100 m, but for a semiconductor laser it is approximately 1 mm. Measurement of the coherence length or coherence time of a laser is accomplished by *intensity fluctuation spectroscopy*, and provides a useful means of ascertaining the emission linewidth (see Sect. 3.6.2).

Chaotic and coherent radiation have quite different *photon statistics*, as illustrated by

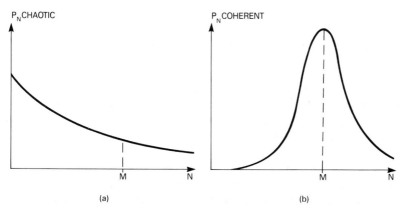

Fig. 1.11 a and b. Photon statistics of (a) thermal radiation and (b) laser radiation of the same mean intensity.

the two graphs in Fig. 1.11. These graphs show the distribution of probability for finding N photons in a volume which contains on a time-average a mean number M. Chaotic light satisfies a Bose-Einstein distribution given by

$$P_N^{chaotic} = M^N/(M + 1)^{N+1},$$ (1.28)

whereas coherent light generally satisfies the Poisson distribution

$$P_N^{coherent} = M^N e^{-M}/N!.$$ (1.29)

Although most processes involving the interaction of light and matter cannot distinguish between the two kinds of light if they have the same mean irradiance (related to M by $M = IV/hc'v$, from Eq. 1.6), this is not the case with multiphoton processes, as we shall see in Sect. 4.6.

There are surprisingly few applications of laser coherence; the main one is *holography*, which is a technique for the production of three-dimensional images. The process involves creating a special type of image, known as a *hologram*, on a plate with a very fine photographic emulsion. Unlike the more usual kind of photographic image, the hologram contains information not only on the intensity, but also on the phase of light reflected from the subject; clearly, such an image cannot be created using a chaotic light source. Subsequent illumination of the image reconstructs a genuinely three-dimensional image. One of the main problems at present is that only monochrome holograms can be made, because phase information is lost if a range of wavelengths is used to create the initial image; although holograms which can be viewed in white light are now quite common, the colours they display are only the result of interference, and not the colours of the original subject.

1.6.4 Monochromaticity

The last of the major characteristics of laser light, and the one which has the most relevance for chemical applications, is its essential monochromaticity, resulting from the fact that all of the photons are emitted as the result of a transition between the same two atomic or molecular energy levels, and hence have almost exactly the same frequency. Nonetheless, as we saw in Sect. 1.5.3, there is always a small spread to the frequency distribution which may cover several discrete frequencies or wavelengths satisfying the standing-wave condition of which we spoke earlier; this situation is illustrated in Fig. 1.12. The result is that a small number of closely separated frequencies may be involved in laser action, so that an additional means of frequency selection has to be built into the laser in order to achieve the optimum monochromaticity. What is generally used is an *etalon*, which is an optical element placed within the laser cavity and arranged so that only one well-defined wavelength can travel back and forth indefinitely between the end-mirrors.

With lasers which have a continuous output, it is quite easy to obtain an emission linewidth as low as 1 cm^{-1}, and in frequency-stabilized lasers the linewidth may be four or five orders of magnitude smaller. One of the important factors which characterize lasers is the quality factor Q, which equals the ratio of the emission frequency v to the linewidth Δv,

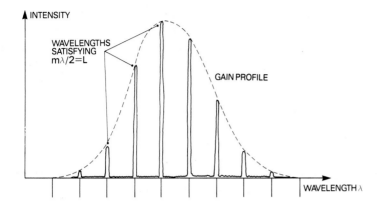

Fig. 1.12. Emission spectrum of a typical laser. Emission occurs only at wavelengths within the fluorescence linewidth of the laser medium which experience gain and also satisfy the standing-wave condition.

$$Q = v/\Delta v. \tag{1.30}$$

The value of the Q-factor can thus easily be as high as 10^8, which is clearly of great significance for *high resolution spectroscopy*, which we shall examine at some length in Chap. 4. The spectroscopist often prefers to have linewidth expressed in terms of wavelength or wavenumber units, where the latter represents the number of wavelengths of radiation per unit length, usually per centimetre ($\tilde{v} = 1/\lambda$). Useful relationships between the *magnitudes* of the corresponding linewidth parameters are as follows;

$$\Delta\lambda = \lambda/Q; \tag{1.31}$$

$$\Delta\tilde{v} = \Delta\lambda/\lambda^2. \tag{1.32}$$

Another important area of technical application associated with the high degree of monochromaticity of a laser source lies in *isotope separation*. Since molecules which differ in isotopic constitution generally have slightly different absorption frequencies, by using a very narrow linewidth laser a mixture of the compounds can be selectively excited, and then separated by other means. Not surprisingly, there is a great deal of interest in this kind of application in the nuclear industry. Again, a full discussion is reserved for later (Sect. 5.3).

2 Laser Sources

Since the construction of the first laser based on ruby, widely ranging materials have been adopted as laser media, and the range is still continually being extended to provide output at new wavelengths: according to Charles H. Townes, one of the pioneers of laser development, 'almost anything works if you hit it hard enough'. As can be seen from a glance at Appendix 1, output from commercial lasers now covers most of the electromagnetic spectrum through from the microwave region to the ultraviolet, and much effort is being concentrated on extending this range to still shorter wavelengths. Amongst a host of tantalising possibilities is the prospect of obtaining holograms of molecules by use of an X-ray laser, for example.

Having discussed the general principles and characteristics of laser sources in the last Chapter, we can now have a look at some of the specific types of laser which are commercially available and commonly used both within and outside of chemistry. Space does not permit a thorough description of the large number of lasers which fall into this category, and the more limited scope of this Chapter is therefore a discussion of representative examples from each of the major types of laser. The last laser to be discussed, the free-electron laser, is different in that it is not commercially available; however, it is included here since it represents the most likely means of approaching the goal of a genuine X-ray laser.

2.1 Solid-state Transition Metal Ion Lasers

With the exception of semiconductor lasers, which we shall deal with separately, the active medium in solid-state lasers is generally a transparent crystal or glass into which a small amount of transition metal is doped; transitions in the transition metal ions are responsible for the laser action. The two most common dopant metals are chromium, in the ruby and alexandrite lasers, and neodymium, in the Nd:YAG and Nd:glass lasers, and the dopant concentration is typically 1 % or less. All such lasers are optically pumped by a broad-band flashlamp source, and can be pulsed to produce the highest available laser intensities. A natural irregularity in the temporal profile of each pulse known as *spiking* makes this kind of laser particularly effective for cutting and drilling applications.

2.1.1 Ruby Laser

The most familiar solid-state laser is the ruby laser, which has an important place in the history of lasers as it was the first type ever constructed, in 1960. In a beautiful example of the ironies of science, the inventor Theodore H. Maiman had his original paper describing this first laser turned down for publication by *Physical Review Letters*, because it was deemed to be of insufficient interest. In construction, the modern ruby laser comprises a rod of commercial ruby (0.05 % Cr_2O_3 in an Al_2O_3 lattice) of between 3 and 25 mm diameter and up to 20 cm length. The chromium ions are excited by the broadband emission from a flashlamp coiled around it, or placed alongside it within an elliptical reflector as shown in Fig. 2.1.

Before we consider the details of the energy level scheme for the ruby laser, it is interesting to note that we can learn something straightaway from the colour of light it emits. The fact that a ruby laser emits red light might not seem too surprising, until we remember that the reason ruby appears red is because it *absorbs* in the green and violet regions of the spectrum; that is why it *transmits* (or reflects) red light. So since the absorption from the flashlamp and the laser emission evidently occur at different wavelengths, it is immediately clear that ruby is a laser involving more than two levels, as we should expect from our discussion in Chap. 1. In fact, the energy level diagram is as shown in Fig. 2.2, and may be regarded as a *pseudo- three-level system*, in the sense that in the course of excitation, decay and emission, only three Cr^{3+} levels are involved; $^4A_2$, $^4T_1$ or $^4T_2$, and ^2E (cf. Fig. 1.7). It is worth noting that these energy levels are quite

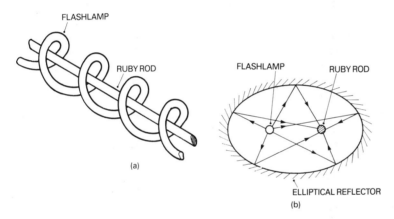

Fig. 2.1a and b. Arrangement of flashlamp tube and ruby rod in a ruby laser; **(a)** illustrates the helical flashlamp option, and **(b)** the highly efficient pumping obtained from a confocal linear flashlamp and laser rod within an elliptical reflector

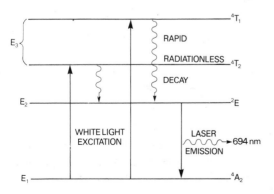

Fig. 2.2. Energetics of the ruby laser

different from those of a free chromium atom. The electrostatic environment created by the surrounding atoms of the host lattice, known as the *crystal field*, produces large splittings between energy levels which in the free atom would normally be degenerate.

The initial flashlamp excitation takes the Cr^{3+} ions up from the ground state E_1 ($^4A_2$) to one of the two E_3 (^4T) levels. Both of these levels have sub-nanosecond life-times, and rapidly decay into the E_2 level (^2E). This decay takes place simply by non-radiative processes which channel energy into lattice vibrations and so result in a heating up of the crystal. Since the E_2 level has a much longer decay time of around 4 ms, a population inversion is created between the E_2 and E_1 states, leading to laser emission at a wavelength of 694.3 nm as the majority of Cr^{3+} ions simultaneously cascade down to the ground state. The lasing process thus generates a single, intense pulse of light of between 0.3 and 3 ms duration, and it is necessary to recharge the flashlamp before the next pulse can be created. The delay between successive pulses usually lasts several seconds, and can be as long as a full minute. A very different kind of pulsing can be created by the technique known as *Q-switching*, which is discussed in Sect. 3.3.2.

One problem with the ruby laser, which is common to all lasers of this type, is the damage caused by the repeated cycle of heating and cooling associated with the generation of each pulse, which ultimately necessitates replacement of the ruby rod. To improve performance, the rod usually needs to be cooled by circulation of water in a jacket around it. Despite its drawbacks, with pulse energies as high as 200 J the ruby laser represents a very powerful source of monochromatic light in the optical region, and it has found many applications in materials processing. The emission bandwidth is typically about 0.5 nm ($10 \, cm^{-1}$), but this can be reduced by a factor of up to 10^4 by introducing intracavity etalons. With such narrow linewidth, the laser is then suitable for holography, and this is where most of its applications are now found. Anther field of application is in lidar (Sect. 3.6.5). The beam diameter of a low-power ruby laser can be as little as 1 mm, with 0.25 milliradian divergence; the most powerful lasers may have beams up to 25 mm in diameter, and a larger divergence of several milliradians.

Recently, another type of solid-state laser has been developed, which although also based on optical transitions in chromium ions embedded in a host ionic crystal, has quite different properties from the ruby laser. This is the *alexandrite* laser, in which the active medium is chromium-doped alexandrite ($BeAl_2O_4$) crystal. The difference between the ruby and alexandrite lasers arises because the Cr^{3+} electronic $^4A_2$ ground state is in the latter case no longer discrete, but as a result of coupling with lattice vibrations consists of a broad continuous band of *vibronic* energy levels. Laser emission, following the usual flashlamp excitation, can in this case take place through downward transitions from the $^4T_2$ state to anywhere in the ground state continuum, and the result is a tunable output covering the region 700–815 nm. For this reason, the alexandrite laser is often referred to as a *vibronic laser*; other vibronic lasers include Ni^{2+}- and Co^{2+}-doped MgF_2. Full consideration of the more general features of such tunable lasers is reserved for the discussion of dye lasers in Sect. 2.5.

2.1.2 Neodymium Lasers

Neodymium lasers are of two types; in one the host lattice for the neodymium ions is yttrium aluminium garnet crystal ($Y_3Al_5O_{12}$), and in the other the host is an amorphous glass. These are referred to as Nd:YAG and Nd:glass, respectively. Although transitions in the neodymium ions are responsible for laser action in both cases, the emission characteristics differ because of the influence of the host lattice on the neodymium energy levels. Also, glass does not have the excellent thermal conductivity properties of YAG crystal, so that it is more suitable for pulsed than for continuous-wave (CW) operation. Like the ruby laser, both types of neodymium laser are optically pumped by a flashlamp arranged confocally alongside a rod of the laser material in an elliptical reflector.

Once again, the energy levels of neodymium ions involved in laser action, which are naturally degenerate in the free state, are split by interaction with the crystal field. In this case, the splitting is readily illustrated in schematic form, as in Fig. 2.3. As a result, transitions between components of the $^4F_{3/2}$ and $^4I_{11/2}$ states, which are forbidden in the free state, become allowed and can give rise to laser emission. The $^4F_{3/2}$ levels are initially populated following non-radiative decay from higher energy levels excited by the flashlamp, and since the terminal $^4I_{11/2}$ laser level lies above the $^4I_{9/2}$ ground state, we thus have a *pseudo- four-level system* (cf. Fig. 1.8).

The principal emission wavelength for both types of neodymium laser is around 1.064 μm, in the near- infra-red; some commercial lasers can also be operated on a different transition producing 1.319 μm output. The YAG and glass host materials impose very different characteristics on the emission, however. Quite apart from the differences in thermal conductivity which determine the question of continuous or pulsed operation, one of the main differences appears in the linewidth. Since glass has an amorphous structure, the electrical environment experienced by different neodymium ions varies, and so the crystal field splitting also varies from ion to ion. Because of this, the linewidth tends to be much broader than in the Nd:YAG laser, where the lattice is much more regular and the field splitting more constant. However, the concentration of neodymium dopant in glass may be as large as 6 %, compared with 1.5 % in a

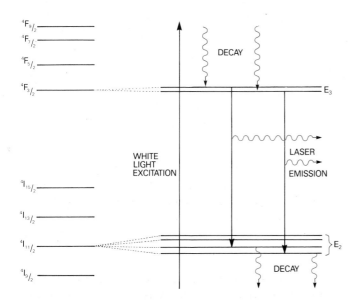

Fig. 2.3. Energetics of a neodymium laser

YAG host, so that a much higher energy output can be obtained. For both these reasons, the Nd:glass laser is ideally suited for the production of extremely high intensity ultrashort pulses by the technique of mode-locking (see Sect. 3.3.3). In fact it is with the neodymium lasers that we find the greatest commercially available laser beam intensities.

The output power of a continuous-wave Nd:YAG laser is typically several Watts, and can exceed 100 W. When operating in a pulsed mode, the energy per pulse varies according to the method of pulsing and the pulse repetition rate, but can be anywhere between a small fraction of a Joule and 100 J for a single pulse. Such powerful sources of infra-red radiation now find many applications in materials processing. In the last few years, Nd:YAG lasers have also made substantial inroads into the surgical laser market. Whilst use is often made of localised thermally induced processes caused through the absorption of intense infra-red radiation, there is much interest in the alternative *breakdown-mode surgery* possible with Q-switched or mode-locked YAG lasers. Here, the enormous electric field (typically 3×10^8 V m^{-1}) associated with each focussed pulse of laser light strips electrons from tissue molecules, and the resultant plasma creates a shock wave which causes mechanical rupture of tissue within a distance of about 1 mm from the focus. This method is proving highly useful for a number of ophthalmic microsurgical applications.

Finally, it is important to note that the majority of Nd:YAG research applications in photochemistry and photobiology make use not of the 1.064 µm radiation as such, but rather the high intensity visible light which can be produced by frequency conversion methods. Particularly important in this respect are the wavelengths of 532, 355 and 266 nm obtained by harmonic generation (see Sect. 3.2).

One other type of solid-state laser in which the active medium is a regular ionic crystal may be mentioned here, although it does not involve transition metal ions. This is the so-called *F-centre* (or colour centre) laser, which operates on optical transitions at defect sites in alkali halide crystals, as for example in KCl doped with thallium; the colour centres are typically excited by a pump Nd:YAG or argon/krypton ion laser. Such lasers produce radiation which is tunable using a grating end mirror over a small range of wavelengths in the overall region 0.8–3.3 µm; different crystals are required for operation over different parts of this range. One disadvantage of this type of laser is that it requires use of liquid nitrogen since the crystals need to be held at cryogenic temperatures.

2.2 Semiconductor Lasers

In the solid-state lasers we have considered so far, the energy levels are associated with dopant atoms in quite low concentrations in a host lattice of a different material. Because under these conditions the dopant atoms are essentially isolated from each other, their energy levels remain discrete, and we obtain the same kind of line spectrum which we generally associate with isolated atoms or molecules. In the case of semiconductor lasers, however, we are dealing with energy levels of an entire lattice, and so it is energy bands, rather than discrete energy levels which we have to consider.

The characteristic properties of a semiconductor arise from the fact that there is a small energy gap between two energy bands known as the valence band and the conduction band. At very low temperatures, the electrons associated with the valence shell of each atom occupy energy levels within the valence band, and the higher energy conduction band is empty. Hence electrons are not able to travel freely through the lattice, and the material has the electrical properties of an insulator. At room temperature, however, thermal energy is sufficient for some electrons to be excited into the conduction band where passage through the lattice is relatively unhindered, and hence the electrical conductivity rises to somewhere inbetween that expected of a conductor and an insulator.

Solid-state electronic devices generally make use of junctions between p-type and n-type semiconductor. The former type has impurity atoms in its lattice which possess fewer valence electrons than the atoms they replace; the latter has impurity atoms possessing more valence electrons than the atoms they replace. The most familiar semiconductor materials are Class IV elements such as silicon and germanium. However, binary compounds between Class III and Class V elements, such as gallium arsenide (GaAs), exhibit similar behaviour. In this case, the lattice consists of two interpenetrating face-centred cubic lattices, and p- and n-type crystal are obtained by varying the stoichiometry from precisely 1:1. In the p-type material, some of the arsenic atoms are replaced by gallium; in n-type crystal, the converse is true.

Semiconductor lasers operate in a broadly similar way to the more familiar light-emitting diodes (LED's) widely used in electronic gadgets. By applying an electrical potential across a simple diode junction between p- and n-type crystal, electrons crossing the semiconductor boundary drop down from a conduction band to a valence band emitting radiation in the process as illustrated in Fig. 2.4. The emission is most often in the infra-red, and the optical properties of the crystal at such wavelengths make it possible for the crystal end-faces to form the confines of the resonant cavity. One advantage of this kind of laser is its extremely small size, which is usually about half a millimetre. However, this does result in very poor collimation; beam divergences of 10° are by no means unusual, and thus create the need for corrective optics in many applications.

There are two main types of semiconductor laser; those which operate at fixed wavelengths, and those which are tunable. The three most common fixed-wavelength types are gallium arsenide, gallium aluminium arsenide, and indium gallium arsenide phosphide. Gallium arsenide lasers emit at a wavelength of around 0.904 μm; the wide range of flexibility in the stoichiometry of the other types makes it possible to produce

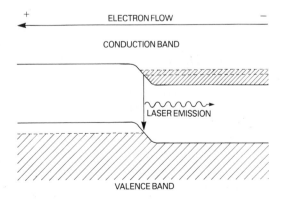

Fig. 2.4. Energetics of a diode laser; the energy levels of the p-type material are shown on the left, and the n-type on the right

a range of lasers operating at various fixed wavelengths in the region 0.8-1.3 μm. The so-called 'lead salt' diode lasers, however, which are derived from non-stoichiometric binary compounds of lead, cadmium and tin with tellurium, selenium and sulphur, emit in the range 2.8-30 μm (3,500-330 cm^{-1}) depending on the exact composition. Although these lasers require a very low operating temperature, typically in the range 15-90 °K, the operating wavelength is very temperature-dependent, and so the wavelength can be tuned by varying the temperature. The tuning range for a lead salt diode laser of a particular composition is typically about 100 cm^{-1}.

Modes in a diode laser are typically separated by 1-2 cm^{-1}, and each individual mode generally has a very narrow linewidth, of 10^{-3} cm^{-1} or less. The output power of continuous semiconductor lasers is generally measured in milliwatts, with some hundreds of milliwatts now available from the most intense devices. By far the most important applications lie in the area of optical communications and fibre optics; not surprisingly, diode lasers currently represent the most rapidly growing area of the laser market. However, diode lasers are also very well suited to high-resolution infra-red spectroscopy, since the linewidth is sufficiently small to enable the rotational structure of vibrational transitions to be resolved for many small molecules (see Sect. 4.2). This method has proved particularly valuable in characterising short-lived intermediates in chemical reactions. More diverse applications can be expected as research presses on towards the goal of a diode laser emitting at the red end of the visible spectrum. When this has been achieved, diode lasers may be expected to oust helium-neon lasers from many of their traditional roles.

2.3 Atomic and Ionic Gas Lasers

The class of lasers in which the active medium is a gas covers a wide variety of devices. Generally, the gas is either monatomic, or else it is composed of very simple molecules. Representative examples of monatomic lasers are discussed in this Section; molecular gas lasers are discussed in Sect. 2.4. In both cases, since laser emission results from optical transitions in *free* atoms or molecules, the emission linewidth can be very small. The gas is often contained in a sealed tube, with the initial excitation provided by an electrical discharge, so that in many cases the innermost part of the laser bears a superficial resemblance to a conventional fluorescent light.

The laser tube can be constructed from various materials, and need not necessarily be transparent. Unfortunately metals are generally ruled out because they would short-circuit the device. Silica is commonly employed, and also beryllium oxide which has the advantage for high power sources of a high thermal conductivity. It is quite a common feature to have the laser tube contain a mixture of two gases, one of which is involved in the pump step, and the other in the laser emission. Such gas lasers are usually very reliable, since there is no possibility of thermal damage to the active medium which there is for solid-state lasers, and for routine purposes they are the most widely used type.

2.3.1 Helium-neon Laser

The helium-neon laser was the first CW laser ever constructed, and was also the first laser to be made available to a commercial market, in 1962. The active medium is a mixture of the two gases contained in a glass tube at low pressure; the partial pressure of helium is approximately 1 mbar, and that of neon 0.1 mbar. The initial excitation is provided by an electrical discharge, and serves primarily to excite helium atoms by electron impact. The excited helium atoms subsequently undergo a process of collisional energy transfer to neon atoms; it so happens that certain levels of helium and neon are very close in energy, so that this transfer takes place with a high degree of efficiency. Because the levels of neon so populated lie above the lowest excited states, a population inversion is created relative to these levels, enabling laser emission to occur as shown in Fig. 2.5. Two points about this should be made in passing. First, note that the usual state designation cannot be given for the energy levels of neon, because Russell-Saunders coupling does not apply. Secondly, each electron configuration gives rise to several closely spaced states, but only those directly involved in the laser operation have been shown on the diagram.

Three distinct wavelengths can be produced in the laser emission stage; there is one visible wavelength, typically of milliwatt power, which appears in the red at 632.8 nm, and there are two infra-red wavelengths of somewhat lower power at 1.152 and 3.391 μm. Obviously, infra-red optics are required for operation of the laser at either of these wavelengths. Following emission, the lasing cycle is completed as the neon undergoes a two-step radiationless decay back down to its ground state. This involves transition to a metastable $2p^53s^1$ level, followed by collisional deactivation at the inner surface of the tube. The last step has to be rapid if the laser is to operate efficiently; for this reason, the surface/volume ratio of the laser tube has to be kept as large as possible, which generally means keeping the tube diameter small. In practice, tubes are

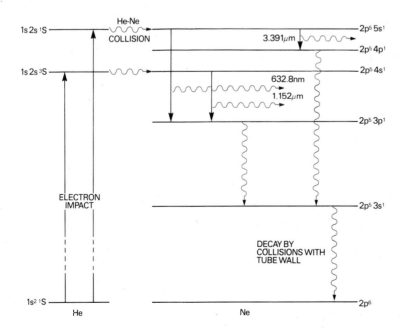

Fig. 2.5. Energetics of the helium-neon laser

commonly only a few millimetres in diameter. One other very weak transition has recently been utilised in the production of a 1 mW helium-neon laser emitting at a wavelength of 543.5 nm in the green; the principal virtue of this laser is that it is substantially cheaper than any other green laser.

Helium-neon lasers operate continuously, and despite their low output power, they have the twin virtues of being both small and relatively inexpensive. Consequently, they can be found in more widely ranging applications than any other kind of laser. The principal applications are those which hinge on the typically narrow laser beamwidth, but where power is not too important. Examples include many optical scanning devices used for quality control and measurement in industry; helium-neon scanners are also used in optical video disc systems, supermarket bar-code readers and optical character recognition equipment. Other uses include electronic printing and optical alignment. One other laser of a similar type is the helium-cadmium laser, in which transitions in free cadmium atoms result in milliwatt emission at 442 nm in the blue, and 325 nm in the ultraviolet. The blue line is particularly well suited to high-resolution applications in the printing and reprographics industry.

2.3.2 Argon Laser

The argon laser is the most common example of a family of ion lasers in which the active medium is a single-component inert gas. The gas, at a pressure of approximately 0.5 mbar, is contained in a plasma tube of 2–3 mm bore, and is excited by a continuous electrical discharge (see Fig. 2.6). Argon atoms are ionised and further excited by electron impact; the nature of the pumping process produces population of several ionic excited states, and those responsible for laser action are on average populated by two successive impacts. The sustainment of a population inversion between these states and others of lower energy results in emission at a series of discrete wavelengths over the range 350–530 nm (see Appendix 1), the two strongest lines appearing at 488.0 and 514.5 nm. These two wavelengths are emitted as the result of transitions from the singly ionised states with electron configuration $3s^2 3p^4 4p^1$ down to the $3s^2 3p^4 4s^1$ state. Further radiative decay to the multiplet associated with the ionic ground configuration $3s^2 3p^5$ then occurs, and the cycle is completed either by electron capture or further impact excitation. Doubly ionised Ar^{2+} ions contribute to the near-ultraviolet laser emission.

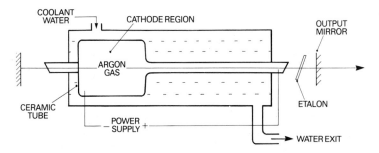

Fig. 2.6. Schematic construction of an argon ion laser. Magnetic coils (not shown) may be placed around the discharge tube to keep the plasma confined to the centre of the bore

Since several wavelengths are produced by this laser, an etalon or dispersing prism is generally placed between the end-mirrors to select one particular wavelength of light for amplification; the output wavelengths can thus be varied by changing its orientation. By selecting a single longitudinal mode, an output linewidth of only 0.0001 cm^{-1} is obtainable. The pumping of ionic levels required for laser action requires a large and continuous input of energy, and the relatively low efficiency of the device means that a large amount of thermal energy has to be dispersed. Cooling therefore has to be a major consideration in the design, and as shown in Fig. 2.6, circulation of water in a jacket around the tube is the most common solution, although air cooled argon lasers of low power have recently become available. The output power of a CW argon laser usually lies in the region running from milliwatts up to about 20 W.

Argon lasers are fairly expensive, comparatively fragile, and have tube lifetimes generally limited to between 1000 and 10,000 hours. One of the principal reasons for the limited lifetime is erosion of the tube walls by the plasma, resulting in deposition of dust on the Brewster output windows. The argon itself is also depleted to a small extent by ions becoming embedded in the tube walls. Despite these drawbacks, such lasers have found widespread research applications in chemistry and physics, particularly in the realm of spectroscopy, where they are often employed to pump dye lasers (see Sect. 2.5). Argon lasers have also made a strong impact in the printing and reprographics industry, and they play an increasingly important role in medical treatment, especially in laser ophthalmological procedures. One other area of application worth mentioning is in entertainment and visual displays. Curiously enough, it is in this somewhat less serious, though undoubtedly exciting application that most people first see laser light.

The other common member of the family of ion lasers is the krypton laser. In most respects it is very similar to the argon laser, emitting wavelengths over the range 350–800 nm, although because of its lower efficiency the output takes place at somewhat lower power levels, up to around 5 W. The strongest emission is at a wavelength of 647.1 nm. In fact, the strong similarity in physical requirements and performance between the argon and krypton lasers makes it possible to construct a laser containing a mixture of the two gases, which provides a very good range of wavelengths across the whole of the visible spectrum, as can be seen from Appendix 1. Such lasers emit many of the wavelengths useful for biomedical applications, the blue-green argon lines being especially useful since they are strongly absorbed by red blood cells.

2.3.3 Copper Vapour Laser

The copper vapour laser is one of the most recently developed lasers, which has yet to make a strong impact on the laser market. However, it does have a number of features which should make it a very attractive competitor for a number of applications. It belongs to the class of *metal-vapour lasers*, in which transitions in free uncharged metal atoms give rise to laser emission.

The copper laser is essentially a three-level system involving the energy levels shown in Fig. 2.7. Electron impact on the ground state copper atoms results in excitation to ^{2}P states belonging to the electron configuration $3d^{10}4p^1$, from which transitions to lower-lying $3d^94s^2$ ^{2}D levels can take place. Laser emission thus occurs at wavelengths of 510.5 nm in the green and 578.2 nm in the yellow. Further collisions of the excited at-

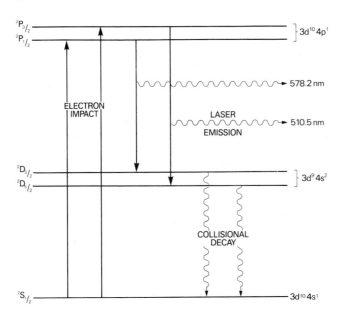

Fig. 2.7. Energetics of the copper vapour laser

oms with electrons or the tube walls subsequently result in decay back to the ground state. One of the problems with this particular scheme is that electron impact on ground state copper atoms not only populates the ^{2}P levels but also the ^{2}D levels associated with the lower end of the laser transitions. For this reason, it is not possible to sustain a population inversion between the ^{2}P and ^{2}D levels, and so the laser naturally operates in a pulsed mode, usually with a pulse repetition frequency of about 5 kHz. Each pulse typically has a duration of 30 ns and an energy in the millijoule range.

The physical design of the laser involves an alumina plasma tube containing small beads or other sources of metallic copper at each end. The tube also contains a low pressure of neon gas (approximately 5 mbar) to sustain an electrical discharge. Passage of a current through the tube creates temperatures of 1400–1500 °C, which heats the copper and produces a partial pressure of Cu atoms of around 0.1 mbar; this then acts as the lasing medium. One of the initial drawbacks of the laser, the long warm-up time of about an hour, has recently been overcome in a variation on this design which operates at room temperature.

The chief advantages of the copper vapour laser are that it emits visible radiation at very high powers (the time average over a complete cycle of pulsed emission and pumping is 10–40 W), and that it is reasonably priced and highly energy-efficient; the most powerful 100 W copper vapour laser consumes only half as much input power as a 20 W argon laser, for example. Applications of the copper laser, which are at present mostly at the research and development stage, include uranium isotope separation (see Sect. 5.3.1). There has also been interest in use of the laser for photography and holography, as well as in underwater illumination, for which the emission wavelengths are particularly well suited to minimise attenuation. Dermatological applications are also

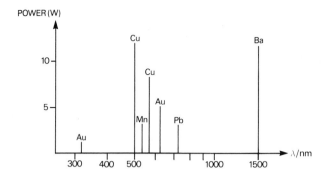

Fig. 2.8. Typical metal vapour output powers. (Reproduced by kind permission of Quentron Optics Pty. Ltd.)

being investigated, since the 578 nm emission lies helpfully close to the haemoglobin absorption peak at 577 nm.

The only other metal-vapour laser yet to graduate from the research to the production stage is the gold laser, emitting a few watts power principally at a wavelength of 628 nm. This has already proved highly effective for cancer phototherapy (see Sect. 5.4.3). The output powers and wavelengths available from the gold and other metal vapour lasers are shown in Fig. 2.8

2.4 Molecular Gas Lasers

2.4.1 Carbon Dioxide Laser

The carbon dioxide laser is our first example of a laser in which the transitions responsible for stimulated emission take place in free molecules. In fact the CO_2 energy levels involved in laser action are not electronic but rotation-vibration levels, and emission therefore occurs at much longer wavelengths, well into the infra-red. The lasing medium consists of a mixture of CO_2, N_2 and He gas in varying proportions, but often in the ratio 1:4:5; the helium is added to improve the lasing efficiency, and the nitrogen plays a role similar to that of helium in the He-Ne laser.

The sequence of excitation is illustrated in Fig. 2.9. The first step is population of the first vibrationally excited level of nitrogen by electron impact. In its ground vibrational state each nitrogen molecule can possess various discrete amounts of rotational energy, and various rotational sub-levels belonging to the vibrationally excited state are populated by the electron collision. These levels are all metastable, since radiative decay back down to the vibrational ground state is forbidden by the normal selection rules for emission. However, one of the vibrationally excited states of carbon dioxide, labelled (001) since it possesses one quantum of energy in the v_3 vibrational mode (the antisymmetric stretch) has almost exactly the same energy as the vibrationally excited nitrogen molecule. Consequently, collision between the two molecules results in a very efficient transfer of energy to the carbon dioxide; once again, it is in fact the rotational sub-levels belonging to the (001) state which are populated.

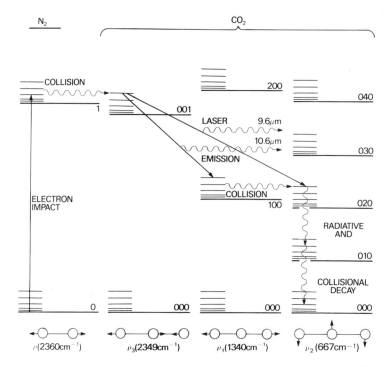

Fig. 2.9. Energetics of the carbon dioxide laser. The rotational structure of the vibrational levels is shown only schematically

Laser emission in the CO_2 then occurs by two routes, involving radiative decay to rotational sub-levels belonging to the (100) and (020) states; the former possess one quantum of vibrational energy in the ν_1 symmetric stretch mode, while the latter have two quanta of energy in the ν_2 bending mode. These states cannot be populated directly by collision with N_2, so that they exist in a population inversion with respect to the (001) levels. The two laser transitions result in emission at wavelengths of around 10.6 μm and 9.6 μm, respectively. Both pathways of decay ultimately lead to the (020) states, as shown in the diagram; subsequent deactivation results both from radiative decay and from collisions with He atoms. Since various rotational sub-levels can be involved in the emissive transitions, use of suitable etalons enable the laser to be operated at various discrete frequencies within the 10.6 μm and 9.6 μm bands as listed in Table 2.1; in wavenumber terms a typical gap between successive emission frequencies is 2 cm⁻¹. Since the positions of the CO_2 energy levels vary for different isotopic species, other wavelengths may be obtained by using isotopically substituted gas.

One of the problems which must be overcome in a carbon dioxide laser is the possibility that some molecules will dissociate into carbon monoxide and oxygen in the course of the excitation process. When a sealed cavity is used, this problem is generally overcome by the admixture of a small amount of water vapour, which reacts with any carbon monoxide and regenerates carbon dioxide. Such a method is unnecessary if cooled carbon dioxide is continuously passed through the discharge tube. This has

Table 2.1: Principal Emission Lines of the Carbon Dioxide Laser.
The emission lines, whose frequencies are expressed in wavenumber units $\tilde{\nu}$ below, result from rotation-vibration transitions in which the rotational quantum number J changes by one unit. Transitions from $J \rightarrow J-1$ are denoted by P(J); transitions from $J \rightarrow J+1$ are denoted by R(J).

10.6 μm band (001 $\rightarrow$ 100 transitions)

Line	$\tilde{\nu}$ (cm^{-1})	Line	$\tilde{\nu}$ (cm^{-1})	Line	$\tilde{\nu}$ (cm^{-1})
P(56)	907.78	P(22)	942.38	R(18)	974.62
P(54)	910.02	P(20)	944.19	R(20)	975.93
P(52)	912.23	P(18)	945.98	R(22)	977.21
P(50)	914.42	P(16)	947.74	R(24)	978.47
P(48)	916.58	P(14)	949.48	R(26)	979.71
P(46)	918.72	P(12)	951.19	R(28)	980.91
P(44)	920.83	P(10)	952.88	R(30)	982.10
P(42)	922.92	P(8)	954.55	R(32)	983.25
P(40)	924.97	P(6)	956.19	R(34)	984.38
P(38)	927.01	P(4)	957.80	R(36)	985.49
P(36)	929.02	R(4)	964.77	R(38)	986.57
P(34)	931.00	R(6)	966.25	R(40)	987.62
P(32)	932.96	R(8)	967.71	R(42)	988.65
P(30)	934.90	R(10)	969.14	R(44)	989.65
P(28)	936.80	R(12)	970.55	R(46)	990.62
P(26)	938.69	R(14)	971.93	R(48)	991.57
P(24)	940.55	R(16)	973.29	R(50)	992.49

9.6 μm band (001 $\rightarrow$ 020 transitions)

Line	$\tilde{\nu}$ (cm^{-1})	Line	$\tilde{\nu}$ (cm^{-1})	Line	$\tilde{\nu}$ (cm^{-1})
P(50)	1016.72	P(20)	1046.85	R(16)	1075.99
P(48)	1018.90	P(18)	1048.66	R(18)	1077.30
P(46)	1021.06	P(16)	1050.44	R(20)	1078.59
P(44)	1023.19	P(14)	1052.20	R(22)	1079.85
P(42)	1025.30	P(12)	1053.92	R(24)	1081.09
P(40)	1027.38	P(10)	1055.63	R(26)	1082.30
P(38)	1029.44	P(8)	1057.30	R(28)	1083.48
P(36)	1031.48	P(6)	1058.95	R(30)	1084.64
P(34)	1033.49	P(4)	1060.57	R(32)	1085.77
P(32)	1035.47	R(4)	1067.54	R(34)	1086.87
P(30)	1037.43	R(6)	1069.01	R(36)	1087.95
P(28)	1039.37	R(8)	1070.46	R(38)	1089.00
P(26)	1041.28	R(10)	1071.88	R(40)	1090.03
P(24)	1043.16	R(12)	1073.28	R(42)	1091.03
P(22)	1045.02	R(14)	1074.65	R(44)	1092.01

the added advantage of increasing the population inversion, thereby further improving the efficiency.

A small carbon dioxide laser, with a discharge tube about half a metre in length, may have an efficiency rating as high as 30 %, and produce a continuous output of 20 W; even a battery-powered hand-held model can produce 8 W cw output. Much higher powers are available from longer tubes; although efficiency drops, outputs in the kilowatt range are obtainable from the largest room-sized devices. Apart from extending the cavity length, another means used to increase the output power in such lasers is to increase the pressure of carbon dioxide inside the discharge tube, so increasing the number of molecules available to undergo stimulated emission. In fact, carbon dioxide lasers can be made to operate at or above atmospheric pressure, although in these cases much stronger electric fields must be applied to sustain the discharge. To produce sufficiently strong fields without using dangerously high voltages, it is necessary to apply a potential across, rather than along the tube, as illustrated in Fig. 2.10. Such a laser is referred to as a transverse excitation atmospheric (TEA) laser. With higher pressures of around 15 atmospheres, pressure broadening results in a quasi-continuum of emission frequencies, enabling the laser to be continuously tuned over the range 910–1100 cm^{-1}.

In passing, it is worth noting that there is one other very unusual way of pumping a carbon dioxide laser, which does not involve the use of electrical excitation. In the *gas dynamic laser*, a mixture of carbon dioxide and nitrogen is heated and compressed, and then injected into a low-pressure laser cavity at supersonic speed. Since the (001) vibration-rotation states have a longer lifetime, the rapid cooling associated with this process depopulates these upper laser levels less rapidly than the lower levels. A population inversion thus ensues, resulting in the usual laser action. Whilst such devices can produce outputs of 100 kW or more, the emission is limited to a few seconds duration, the construction is necessarily unwieldy, and the device has the dubious distinction of being the only laser which is literally noisy.

Carbon dioxide lasers are extensively used in the area of laser-induced chemical reactions, as we shall see in Chap. 5. However, the majority of their *industrial* applications are at present to be found in material processing, such as hole drilling, welding, cutting and surface treatment. Despite the fact that metals in particular tend to be quite reflective in the wavelength region where these lasers operate, the enormous intensities of about 10^{10} W m^{-2} which they produce at focus more than compensate. It is also important to note that the *total* amount of heat transferred to the metal by the laser beam is minimal. Applications of this kind account for the fact that carbon dioxide lasers represent the largest sector of commercial laser sales.

One other important area of application for CO_2 lasers lies in surgical procedures. The cells of which biological tissue is composed largely consist of water, and as such

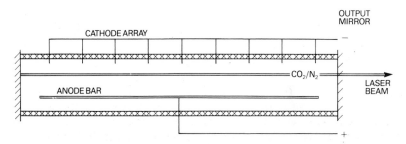

Fig. 2.10. Electrode arrangement in a transverse excitation atmospheric (TEA) carbon dioxide laser

can be instantly vapourised by any powerful CO_2 laser beam; moreover, the heat supplied to the surrounding tissue cauterises the wound and prevents the bleeding normally associated with conventional surgery. For these reasons there are an increasing number of operations where CO_2 laser radiation is proving a more acceptable alternative to the scalpel. The radiation not only provides a very neat method of incision, but in other cases can be used for completely removing large areas of tissue. The recent development of flexible catheter waveguides for CO_2 laser radiation should increase the scope for applications of this type.

A related gas laser is the carbon monoxide laser. This operates in many respects in a similar way to the carbon dioxide system, with an electrical discharge initially exciting nitrogen molecules. This leads to collisional activation of the carbon monoxide, and subsequent laser emission in the range 4.97-8.26 μm. The main difference is that being a diatomic species, CO has only one mode of vibration. This laser has no particular advantages over other kinds, and very few are commercially produced.

2.4.2 Nitrogen Laser

Another gas laser based on a simple chemically stable molecular species is the nitrogen laser. This laser has three main differences from the carbon dioxide laser. Firstly, it operates on electronic, rather than vibrational transitions. The gas is excited by a high-voltage electrical discharge which populates the $C^3\pi_u$ triplet electronic excited state, and the laser transition is to the lower energy metastable $B^3\pi_g$ state. The second difference results from the fact that the upper laser level, with a lifetime of only 40 ns, has a much shorter lifetime than the lower level, and consequently a population inversion cannot be sustained. The third difference is that essentially all the excited nitrogen molecules undergo radiative decay together over a very short period of time, effectively emptying the cavity of all its energy. This kind of process is known as *superradiant emission*, and is sufficiently powerful that a highly intense pulse is produced without the need for repeatedly passing the light back and forth between end-mirrors. Indeed, the nitrogen laser can be successfully operated without any mirrors, though in practice a mirror is placed at one end of the cavity so as to direct the output.

The configuration of the nitrogen laser is thus similar to that shown in Fig. 2.10, except that the output mirror is absent. The laser therefore automatically operates in a pulsed mode, producing pulses of about 10 ns or less duration at a wavelength of 337.1 nm. Bandwidth is approximately 0.1 nm, and pulse repetition frequency 1-200 Hz. Pulses can tend to have a relatively unstable temporal profile as a result of the low residence time of photons in the laser cavity. Since the laser can produce peak intensities in the 10^{10} W m^{-2} range, the nitrogen laser is one of the most powerful commercial sources of ultraviolet radiation, and it is frequently used in photochemical studies. It is also a popular choice as a pump for dye lasers.

2.4.3 Chemical Lasers

In all of the lasers we have examined so far, the pumping mechanism used to initiate the population inversion has involved an external source of power. In a chemical laser, by contrast, population inversion is created directly through an exothermic chemical reaction or other chemical means. To be more precise, we can define a chemical laser as one in which irreversible chemical reaction accompanies the laser cycle. This definition is somewhat more restrictive than that used by some other authors, and thereby excludes the iodine and exciplex lasers dealt with in the following Sections. The concept of a chemical laser is a very appealing one since large amounts of energy can be released by chemical reaction, so that if the laser is efficient in its operation it can produce a very substantial output of energy in the form of light. One of the earliest examples of a chemical laser was the hydrogen chloride laser, whose operation is based on the chemical reaction between hydrogen and chlorine gas, according to the following sequence of reactions;

$$Cl_2 + h\nu_P \rightarrow 2Cl; \tag{2.1}$$

$$Cl + H_2 \rightarrow HCl^* + H; \tag{2.2}$$

$$H + Cl_2 \rightarrow HCl^* + Cl; \tag{2.3}$$

Where $h\nu_P$ is a pump ultraviolet photon provided by a flashlamp. The appropriate laser transition is provided by the subsequent radiative decay of vibrationally excited hydrogen chloride molecules. Although the free radical propagation reactions (2.2) and (2.3) leading to the production of excited HCl are exothermic, the initiation stage (2.1) requires the input of radiation, and hence an external source of power is still necessary.

A modification of this scheme involving hydrogen fluoride is much more popular, and is in commercial production. The only major difference is that in the initiation stage free fluorine radicals are generally produced by an electrical discharge, for example by the electron impact dissociation of a species such as SF_6, which is appreciably less hazardous than the alternative F_2. Oxygen gas is also included in the reaction mixture to scavenge sulphur from the decomposition of SF_6 by reaction to form SO_2. The essential features of the laser are illustrated in Fig. 2.11, where the resonant cavity is anything up to a metre in length; the output beam is typically 2–3 mm in diameter, and has a divergence of about 2 milliradians.

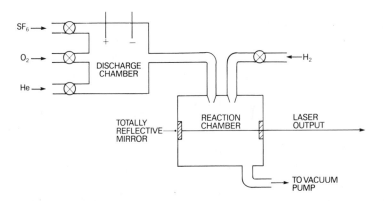

Fig. 2.11. Essential features of a hydrogen fluoride laser

Since only about 1 % of the reactant gases are expended on passage through the laser, removal of HF and replenishment of SF_6 and H_2 enables the exhaust gas to be recycled if required. The HF laser produces output in the region 2.6–3.0 µm, consisting of a series of discrete wavelengths associated with rotation-vibration transitions; its deuterium counterpart based on DF transitions emits wavelengths a factor of approximately $2^{1/2}$ longer, i. e. in the range 3.6–4.0 µm. Continuous-wave power ranges from a few watts to 150 W. Amongst other research applications, the hydrogen fluoride laser is uniquely suited to absorption-based measurements of HF atmospheric pollution in the neighbourhood of industrial plants. Another example of a chemical laser is afforded by the following scheme utilizing a mixture of nitric oxide, fluorine, deuterium and carbon dioxide gases;

$$NO + F_2 \quad \rightarrow \quad ONF + F; \tag{2.4}$$

$$F + D_2 \quad \rightarrow \quad DF^* + D; \tag{2.5}$$

$$DF^* + CO_2 \quad \rightarrow \quad DF + CO_2^*. \tag{2.6}$$

The result of this sequence of reactions is the collisional transfer of vibrational energy to carbon dioxide molecules, thus creating the conditions under which laser emission can occur exactly as in the carbon dioxide laser discussed earlier. Such a laser essentially has its own built-in power supply, and can be turned on simply by opening a valve to allow mixture of the reactant gases. As such, it obviates the need for an external power supply, and thus holds an important advantage over other kinds of laser for many field applications. For higher power continuous wave chemical lasers, relatively large volumes of reactant gases have to be rapidly mixed at supersonic velocities, and a high degree of spatial homogeneity is required in the mixing region for a stable output.

2.4.4 Iodine Laser

Another type of laser operating on broadly similar principles is the iodine laser. This particular laser refuses to fit squarely into most classification schemes; as we shall see, whilst polyatomic chemistry is involved in its operation, the crucial laser transition actually takes place in free atomic iodine, so that it does not strictly deserve to be classed as a molecular laser. Also, it does not involve irreversible chemistry in the laser cycle, only in side-reactions, so that from this point of view it is equally wrong to class it as a chemical laser. Nonetheless, the iodine laser does share many features in common with the chemical lasers discussed in Sect. 2.4.3, and hence it seems most appropriate to consider it here.

The driving principle involved in the iodine laser, more fully termed the *atomic iodine photodissociation laser*, is the photolysis of iodohydrocarbon or iodofluorocarbon gas by ultraviolet light from a flashlamp. One of the gases typically used for this purpose is 1-iodoheptafluoropropane, C_3F_7I, which is stored in an ampoule and introduced into the silica laser tube at a pressure of between 30 and 300 mbar. The reaction sequence is then as follows;

$$C_3F_7I + h\nu_P \quad \rightarrow \quad C_3F_7 + I^*; \tag{2.7}$$

$$I^* \rightarrow I + h\nu_L; \tag{2.8}$$

$$C_3F_7 + I + M \rightarrow C_3F_7I + M; \tag{2.9}$$

where $h\nu_P$ is a pump photon provided by the flashlamp, and $h\nu_L$ is a laser emission photon; the corresponding energy-level diagram is schematically illustrated in Fig. 2.12. The laser emission stage involves transition between an excited metastable $^2P_{1/2}$ state and the ground $^2P_{3/2}$ state of atomic iodine; this results in narrow linewidth output at a wavelength of 1.315 μm (7605 cm^{-1}), consisting of six very closely spaced hyperfine components encompassing a range of less than 1 cm^{-1}.

Despite the fact that the sequence of reactions (2.7) to (2.9) in principle represents a repeatable cycle, there are competing irreversible side-reactions which destroy approximately 10 % of the active medium during each laser cycle. The main side-reactions are

$$C_3F_7 + C_3F_7 + M \rightarrow C_6F_{14} + M; \tag{2.10}$$

$$I + I + M \rightarrow I_2 + M. \tag{2.11}$$

Any build-up of molecular iodine through the mechanism of equation (2.11) further reduces the efficiency of the laser process, because it very effectively quenches the upper laser level by the reaction

$$I_2 + I^* \rightarrow I_2^* + I, \tag{2.12}$$

and so diminishes the extent of population inversion. For this reason, the photolysed gas must be exhausted from the laser after emission has taken place, and the tube recharged with fresh gas for the next pulse. The exhaust gas can in fact be recycled provided the molecular iodine is trapped, for example in a low-temperature alkyl iodide solution.

One of the chief advantages of the iodine laser lies in the fact that the active medium is comparatively cheap, and hence available in large quantities. In the absence of

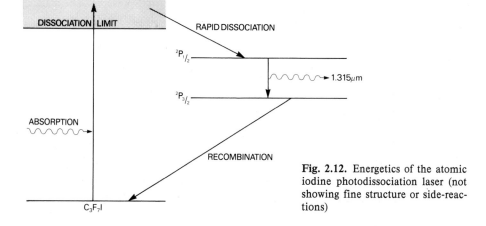

Fig. 2.12. Energetics of the atomic iodine photodissociation laser (not showing fine structure or side-reactions)

any other pulsing mechanism, the laser typically produces pulses of microsecond duration, and pulse energies of several joules; however the output is often modified by Q-switching or mode-locking (see Sect. 3.3) to produce pulse trains of nanosecond or sub-nanosecond duration. One application of the laser of particular interest to chemists is the ability to produce a rapid increase of temperature in aqueous media. Since water absorbs very strongly at 1.315 μm with an efficiency of about 30% per centimetre, the iodine laser can induce nanosecond temperature jumps of several degrees Celsius, thus offering a wide range of possibilities for the study of fast chemical and biological solution kinetics. Other uses include lidar and fibre-optical applications.

2.4.5 Exciplex Lasers

The next category of commonly available lasers consists of those in which the active medium is an *exciplex*, or excited diatomic complex. The crucial feature of an exciplex is that only when it is electronically excited does it exist in a bound state with a well-defined potential energy minimum; the ground electronic state generally has no potential energy minimum, or else only a very shallow one. Examples are generally found in the inert gas halides such as KrF, whose potential energy curves are illustrated in Fig. 2.13. Other homonuclear diatomic species which fall into the same category, such as Xe_2, are termed *excimers*, although the term is often found incorrectly applied to heteronuclear exciplexes.

The exciplex is generally formed by chemical reaction between inert gas and halide ions produced by an electrical discharge. For KrF, the exciplex is produced by the following series of reactions;

$$Kr + e \rightarrow Kr^+ + 2e;$$
                                                                              (2.13)

$$F_2 + e \rightarrow F^- + F;$$
                                                                              (2.14)

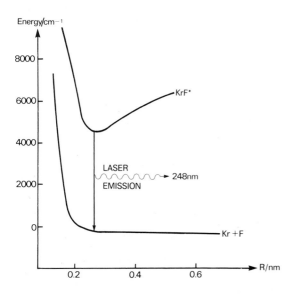

Fig. 2.13. Energetics of a krypton fluoride exciplex laser. Only the electronic states directly involved in laser action are shown

$$F^- + Kr^+ + He \rightarrow KrF^* + He. \tag{2.15}$$

In the three-body reaction (2.15), the helium simply acts as a buffer. Because the krypton fluoride so produced is electronically excited and has a short lifetime (about 2.5 ns), it rapidly decays by photon emission to the lower energy state as shown on the diagram. Since this is an unbound state, in which the force between the atoms is always repulsive, the exciplex molecule then immediately dissociates into its constituent atoms. Consequently, this state never attains a large population, and a population inversion therefore exists between it and the higher energy bound exciplex state. The decay transition can thus be stimulated to produce laser emission with a high efficiency, typically around 20 %. One noteworthy feature of this particular laser scheme is that it represents a rare example of a genuinely two-level laser.

Although the exciplex laser may once again in a sense be regarded as a type of chemical laser, it is worth noting that at the end of a complete cycle of laser transitions the two processes shown below regenerate the starting materials, i. e. krypton and fluorine gas;

$$KrF^* \rightarrow Kr + F + h\nu_L; \tag{2.16}$$

$$F + F \rightarrow F_2. \tag{2.17}$$

Hence the laser can be operated continuously without direct consumption of the active medium, in contrast to the chemical lasers examined earlier. Thus a sealed cavity can be used, in which for the case of KrF the mix of gases is typically 2 % krypton, 0.2 % fluorine and 97.8 % helium, to an overall pressure of 2.5–3.0 atmospheres. Of course, the cavity wall material has to be carefully chosen in view of the highly corrosive halogen gas used. Moreover, since the walls are rapidly poisoned by the gas, it is not possible to use the same laser tube for different halogens.

Exciplex lasers are superradiant, and produce pulsed radiation with pulse durations of 10–20 ns, with pulse repetition frequencies in the 1–500 Hz range. Pulse energies can be up to 1 J, and the average power ranges from 20–100 W. The emission wavelengths of commercially available systems are as follows; ArF 193 nm; KrF 248 nm; KrCl 222 nm; XeF 351 and 353 nm; XeCl 308 nm. The combination of high output power and short wavelength have made these lasers increasingly attractive for applications in photochemistry. It has also been demonstrated that there may be significant surgical applications; tissue irradiated with such short wavelengths will undergo molecular fragmentation and volatilisation without the degree of thermal damage associated with the more common medical lasers which operate in the infra-red. In conjunction with an optical fibre catheter, for example, such lasers have proved highly effective at removing plaque from clogged arteries. There is still a great deal of development work in progress both on the inert gas halide systems mentioned above, and on other exciplex and excimer lasers.

2.5 Dye Lasers

The last category of commonly available lasers consists of dye lasers. These are radically different from each of the types of laser we have discussed so far. All of the differences can be traced back to the unusual nature of the active medium, which is a solution of an organic dye. A wide range of dyes can be used for this purpose; the only general requirements are an absorption band in the visible, and a broad fluorescence spectrum. The kind of compounds which satisfy these criteria best consist of comparatively large polyatomic molecules with extensive electron delocalisation; the most widely used example is the dye commonly known as Rhodamine 6G ($C_{28}H_{31}N_2O_3Cl$), a chloride of the cation whose structure is shown in Fig. 2.14. With 64 atoms, this species

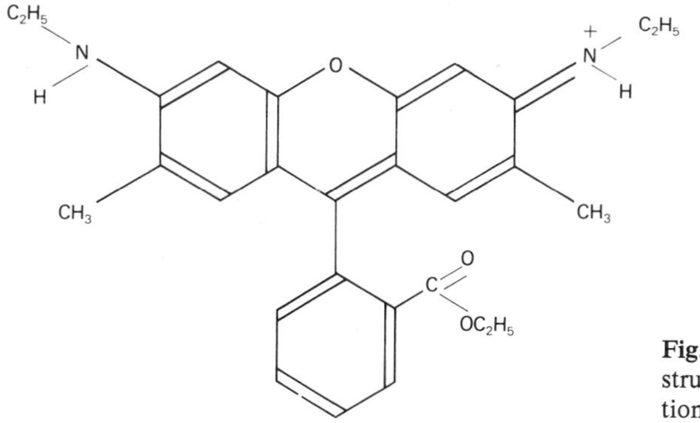

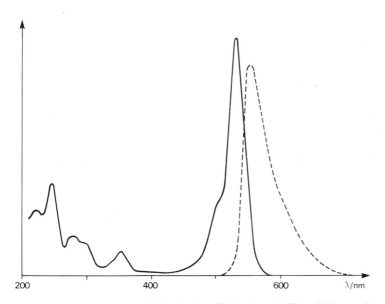

Fig. 2.14. One of the resonance structures of the Rhodamine 6G cation

Fig. 2.15. Solution spectra of Rhodamine 6G in ethanol. The solid curve shows the absorption, and the dotted curve the fluorescence at longer wavelength

has 186, (3N-6), distinct modes of vibration. In solution, the corresponding energy levels are of course broadened due to the strong molecular interactions of the liquid state, and they overlap to such an extent that an energy continuum is formed for each electronic state. This results in broad band absorption and fluorescence, as illustrated by the Rhodamine 6G spectra in Fig. 2.15.

Generally speaking, the absorption of visible light by polyatomic dyes first of all results in a transition from the ground singlet state S_0 to the energy continuum belonging to the first excited singlet state S_1. The singlet designation arises from the fact that a state with no unpaired electron spin is non-degenerate (i. e. $2S + 1 = 1$ if $S = 0$). This is immediately followed by a rapid radiationless decay to the lowest energy level within the S_1 continuum, as illustrated in Fig. 2.16; in the case of Rhodamine 6G, this process is known to be complete within 20 ps of the initial excitation. Fluorescent emission then results in a downward transition to levels within the S_0 continuum, followed by further radiationless decay. It is the fluorescent emission process which can be made the basis of laser action, provided a population inversion is set up between the upper and lower levels involved in the transition; we are therefore essentially dealing with a four-level laser (compare Fig. 1.8). Clearly, since the emitted photons have less energy, the fluorescence must always occur at longer wavelengths than the initial excitation.

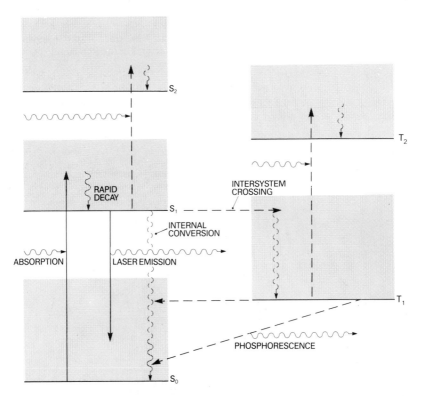

Fig. 2.16. Jablonski diagram for a laser dye. The solid vertical arrows show the transitions involved in laser action, and the dotted arrows illustrate some of the competing transitions

However, as the dotted arrows on the diagram show, there are usually several other processes going on which complicate the picture. One of the most important of these competing transitions is known as *intersystem crossing*, since the molecule changes over from one of the system of singlet states to a triplet state, in which two electron spins are parallel ($2S + 1 = 3$ if $S = 1$). Such singlet-to-triplet transitions which are formally forbidden nonetheless take place at a small but significant rate. The T_1 state is to a small extent depopulated by the comparatively slow process of phosphorescence, which is also spin-forbidden and returns the molecule to its ground electronic state S_0. It is also depopulated through non-radiative intersystem crossing to S_0, and through further absorption of radiation which populates higher triplet states. In addition to all these processes, a molecule in the singlet state S_1 can also undergo radiationless internal conversion to the S_0 state, or absorb further radiation and undergo a transition to a higher singlet state. Together, these processes contribute to a depopulation of the upper S_1 laser level, an increase in the population of the lower S_0^* laser levels, and a reduction in the output intensity, all of which contribute to a decrease of laser efficiency.

Quenching of the excited states through interaction with other molecules also occurs, however, and this effect is particularly important if the dye solution contains dissolved oxygen. In some cases triplet quenchers such as dimethylsulphoxide (DMSO) are specially added to the dye solution to increase output power by repopulating singlet states. The photochemical and thermal stability of the dyes used in dye lasers is another clearly important factor. The heat produced by the radiationless decay transitions can very rapidly degrade a dye, and for this reason it is common practice to continuously circulate the dye solution, to allow cooling to take place. A common set-up is shown in Fig. 2.17. The pump radiation from a flashlamp or a primary laser source with emission in the visible or near-ultraviolet range is focussed at a point traversed by a jet of dye solution, which typically has a concentration in the range 10^{-2}–10^{-4} mol 1^{-1}. The dye solvent is usually based on ethylene glycol, which provides the viscosity needed to maintain an optically flat jet stream.

The fluorescent emission from the dye jet is stimulated by placing two parallel cavity-end mirrors on either side of the jet. However, it is at this stage that the unique properties of the dye laser become evident. Since the fluorescence occurs over a range of

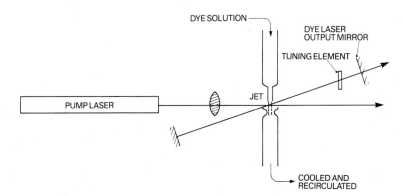

Fig. 2.17. Schematic diagram of a laser-pumped dye laser

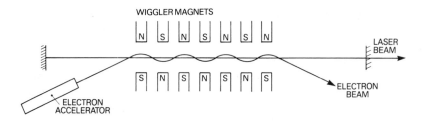

Fig. 2.18. Configuration of a free-electron laser. The electron accelerator is in fact a massive piece of equipment on a far different scale to the ancillary equipment associated with most lasers

wavelengths, monochromatic laser emission can only be obtained by the use of additional dispersive optics such as a diffraction grating or etalon within the cavity. However, rotation of this element obviously changes the wavelength amplified within the dye laser, so that *tunable emission* is obtained. A dye laser based on a solution of Rhodamine 6G in methanol, for example, is continuously tunable over the range 570-660 nm. The full range of commercially available dye lasers provides complete coverage of the range 200 nm – 1 μm. A thoroughly comprehensive listing of laser dyes and tuning ranges can be found in the book by Maeda[1].

The efficiency of a dye laser is often around 5 %, and the output power depends principally on the source of the pumping radiation. For CW output, the usual pump is an inert gas ion laser; other commonly used laser pumps such as the nitrogen, exciplex or solid-state transition-metal ion lasers, or else xenon flashlamps, generally result in pulsed output. Use of a dye laser as a means of laser frequency conversion is discussed further in Sect. 3.2. The highest powers are obtained by pumping with harmonics of a Nd laser. CW dye lasers produce emission with linewidth in the range 20-40 GHz (around 1 cm^{-1}), although with suitable optics this can be reduced to about 1 GHz. The combination of narrow linewidth, good frequency stability and tunability is particularly attractive for many spectroscopic applications. One disadvantage of the dye laser is that it tends to have rather poorer amplitude stability than a gas laser, so that indirect spectroscopic methods such as fluorescence or the photoacoustic effect are often most appropriate.

One interesting variation on the dye laser concept is the *ring dye laser*, in which the laser radiation travels around a cyclic route between a series of mirrors, rather than simply back and forth between two. When both clockwise and anticlockwise travelling waves are present, their frequencies are normally identical. However, any rotation of the laser itself results in a small difference between these two frequencies, and detection of this difference can be used as the basis for very accurate measurement of the rotation; this is the principle behind the ring laser gyroscope. Alternatively, an optical element which behaves somewhat like an optical counterpart to an electronic diode can be placed within the cavity to select one particular direction of propagation (clockwise or anticlockwise). In this case the ring laser has an emission linewidth typically at least a factor of ten smaller than a conventional dye laser, but in optimum actively stabilised cases the linewidth may be as small as 4×10^{-6} cm^{-1}.

[1] M. Maeda, 'Laser Dyes' (Academic, New York, 1984).

2.6 Free-electron Laser

The last type of laser we shall consider is one which is radically different from any of those mentioned so far. This is a laser in which the active medium consists purely and simply of a beam of free electrons, and the optical transitions on which laser action is based result from the acceleration and deceleration of these electrons in a magnetic field. One of the most common experimental arrangements is illustrated in Fig. 2.18, and involves passing a beam of very high-energy electrons from an accelerator between the poles of a series of regularly spaced magnets of alternating polarity. The electrons typically need to have energies in the 10^7–10^8 MeV range, with the magnet spacing a few centimetres. The result is that the electrons are repeatedly accelerated and decelerated in a direction perpendicular to their direction of travel, resulting in an oscillatory path, as shown. For this reason, the magnets are generally referred to as *wiggler magnets*. The effect of this process is to produce emission of *bremsstrahlung* radiation along the axis of the laser, which is then trapped between parallel mirrors and stimulates further emission in the usual way.

The wavelength of light emitted by the free-electron laser is determined by the energy of the electrons and the spacing of the magnets. High-energy electrons travel at an appreciable fraction of the speed of light, and can be characterised by a parameter f, denoting the ratio of their relativistic total energy to their rest-mass energy. If we denote the magnet separation by d, then the laser emission wavelength is given by the simple formula;

$$\lambda = d/2f^2. \tag{2.18}$$

Since the energy of the electrons emerging from the accelerator can be continuously varied, the result is once again a laser with a tuning capability.

In contrast to the dye laser, however, there is in principle no limit to the range of tuning right across the infrared, visible and ultraviolet regions of the electromagnetic spectrum. Moreover, this kind of laser has been shown to produce high average powers and to be capable of a reasonable efficiency. For example a setup in the Naval Research Laboratory in the U.S.A. can produce 75 MW pulses of 4 mm radiation with an efficiency of about 6%. The efficiency in the visible and ultraviolet is, nonetheless, generally much lower, and a great deal of research effort is currently being directed towards this problem. For many other reasons, not least the high power requirement and large bulk of a suitable electron accelerator, the free-electron laser is not in commercial production, and is likely to remain in the province of highly specialised research equipment for the time being. Nevertheless, for applications which require tunable radiation at very high power levels, the free-electron laser will prove hard to beat.

3 Laser Instrumentation in Chemistry

In the first two Chapters of this book, we have considered the chemical and physical principles underlying the operation of various types of laser, and the characteristic properties of the light which they emit. In the remainder of the book, we shall be concerned with chemical applications of lasers, paying particular attention to the ways in which each application has been developed to take the fullest advantage of the unique properties of laser light.

Laser instrumentation is now widely used to probe systems of chemical interest. These systems may be chemically stable, or they may be in the process of chemical reaction, but in either case the laser can be utilised as a powerful analytical device. There are numerous and widely diverse ways in which the laser is now used in chemical laboratories, and this Chapter provides an introduction to some of the general instrumental principles concerned. By far the most widely used methods are spectroscopic in nature, and a more detailed discussion of the enormously varied chemical applications in this area is reserved for the following Chapter. The quite distinct area of chemistry in which the laser is used to *promote* chemical reaction is dealt with in Chap. 5. Other chemical applications which fall outside the province of Chaps. 4 and 5 will be examined during the course of this Chapter.

We begin with a discussion of some of the ways in which laser output can be modified. We have seen that most of the important emission characteristics of any particular type of laser are determined by the energy levels and kinetics associated with the cycle of laser transitions in the active medium; the physical properties of the laser cavity and the nature of the pump are also involved, as discussed earlier. However with any laser there are certain applications for which the optimum specifications are somewhat different, and it is therefore desirable to tailor the laser output in certain ways. Three particularly important and widely used modifications involve polarisation modification, frequency conversion, and pulsing.

3.1 Polarising Optics

Every photon has an associated electric and magnetic field which oscillate perpendicularly to its direction of propagation. The polarisation of a monochromatic beam of photons is one measure of the extent of correlation between these oscillations. Some important types of beam polarisation are illustrated in Fig. 3.1. If the resultant of the photon oscillation lies in a single plane, as in (a), then we have *plane polarisation*, also known as linear polarisation. If the resultant describes a helix about the direction of propagation, as in (b) and (c), then we have left- or right-handed *circular polarisation*. Finally, there is the case of *elliptical polarisation*, illustrated by (d), which is essentially

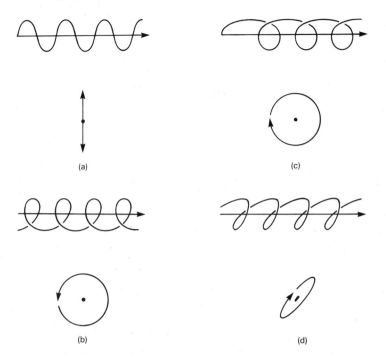

Fig. 3.1 a–d. Representations of (**a**) plane polarised, (**b**) left-circularly polarised, (**c**) right-circularly polarised, and (**d**) elliptically polarised light. The lower diagrams show the apparent motion of the electric field vector as the beam travels towards the observer

an intermediate between (a) and (c). If there is no correlation between the photons, then the state of the beam is regarded as *unpolarised*, and may be represented by two orthogonal plane polarisations with no phase relationship between them.

Laser operation does not by itself automatically produce polarised light. The simple diagram of Fig. 1.4 shows a cavity mode which is plane polarised in the plane of the diagram; however, plane polarisation perpendicular to the plane of the diagram is equally possible. In the absence of any discriminatory optics both will be present in the cavity, and will lead to an unpolarised laser output. For a number of laser applications, however, it is desirable to obtain polarised emission, and it is usual to incorporate optics into the laser cavity to accomplish this. The standard method is to use a *Brewster angle window*. This is a parallel-sided piece of glass set at the Brewster angle, defined as

$$\theta_B = \tan^{-1} n, \tag{3.1}$$

where n is the refractive index of the glass for the appropriate wavelength. The Brewster angle is the angle at which reflected light is completely plane polarised. When such elements are used for the end-windows of a gas laser tube, for example, then only one selected polarisation experiences gain and is amplified within the cavity. Polarised

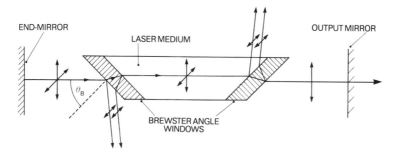

END-MIRROR

LASER MEDIUM

OUTPUT MIRROR

θ_B

BREWSTER ANGLE
WINDOWS

Fig. 3.2. Schematic diagram illustrating the use of Brewster angle windows in a gas laser. Each window reflection results in loss of the same polarisation component from the cavity. Radiation which is unpolarised as it leaves the left-hand end-mirror thus becomes nearly plane polarised before it reaches the output mirror. The windows thus enable only one polarisation component to be amplified as light passes back and forth between the end-mirrors

emission thus ensues as illustrated in Fig. 3.2. An alternative method of achieving the same result is to make use of a *birefringent* crystal such as calcite ($CaCO_3$), in which orthogonal polarisations propagate in slightly different directions and can thus be separated.

Calcite is a good example of an optically anisotropic material, i. e. one whose optical properties such as refractive index are not the same in all directions. Crystals of this type have a number of other important uses in controlling laser polarisation. Mostly, these stem from the fact that in such birefringent media different polarisations propagate at slightly different velocities. By passing *polarised* light in a suitable direction through such a crystal, the birefringence effects a change in the polarisation state, the extent of which depends on the distance the beam travels through the crystal. The changing state of polarisation of an initially plane polarised beam as it traverses an optically anisotropic crystal is shown at the top of Fig. 3.3. As shown beneath, with a slice of crystal of the correct thickness d, plane polarised light thus emerges circularly polarised (a), or *vice versa* (b), and with a crystal of twice the thickness, plane polarised light emerges with a rotated plane of polarisation (c), or circularly polarised light reverses its handedness (d). The two optical elements based on these principles are known as a *quarter-wave plate* and a *half-wave plate* respectively.

One other type of polarising optic which is particularly useful in many laser applications is the *Pockels cell*. This is a cell consisting of a crystalline material such as potassium dihydrogen phosphate (KDP) which exhibits the Pockels effect, essentially a proportional change in refractive index on application of an electric field. (There is also in general a change in refractive index proportional to the *square* of the electric field; this is known as the *Kerr effect*.) A Pockels cell can once again be used to rotate the plane of polarisation of light passing through it, or to reverse the handedness of circularly polarised light. However, its virtue is that because the Pockels effect only takes place whilst the electric field is applied, modulation of the field at a suitable frequency results in a corresponding modulation of the beam polarisation. This is a particularly useful way of increasing detection sensitivity in many techniques based on optical activity, such as polarimetry (Sect. 3.7) and circular differential Raman scattering (Sect. 4.5.6).

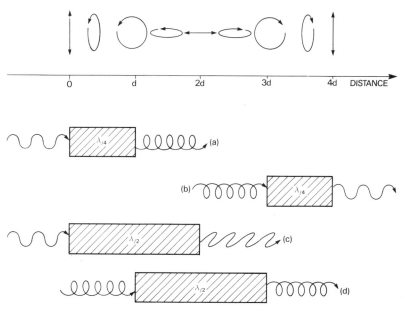

Fig. 3.3 a-d. Illustration of the changing state of polarisation of a polarised beam of light propagating through an optically anisotropic medium; (**a**) and (**b**) represent the use of a quarter-wave plate; (**c**) and (**d**) the use of a half-wave plate; see text for details

3.2 Frequency Conversion

We now turn to a consideration of another more important way in which laser output can be modified. Since for a given laser the output frequencies are determined by the nature of the lasing material, it is often helpful to be able to convert the output to a different frequency more suited to a particular application. The two most widely adopted methods of frequency conversion involve dye lasers and frequency-doubling crystals.

The principle behind the operation of the dye laser has already been discussed in some detail in Sect. 2.5. Its use as a frequency-conversion device follows a similar line, using a primary laser source (often an inert gas or nitrogen laser) rather than the broadband emission from a flashlamp as the pump. Thus the excitation is often essentially monochromatic, and the excitation and emission curves are as shown in Fig. 3.4. The most important feature of this method of frequency conversion is that the dye laser output has to occur at a lower frequency, and hence a *longer wavelength*, than the pump laser. Nevertheless, the exact range of output wavelengths depends on the choice of dye, and the facility for tuning within this range provides a very useful method of obtaining laser emission at a chosen wavelength. For example by using various lines from a krypton/argon ion laser as pump, coherent radiation can be obtained right across and well beyond the visible spectrum, as shown in Fig. 3.5. Conversion efficiencies are generally somewhat low; 5-10 % is not unusual, although with certain dyes such as Rhodamine 6G, figures of up to 20 % are possible.

The second commonly used method of frequency conversion is known as *frequency*

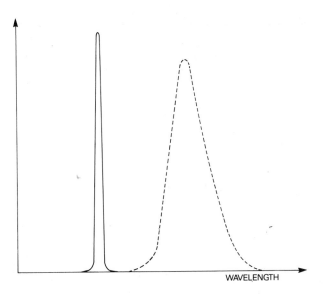

Fig. 3.4. Incident radiation (solid curve) and emission (dotted curve) in a dye laser pumped by a single line from an ion laser

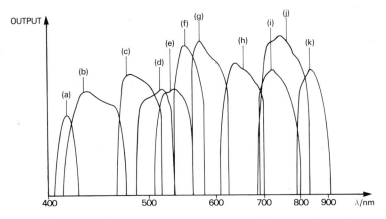

Fig. 3.5. Tuning curves for a dye laser pumped by various lines from a krypton/argon laser. The dye and pump wavelength for each curve are given in Table 3.1. (Reproduced by kind permission of Coherent Radiation Ltd.)

doubling, or to use the physicists' term, *second harmonic generation*. The energetics of this process are illustrated in Fig. 3.6. Two photons of laser light with frequency ν are absorbed by a substance in its ground state, and a single photon of frequency ν′ is emitted through a transition back to the ground state. Note that the entire process is a concerted one, and there is no intermediate excited state with any measurable lifetime; hence the energy-time uncertainty principle allows the process to take place regardless of whether there are energy levels of the substance at energies hν or 2hν above the

Table 3.1   Dyes and Pump Wavelengths for the Tuning Curves of Figure 3.5

Curve	Dye	Pump laser; wavelength λ/nm
(a)	Stilbene 1	Argon; 333.6, 351.1, 363.8
(b)	Stilbene 3	Argon; 333.6, 351.1, 363.8
(c)	Coumarin 102	Krypton; 406.7, 413.1, 415.4
(d)	Coumarin 30	Krypton; 406.7, 413.1, 415.4
(e)	Coumarin 6[a]	Argon; 488.0
(f)	Rhodamine 110	Argon; 514.5
(g)	Rhodamine 6G	Argon; 514.5
(h)	DCM[b]	Argon; 488.0
(i)	Pyridine 2	Argon; 488.0, 496.5, 501.7, 514.5
(j)	Rhodamine 700	Krypton; 647.1, 657.0, 676.5
(k)	Styryl 9M	Argon; 514.5

[a] With cyclo-octatetraene and 9-methylanthracene;
[b] 4-dicyanomethylene 2-methyl 6-(p-dimethylaminostyrol) 4H-pyran

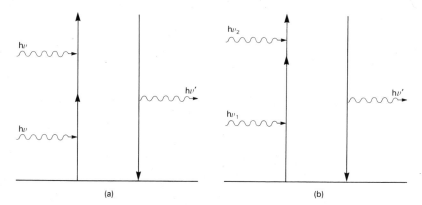

Fig. 3.6 a and b. Energetics of (a) frequency doubling, where the emitted photon has frequency $\nu' = 2\nu$, and (b) frequency mixing, where the emitted photon has frequency $\nu' = \nu_1 + \nu_2$

ground state. Indeed, it is better if there are not, since the presence of such levels can lead to competing absorption processes.

Symmetry considerations reveal that the frequency-doubling process can take place only in media which lack a centre of symmetry. Hence gases and liquids are automatically ruled out, and it is necessary to use a non-centrosymmetric crystal. It is important to note that the process is a *coherent* one, so that with laser input the emission also has typically laser-like characteristics. In order that the frequency-doubled output does not interfere destructively with the input, the orientation of the crystal is often crucial; also, the temperature-dependence of the refractive index means that certain crystals have to be enclosed in a temperature-controlled heating cell. Either the crystal orientation or its temperature can then be varied for any particular incident wavelength to obtain maximum efficiency. Under optimum conditions, conversion efficiencies can often be as high as 20–30 %. Two of the most widely used crystals are KDP (KH_2PO_4) and lithium niobate (Li$_3$NbO$_4$). Frequency doubling is particularly useful for generating powerful visible (532 nm) radiation using a Nd laser pump.

The high conversion efficiency of the frequency-doubling process makes it possible to use a series of crystals to produce 4×, 8×, 16× ..., the pump frequency, and so to obtain coherent radiation at very short wavelengths. More commonly, one doubling crystal is coupled with a dye laser so as to obtain tunable emission over a range of wavelengths below the operating wavelength of the pump. Another technique which has recently passed from the research stage to incorporation in a commercial device is *third harmonic generation*, which can take place in a variety of either centrosymmetric or non-centrosymmetric crystals, and is now used to generate 355 nm radiation from Nd: YAG pump lasers.

There are other means of frequency conversion also worth mentioning. One is *fre-*

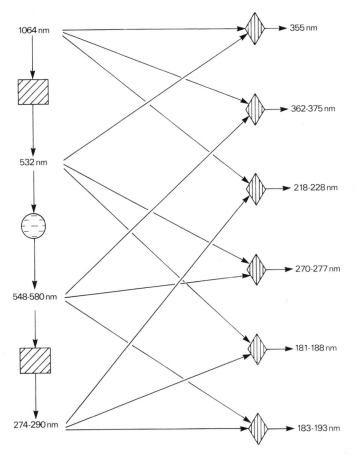

Fig. 3.7. Schematic illustration of *some* of the frequencies which can be obtained by conversion of the 1.064 µm output from a Nd:YAG laser. The rectangles represent frequency doubling, the circle represents dye laser conversion (the range of wavelengths given corresponds to use of Rhodamine 6G solution), and the diamonds represent frequency mixing. Raman shifting, not included here, can also be employed (see text)

quency mixing, a variation of the frequency-doubling process in which two beams of laser radiation with different frequencies are coupled in a suitable crystal (again, one which is of non-centrosymmetric structure) to produce output at the sum frequency, as illustrated in Fig. 3.6 b. Since wavelength is inversely proportional to frequency, the emitted wavelength is given by the formula

$$1/\lambda' = 1/\lambda_1 + 1/\lambda_2, \tag{3.2}$$

and hence is always shorter than either of the input wavelengths. Together with dye lasers and frequency-doubling, this method of frequency conversion enables a single laser to be employed for the production of coherent radiation at a variety of wavelengths, as shown in Fig. 3.7.

One other alternative for frequency conversion is *Raman shifting*, in which the stimulated Raman effect is employed for conversion to either shorter or longer wavelength. The principle of the Raman effect, which is more widely used in a spectroscopic context, will be discussed in detail in Chap. 4. For the present, we simply note that it enables laser frequency to be modified by discrete increments (Stokes and anti-Stokes shifts). For the purpose of frequency conversion, the process is usually accomplished by passage of the laser light through a stainless steel cell containing gas at a pressure of several atmospheres. Conversion efficiency for the principal Stokes shift to longer wavelength can be as high as 35%. The nature of the gas determines the frequency increment, and the most commonly used gases H_2, D_2 and CH_4, produce shifts of 4155, 2987 and 2917 cm^{-1} respectively. Atomic vapours can also be useful for this purpose; for example 455 nm radiation in the blue-green region best suited to submarine-satellite optical communication can be obtained from the 308 nm output of a xenon chloride exciplex laser by Raman conversion in lead vapour.

3.3 Pulsing Techniques

In Chap. 2, we encountered several lasers which can naturally operate on a continuous-wave (CW) basis, whilst others are inherently pulsed, depending on kinetics considerations discussed in Sect. 1.4.3. Nonetheless, there are several reasons why it is common practice to make use of a pulsing device, either to convert a CW laser to pulsed output, or else to shorten the duration of the pulses emitted by a naturally pulsed laser. One motive is to obtain high peak intensities, since if in any period the emission energy can be stored up and then emitted over a much shorter period of time, the result is an increase in the instantaneous irradiance. Another motive is to produce pulses of very short duration in order to make measurements of processes which occur on the same kind of timescale. There are three widely used pulsing methods, which we now consider in detail.

3.3.1 Cavity Dumping

As the name implies, cavity dumping refers to a method of rapidly emptying the laser cavity of the energy stored within it. The simplest method of accomplishing this with a continuous-wave laser would be to have both of the end-mirrors fully reflective, but

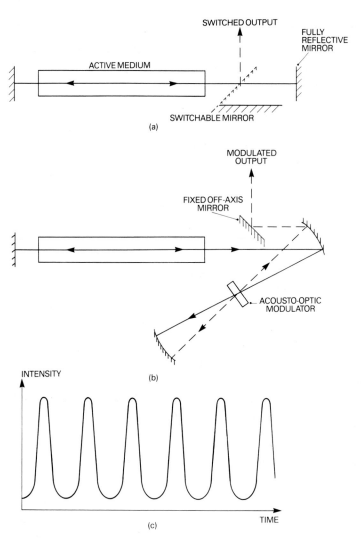

Fig. 3.8 a–c. Cavity dumping, (**a**) to produce a single pulse of output, and (**b**) for a radio-frequency modulated output with temporal profile as in (**c**)

with a third coupling mirror able to be switched into the beam to reflect light out of the cavity, as shown in Fig. 3.8 a. This configuration would deliver a single pulse of light, terminating laser action until the coupling mirror was again switched out of the beam. In practice, an acousto-optic modulator is usually employed, as illustrated in Fig. 3.8b. This double-pass device is driven by a radio-frequency electric field, and generates an acoustic wave which produces off-axis diffraction of the laser beam. The cavity-dumped output of the laser in this case consists of an essentially sinusoidal temporal profile as shown in Fig. 3.8 c; the frequency of oscillation is twice the acoustic frequency of the cavity dumper, typically of the order of MHz. This method of pulsing is often used in conjunction with mode-locking, as we shall see in Sect. 3.3.3.

3.3.2 Q-Switching

In physical terms, the reasoning behind this method of achieving pulsed operation is as follows. If a shutter or some other device is placed within the laser cavity so as to increase the loss per round-trip of the radiation, a sizeable population inversion can be established in the active medium without any appreciable amount of stimulated emission taking place. If the shutter is then opened up so that the cavity can properly act as a resonator, the energy stored by the medium can be released in a single pulse of highly intense light. The term *Q-switching* refers to the fact that such a method essentially involves first reducing, and then swiftly increasing the quality, or Q-factor of the laser.

In Eq. (1.30), Q was expressed in terms of the resonant frequency and linewidth of the emission. However, that equation is formally derived from another defining equation which is as follows;

$$Q = 2\pi \times \text{ energy stored in cavity/energy loss per optical cycle} \qquad (3.3)$$

in which the optical cycle length is simply the inverse of the resonant frequency. Q-switching thus represents the effect of suddenly reducing the rate of energy loss within the laser cavity. In practice, the pumping rate has to exceed the rate of spontaneous decay in order to build up the required population inversion, and the time taken for Q-switching has to be sufficiently short to produce a single pulse.

There are three commonly used methods of Q-switching a laser. The first and simplest method involves replacing one of the cavity end-mirrors with one which revolves at a high speed (typically 500 revolutions per second), as illustrated in Fig. 3.9 a. In this situation, cavity losses are kept high except for the brief periods when this mirror is virtually parallel with the static output mirror. The pulse repetition rate is determined by constraints on the rate of pumping, however.

In the second method, there is a genuine shutter action within the cavity, but one which is usually based on electro-optical rather than mechanical principles. One of the devices commonly used is the Pockels cell described in Sect. 3.1. In this case the cell is designed such that when a suitable potential difference is applied along it, plane polarised light traversing the cell and reflecting back through it suffers an overall 90° rotation of the polarisation plane. Hence with both a polariser and a Pockels cell within the laser cavity, as shown in Fig. 3.9 b, there is an effective shutter action when the voltage is applied, since the polariser cuts out all the light with rotated plane of polarisation. By switching the voltage off, the cavity again becomes effectively transparent, and Q-switched emission can take place. With this method, the switching time is synchronised with the pumping, and is typically on the nanosecond scale. There are several variations on this theme based on other electro-optic, magneto-optic and acousto-optic principles.

The third method of Q-switching involves use a solution of a *saturable absorber* dye. This is a compound, which must display strong absorption at the laser emission wavelength, for which the rate of absorption nonetheless *decreases* with intensity; an example used for the ruby laser is cryptocyanine. In a certain sense, then, such a dye undergoes a reversible bleaching process due to depopulation of the ground state. Thus if a cell containing a saturable absorber solution is placed within the laser cavity as shown in Fig. 3.9c, then in the initial stages of pumping the strong absorption effectively ren-

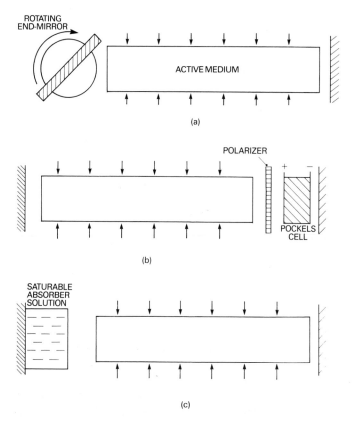

ROTATING
END-MIRROR

ACTIVE MEDIUM

(a)

POLARIZER

POCKELS
CELL

(b)

SATURABLE
ABSORBER
SOLUTION

(c)

Fig. 3.9a–c. Three schemes for obtaining Q-switched laser emission; (a) using a rotating end-mirror, (b) an electro-optic switch, and (c) a saturable absorber cell

ders the cell opaque (in other words it has negligible transmittance), and losses are high. However, once a large population inversion has been created in the active medium, the intensity of emission increases to the point where the dye is bleached, and stimulated emission can result in emission of a pulse of light. This last method of Q-switching is known as *passive*, since the switching time is not determined by external constraints.

Q-switching generally produces pulses which have a duration of the order of 10^{-9}–10^{-8} s. The pulse repetition rate is determined by the time taken to re-establish population inversion, and thus depends amongst other factors on the pumping rate, but the interval between pulses is typically a few seconds. In the case of a laser such as the ruby laser which is inherently pulsed, it is worth noting that Q-switching does somewhat reduce the energy content of each pulse. However, this disadvantage is more than outweighed by the enormous reduction in pulse duration, which thus results in a sizeable increase in the peak irradiance. If we ignore the energy loss, a rough calculation shows that for a laser producing 1 ms pulses of energy 10 J, in other words a mean output power per pulse of 10 kW, then Q-switched operation which results in compression of the pulse to 10 ns increases the mean pulse power to 1 GW. Note that such cal-

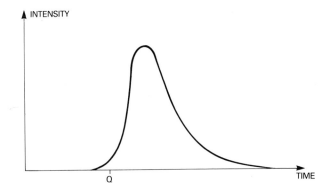

Fig. 3.10. The temporal profile of a Q-switched laser
pulse; the point Q on the
time-axis denotes the onset
of Q-switching

culations are necessarily crude because the intensity of emission does not remain constant throughout the pulse duration; the temporal profile of a typical Q-switched laser
pulse is illustrated in Fig. 3.10. Hence the *peak* intensity may be far greater than the
computed mean.

3.3.3 Mode-locking

The third widely used method of producing laser pulses is known as mode-locking.
This method produces pulses of much shorter duration than either cavity dumping or
Q-switching, typically measured on the picosecond (10^{-12} s) scale and referred to as *ultrashort*. The technique has very important applications in chemistry, for example in
the study of ultrafast reactions (see Sect. 5.2.5), and it is therefore worth examining in
somewhat more detail. The mechanism of mode-locking is as follows.

We have previously noted that laser emission usually occurs over a small range of
wavelengths which fall within the emission linewidth of the active medium (see
Fig. 1.12), and satisfy the standing wave condition

$$m\lambda/2 = L, \tag{3.4}$$

where m is an integer. Hence the emission frequencies are given by

$$\nu = mc'/2L, \tag{3.5}$$

and they are separated by the amount

$$\Delta\nu = c'/2L, \tag{3.6}$$

which is typically around 10^8 Hz. In the case of solid state or dye lasers, where the fluorescence linewidth is quite broad, the number of longitudinal modes N present within
the laser cavity may thus be as many as $10^3 - 10^4$ if frequency-selective elements are
removed.

In such a situation, however, there is not generally any correlation in phase between
the various modes, so that the effect of interference usually results in a continual, and

essentially random fluctuation in the intensity of light within the cavity. The technique of *mode-locking* consists of creating the phase relationship which results in completely constructive interference between all the modes at just one point, with destructive interference everywhere else. Consequently, a pulse of light is obtained which travels back and forth between the end-mirrors, giving a pulse of output each time it is incident upon the semi-transparent output mirror.

It can be shown that the pulse duration, expressed as the width of the pulse at half power, is given by

$$\Delta t = 4\pi L/(2N + 1)c', \tag{3.7}$$

and the interval between successive pulses is obviously the round-trip time given by Eq. (1.21), i.e $\tau = 2L/c'$. Hence mode-locked pulses are typically separated by an interval of around 10 ns, and have a duration of 1–10 ps. The temporal profile of the emission from mode-locked CW and pulsed lasers is thus as shown in Fig. 3.11, and in each

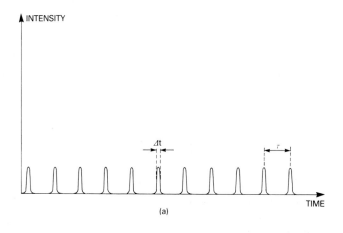

(a)

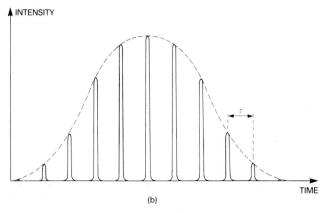

(b)

Fig. 3.11 a and b. The temporal profile of mode-locked emission from **(a)** a CW laser, and **(b)** a pulsed laser. In the second case, very many more mode-locked pulses than illustrated would normally be found within the overall pulse envelope

case consists of a pulse train. The instantaneous intensity of a mode-locked laser pulse can be readily computed on the essentially correct assumption that the overall energy emitted within the duration of one laser round-trip is the same before and after mode-locking. Thus with a pulse-length reduction from 10 ns to 1 ps, the pulse power can be increased by a factor of approximately 10^4, so increasing the power of gigawatt pulses up to the region of 10^{13} W.

As with Q-switching, there are several ways to accomplish mode-locking, involving both active and passive methods. The active methods generally involve modulating the gain of the laser cavity using an electro-optic or acousto-optic switch driven at the frequency $c'/2L$. In much the same way as Q-switching, such elements placed inside the cavity can effectively act as shutters, and for mode-locked operation become transparent only for brief intervals separated by an interval of $2L/c'$. Hence only light which is travelling back and forth in phase with the modulation of the shutter can experience amplification, which naturally results in mode-locking.

Passive mode-locking is generally accomplished using a saturable absorber cell within the laser cavity as in Fig. 3.9c. Here, one of the most widely used saturable absorbers is 3,3'-diethyloxadicarbocyanine iodide, usually known as DODCI. Figure 3.12

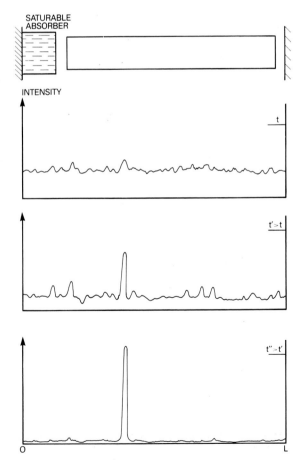

Fig. 3.12. Development in time of a single pulse in a passively mode-locked laser. The times t, t' and t" are separated by an integer number of round-trips in the cavity

illustrates how use of a saturable absorber results in mode-locked operation. At time t, the graph shows the initially random fluctuation of intensity along the length of the laser cavity, resulting from interference between different longitudinal modes. After a few round-trips back and forth, at time t, the intensity pattern is as shown in the second graph. The effect of each passage through the saturable absorber cell is to reduce the intensity of the high amplitude fluctuations to a far lesser extent than the small fluctuations. The loss of intensity by the highest amplitude fluctuations is more than compensated by the amplification they experience on passage through the gain medium; however, the low amplitude fluctuations experience a net loss. Hence in the course of each round-trip, the high intensity fluctuations essentially grow at the expense of the small fluctuations. Thus after several more round-trips, at time t″, the intensity pattern develops to the form shown in the third graph. Here, only one pulse of intensity remains, having grown from the initial fluctuation of the largest amplitude; all the other fluctuations have died away. Again, the result is mode-locked operation.

One other mode-locking technique principally employed with tunable dye lasers is *synchronous pumping*. Here a mode-locked primary laser pumps a dye laser whose optical cavity length is adjusted to equal that of the pump, so that the round-trip time of the dye laser matches the interval between pump pulses. In this way, frequency-converted pulses propagating within the dye laser cavity arrive at the the dye jet at the same rate as the pump pulses. The period over which amplification is effective in the dye laser is thus very short, and once again picosecond pulses emerge. In practice, the dye laser cavity length needs to be kept constant to within a few microns, so that special mounting and temperature control are necessary.

Several further points concerning mode-locked pulses are worth noting. First, it should be noted that although the method produces a train of pulses, it is possible to select out single pulses by using a Pockels cell and polariser combination as discussed earlier (p. 55). Here, however, both are placed outside the laser cavity, and the Pockels cell is activated by an electrical spark of shorter duration than the pulse separation τ; hence only a single pulse from the mode-locked train can pass through. It is also quite common to employ mode-locking in conjunction with radio-frequency cavity dumping, so as to obtain variable repetition-rate trains of picosecond pulses. Because cavity dumping releases large amounts of cavity energy, the power levels of cavity-dumped modelocked pulses are appreciably higher than those obtained by mode-locking alone, often by a factor of about 30.

By using other techniques in conjunction with mode-locking, commercially available lasers can now provide ultrashort pulses of sub-picosecond duration, and at the research level it has proved possible to reduce pulse duration down into the femtosecond (10^{-15} s) region. This, indeed, is one of the most rapidly developing areas of laser physics. In 1980, the shortest recorded laser pulse was of 150 fs duration; at the time of writing five years later, this has already been reduced by more than a factor of 10, to a large extent as a result of the pioneering work of Charles V. Shank and others at the AT & T Bell Laboratories. Indeed, the duration of such pulses less than 10^{-14} s long can now be meaningfully described in terms of a few optical cycles (optical frequencies being in the 10^{15} Hz range). The time-frequency uncertainty principle $\Delta v \Delta t \geq 1/2\pi$ demonstrates that a very broad span of frequencies is present in such pulses, so that further pulse compression has to be at the expense of monochromaticity.

This brings us to another key aspect of ultrashort laser pulses. In 1970[1] Alfano and Shapiro discovered that on passing picosecond laser pulses through certain media, a broad continuum of light was generated, typically covering a range of about $10,000 \ cm^{-1}$ about the laser frequency; moreover, the duration of the continuum pulse was of the order of 100 fs. The effect turned out to be surprisingly easy to produce; even focussing a mode-locked laser into a beaker of water will do the trick. The facility to produce such a broad continuum of light with such short duration has a number of distinctive uses, as for example in the flash photolysis study of ultrafast reactions (see Sect. 5.2.5). The light produced by this method, which results from a mechanism known as self-phase modulation, is referred to as an *ultrafast supercontinuum laser source,* or *picosecond continuum* for short.

Quite apart from the chemical applications of ultrashort laser pulses, there are also some very significant applications in physics, which it is worth drawing attention to before leaving the subject. One is the possibility of inducing nuclear fusion reactions by making use of the enormously high intensities available. If pellets of lithium deuteride, for example, are encapsulated in thin spherical glass shells and irradiated by pulses of say $10^{21} \ W \ m^{-2}$ intensity, instantaneous pressures of about 10^{12} atmospheres can be created which are sufficient to compress the nuclei together and induce nuclear fusion. A great deal of effort is being directed towards the development of this method as a means of producing clean, cheap energy in a controllable way. Other uses have been found in ultra-fast photography, and in the field of optical communications, where it has been estimated that by using a simple pulse code modulation (PCM) technique to switch on or off each pulse in a mode-locked train, at least a million simultaneous telephone conversations can in principle be transmitted by a single laser.

3.4 Detectors

In virtually all laser applications involving measurement of a physical or chemical property, an electrical signal is produced by the response of some photodetector to light incident upon it. There are three main types of detector used for this purpose; the exact choice for any particular application depends on the region of the electromagnetic spectrum and the power level involved.

First we consider *thermal* photodetectors. These operate by electrical detection of the variation in a physical property caused by heating. For example, electrical resistivity, gas pressure and thermoelectric emission can all be monitored for this purpose. Thermal detectors suffer limitations due to the nature of the primary mechanism whereby the energy of absorbed light is converted into thermal energy. Consequently they often have comparatively slow response times; also the wavelength range over which they operate may be restricted by the absorption properties of the material used for the window. The most common type of thermal photodetector is the *pyroelectric* detector whose response is based on measurement of thermally-induced changes in electrical properties. Devices of this kind cover the entire wavelength range 100 nm— 1 mm, and respond to pulse frequencies of up to 10^{10} Hz.

---

[1] Alfano, R. R. and Shapiro, S. L.: Phys. Rev. Letts **24**, 584, (1970)

The second class of laser detectors is known as *photoemissive* and operates on the photoelectric effect. All such detectors have a long-wavelength cut-off, in other words a wavelength beyond which the photon energies are insufficient to release electrons, and therefore induce no response. The simplest device of this kind is a *vacuum photodiode,* in which electrons are released as light impinges upon a photosensitive cathode inside a vacuum tube, producing an electrical signal as they are received by an anode plate. The precise wavelength range over which such a device is effective depends on the nature of the photocathode material, and the wavelength for maximum sensitivity ranges from 240 nm for non-stoichiometric Rb/Te to 1,500 nm for germanium. Since solid-state electronics is involved in the detection process, the response times of semiconductor detectors are very fast, and are generally limited only by the associated circuitry.

Photomultiplier tubes are based on the same principle, but incorporate electronics which multiply the number of electrons received at the anode by a factor of between 10^6 and 10^8; thus in certain designs the absorption of even a single photon generates sufficient current (up to 100 mA) to be easily detected. This sensitivity is obviously highly significant for applications to the measurement of weak fluorescence or light scattering. Photomultiplier devices are most sensitive for the uv/visible region 185–650 nm, but generally have the drawback of a comparatively slow response time on the nanosecond scale. Nonetheless, the fastest devices can respond to pulse repetition frequencies of up to 10^{11} Hz. One other factor which limits the sensitivity of photoemissive devices in general is the so-called 'dark current', the residual current due to emission in the absence of irradiation. This is a strongly temperature-dependent phenomenon, which can be ameliorated by cooling the photomultiplier housing.

The last major category of photodetectors involves *photoconductive* semiconductor sensors, where the underlying principle involved is the excitation of valence electrons into a *conduction* band by the absorption of light. Once again, such detectors have a long-wavelength limit beyond which they are unusable. Here a very wide range of ma-

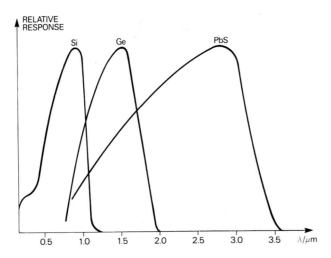

Fig. 3.13. Relative spectral responses of three semiconductor detector materials normalised to the same maximum. (Reproduced by kind permission of Optronic Laboratories Inc.)

terials is employed, many of which are responsive over infra-red regions only. However trialkali antimonides such as Cs_3Sb and Na_2KSb provide a good coverage of much of the visible spectrum, and some silicon devices are sensitive down to 180 nm. Most sensors have a non-stoichiometric composition tailored to provide response over a particular range. It is often necessary, when detecting infra-red radiation, to cool such detectors in order to eliminate the thermal excitation which would otherwise take place even in the absence of radiation. Occasionally, dry ice is sufficient, but frequently liquid nitrogen or even liquid helium may be required. Some photoconductive sensors can respond to pulse repetition rates of up to 10^{10} Hz, but somewhat lower frequency limits are more common; once again, response times are primarily determined by the signal processing circuitry. The spectral response curves for three common semiconductor detectors are shown in Fig. 3.13. Many detectors are now available linked to amplification and signal-processing electronics, and are known as power or energy meters.

3.5 Pulse Detection Systems

In chemistry, laser instrumentation is usually based on detection of the response of a material to laser irradiation. The response is monitored with any suitable detector such as a photodiode or photomultiplier tube which produces an electronic signal, and the signal is then amplified and fed to a display device. For example in many spectroscopic applications a pen recorder is used, so that the response can be plotted against wavelength. It is outside the scope of this book to give fuller details of the electronics generally involved in detection instrumentation. However, there are certain additional features which have to be employed where pulsed irradiation is concerned, and it is worthwhile outlining these before proceeding to a discussion of specific applications.

3.5.1 Lock-in Amplifiers

One of the methods of increasing sensitivity in certain applications of *continuous-wave* lasers is to use a beam chopper and a lock-in amplifier (also known as a *phase-sensitive detector*) as shown in Fig. 3.14. Rotation of the chopper imposes a square-wave modulation in the intensity of light reaching the sample; the frequency of modulation is generally in the kilohertz region, but can lie anywhere between 5 Hz and 20 kHz, depending on the nature of the application. The lock-in amplifier is fed two signals; one is a con-

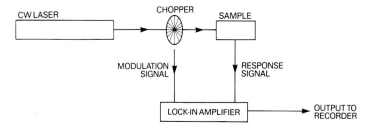

Fig. 3.14. Improvement of signal sensitivity by use of a chopper and lock-in amplifier

trol signal from a detector connected to the chopper, and the other is a response signal from the sample monitoring device. The job of the lock-in amplifier is then to selectively amplify only components of the response signal which oscillate at the same frequency as the control signal.

The result is a very effective discrimination against noise, and hence a large increase in the signal-to-noise ratio. Use of a lock-in amplifier with a dynamic range of 10^3 enables weak signals to be detected even when the level of noise is one thousand times greater. In fact, by electronically pre-filtering the input to discriminate against noise frequencies away from the frequency of interest, the dynamic range can be improved to as much as 10^7. Hence uses arise, for example, in the detection of very weak laser-induced fluorescence.

The lock-in method is also commonly used when an inherently weak *effect* is monitored, as for example in optogalvanic spectroscopy (Sect. 4.2.5). One other more general kind of spectroscopic application lies in the *ratiometric recording* of absorption spectra. Here, the laser beam is split into two paths, one of which is modulated with a chopper and passes through the sample, and the other of which is separately modulated and used as a reference. Both beams are then detected, and produce signals which can be ratioed in conjunction with a lock-in amplifier. In this way the sample spectrum is automatically corrected for the variation of source output and detector efficiency with wavelength.

3.5.2 Boxcar Integrators

A boxcar integrator is a device widely used in conjunction with a *pulsed* laser producing a regular train of pulses, as for example in the case of a mode-locked source. The

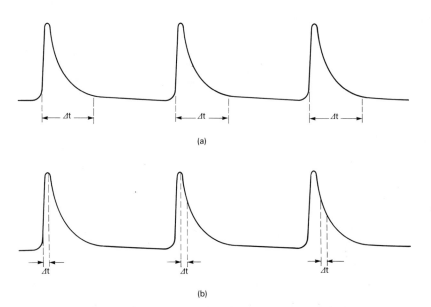

Fig. 3.15a and b. Use of a boxcar integrator: (a) to average the response from a series of laser pulses, and (b) to sample the response at various points during the decay time of a single pulse

response signal from the sample is then also a periodic function; a typical response is represented in Fig. 3.15. There are two different modes of operation of the boxcar. First, as shown in (a), the device can be used as an electronic gate which responds only to signals received during a period Δt comparable to the duration of the pulse response. The signal received during this period is integrated and averaged over many pulses to produce the boxcar output. This method is the simplest to use, and is appropriate when the time-dependence of the response is not of interest. It results in a better signal-to-noise ratio than direct signal integration since the noise usually present during the intervals between pulses is completely eliminated.

The second boxcar method, represented by Fig. 3.15b, involves sampling the response signal at a rate which is slightly different from the pulse repetition rate, with integration over comparatively short time intervals Δt. Thus different portions of the response curve are sampled from each successive pulse, and after a sufficiently large number of pulses the boxcar produces an averaged output which registers the signal level at various points through a typical pulse response. This approach is particularly useful in the determination of fluorescence decay lifetimes.

3.5.3 Single-pulse Systems

In experiments involving the study of laser-induced physical or chemical change in a sample, it is often necessary to monitor the effect produced by one single laser pulse. One method is to use a *transient recorder,* an electronic device which retains a temporal representation of the response signal. This is accomplished by digitally storing in a memory the signal level at a large number of closely spaced but equidistant time intervals. Time resolutions down to 100 ps are obtainable with this type of device. However for the measurement of ultrafast processes such as the primary events in photosynthesis, resolution needs to be much better than this. The device which is capable of the best time resolution is known as the *streak camera.*

Early streak cameras operated by focussing light onto a photographic film shot at very high speeds through a camera. The lateral displacement along the film thus represented a measure of the time elapsed, and by using a microdensitometer on the recorded image the intensity could be plotted as a function of time. As the duration of the shortest laser pulses has been reduced further and further, new methods have been required for their measurement. Today, the name 'streak camera' is usually applied to an electronic instrument operating on the same basic principle, but capable of resolutions down into the subpicosecond range. As illustrated in Fig. 3.16, the device is a

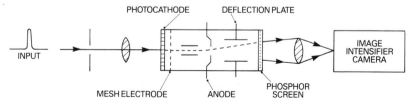

Fig. 3.16. The streak camera. Electrons released as the laser pulse strikes the photocathode are swept out electrically into an elongated image on the screen, whose length depends on the pulse duration

modified cathode-ray tube, which by application of a sweep voltage to the deflection plates allows pulse duration and intensity to be calculated from the spatial extent of the image and its density on the screen.

Whilst streak camera methods are appropriate for the study of ultrashort laser pulses *per se*, or in some cases for laser measurements of fluorescence kinetics, other techniques can be adopted for the study of ultrafast processes directly connected with the *absorption* of ultrashort pulses. A common setup is illustrated in Fig. 3.17, and involves use of a beam-splitter to create two pulses from one original ultrashort pulse; one of these is used to excite the sample, and the other is used as a probe of the excitation. A variable delay is introduced between the two separated pulses using a stepping motor, allowing the difference in optical path length to be changed in small increments. Since 1 ps corresponds to distance of 0.3 mm, and 1 ns to 30 cm, the distances involved are in just the right range for very precise and yet very simple experimental control. The absorption from the probe pulse can thus be monitored at various ultrashort time intervals, so providing information on the kinetics of the processes induced by the initial excitation. The delayed pulse can also be used to generate an ultrafast supercontinuum, as discussed earlier (Sect. 3.3.3), making it possible to probe the response of the sample over a wide range of wavelengths.

Having discussed some of the instrumentation for modification and detection of laser radiation, we now move on to consider the more specific instrumental considerations for particular applications. The techniques discussed in the following Sections of this Chapter are not explicitly spectroscopic in the sense that a spectrum is necessarily recorded, although spectroscopic principles may be involved. Genuinely spectroscopic applications are discussed in Chap. 4.

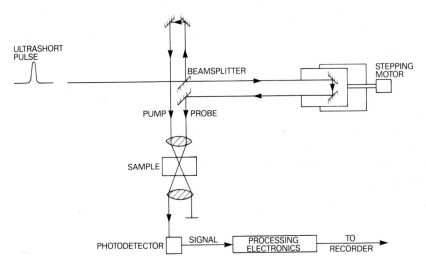

Fig. 3.17. Instrumentation for kinetics measurements over picosecond and sub-picosecond time intervals

3.6 Light Scattering Instrumentation

Many chemical applications of lasers involve irradiation of a sample and detection of its absorption or subsequent fluorescence. Nonetheless, a great deal can be learned from analysis of the light which a sample simply scatters. Most of the light which is scattered has essentially the same frequency as the laser light itself; this is known as *Rayleigh scattering,* and is itself a weak effect, seldom accounting for more than one photon in 10^4. The very small frequency shifts which are associated with Rayleigh scattering result from *Doppler shifting* due to the motion of sample molecules.

A much smaller amount of light is nevertheless inelastically scattered with somewhat larger shifts in frequency, and the monochromaticity of the laser radiation facilitates its detection. Inelastic light scattering comes under two main headings; *Brillouin scattering,* in which energy is transferred to or from acoustic vibrations in bulk materials, and *Raman scattering,* in which it is generally transitions in individual molecules which are responsible for the loss or gain of photon energy. In this Section we examine some of the analytical applications of these light scattering processes; we return to a much more detailed look at the principles of Raman *spectroscopy* in the next Chapter.

3.6.1 Nephelometry

The elastic scattering of light without change in frequency occurs both from free molecules, and from suspensions of larger aggregate particles. The term 'Rayleigh scattering' is more commonly applied to the former case, or to particles of diameter not exceeding roughly a tenth of the wavelength of light involved. Such scattering, whilst not isotropic, nonetheless occurs over all directions, and the angular distribution of scattering intensity provides a crude 'fingerprint' of the medium. This method has, for example, been suggested as a simple but objective means of characterising wines. The technique does not, however, provide much useful chemical information that cannot be more easily obtained by other means.

The more intense scattering of light observed from larger particles has different characteristics, and is especially effective close to the forward-scattering direction. When the size of each particle is sufficiently large that it is no longer transparent, the nature of the effect changes once more as surface reflection takes over. *Nephelometry,* which actually means the measurement of cloudiness, is a quantitative measure of the scattering of light from aggregate particles such as those in colloidal suspension. The instrumentation simply involves placement of a photodetector beside the sample, but out of the direct pathway of the laser radiation. The technique can be especially useful in conjunction with liquid chromatography, as discussed in Sect. 3.8.

3.6.2 Photon Correlation Measurements

Whilst measurement of the gross intensity of Rayleigh scattering provides a certain amount of data on a sample, much more detailed information on solutions and liquid suspensions can be obtained by monitoring the intensity *fluctuations* in the scattering from a continuous-wave laser beam. Assuming that the laser has good amplitude stability, such fluctuations primarily result from the essentially random interference of light scattered by different molecules or molecular aggregates at the focal point in the sample. Typically the scattered light is detected by a photomultiplier, and the signal is processed in a *correlator,* which determines the timescale over which fluctuations take place. The results can ultimately be used to characterise various dynamic properties of the sample.

The quantity actually measured by a correlator fed two time-varying input signals I(t) and J(t) is the *correlation function* G(τ), defined by the formula;

$$G(\tau) = \lim_{T \to \infty} (1/2T) \int_{-T}^{T} I(t)J(t + \tau) \, dt. \tag{3.8}$$

Whilst some *heterodyne* measurements do indeed feed two different signals to the correlator, one derived directly from the laser and the other from the scattered light, *homodyne* measurements are more common, in which I and J are the same signal and G represents the *autocorrelation function* of the scattered light. The technique by which this autocorrelation function is derived is variously known as *quasi-elastic light scattering, dynamic light scattering, intensity fluctuation spectroscopy,* and *photon correlation spectroscopy.* The term 'spectroscopy' is basically a misnomer, since it is not the frequency distribution of the scattered light which is directly measured; nonetheless, the spectral distribution of the scattered light can if desired be obtained from the correlation function by Fourier transform. This method is particularly useful for measurement of very small frequency shifts of less than 1 MHz.

The laser used for photon correlation measurements must operate in the TEM_{00} mode, have good amplitude stability, and emit a suitable wavelength which is not absorbed to any large extent by the sample. For low levels of scattering, the principal 488.0 and 514.5 nm lines from an argon laser are frequently employed; for stronger scattering, 632.8 nm radiation from a He-Ne laser is more commonly used. Typical correlator apparatus can measure intensity fluctuations with a time-resolution of down to 100 ns. It is worth noting that where the scattering intensity is especially weak, data collection times of several hours may be necessary. In such cases, it is of course imperative that the liquid sample is scrupulously clean, since dust, bacteria, and any other foreign matter can produce strong spurious fluctuations in the scattered light.

Various kinds of data are available by use of this technique. For example the rotational and translational diffusion coefficients of macromolecules in solution can be ascertained, which in turn provide valuable information on molecular shape and dimensions. Such information is especially useful in the study of polymers, since it can be related to their degree of polymerisation and mean molecular weight. Similar measurements on micellar systems (colloids) provide data on micellar shape, aggregation number, and micellar interactions. This facilitates the characterisation of biological micelles and synthetic aqueous detergents, for example. Finally, there are a wide range of medical and physiological applications, ranging from immunoassay to determination of the motility of bacteria and spermatozoa.

3.6.3 Brillouin Scattering

As mentioned earlier, Brillouin scattering results in a shift in scattering frequency due to coupling between the radiation and acoustic modes of the sample. The effect is strongest in solids and liquids, but can be observed weakly in gases. Since acoustic frequencies are very low compared to optical frequencies, the resultant shifts are correspondingly small, typically being of the order of 1 cm^{-1} in wavenumber units. Brillouin scattering cannot therefore be resolved from Rayleigh scattering using conventional filters or monochromators; instead, a *Fabry-Perot interferometer* has to be employed. This generally consists of two parallel glass plates each coated with a thin film of silver, one of which is stationary and the other of which can be moved towards or away from it on a carriage. Light travelling back and forth between the two plates undergoes multiple reflections rather like those within a laser cavity, and for any particular separation only a single wavelength can propagate through the system without suffering destructive interference.

Typical apparatus for the measurement of Brillouin scattering is illustrated in Fig. 3.18 a. The source is generally a highly monochromatic ion laser, in which the effects of Doppler broadening are obviated by an intracavity etalon restricting emission to a single mode. Light scattered from the sample is resolved in the interferometer and gives rise to a spectrum such as that shown in Fig. 3.18 b. Various types of information can be deduced from the results. In liquids, Brillouin scattering provides one of the few methods of obtaining information on the low-energy interactions which determine local structure; energy transport mechanisms in gases can also be studied in the same way. In the case of crystalline solids, each Brillouin feature represents coupling with one of the acoustic-frequency lattice vibrations known as *phonons,* and yields informa-

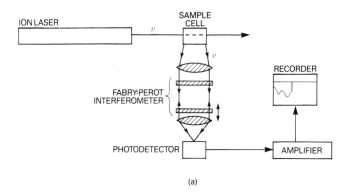

(a)

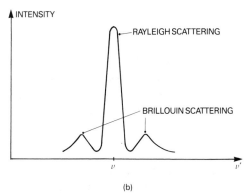

(b)

Fig. 3.18 a and b. Measurement of Brillouin scattering: (a) typical instrumentation using a Fabry-Perot interferometer; (b) a typical spectrum

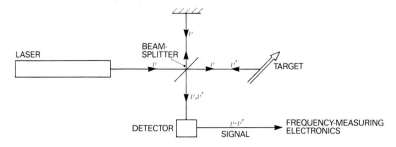

Fig. 3.19. Instrumentation for Doppler velocimetry

tion on structure and binding forces; such measurements are important in the semi-conductor device industry, for example. Other quite different kinds of solid usefully investigated by this method include biological macromolecules, and in particular fibrous materials such as collagen. Here, data on the elastic properties can be interpreted in terms of chemical properties such as the extent of hydration.

3.6.4 Doppler Velocimetry

In Chap. 1, we examined one manifestation of the Doppler effect, in the shift in the frequency of light emitted by molecules due to their Brownian motion. The same principle applies to the scattering of light in any direction other than forward scattering. However, bulk flow of the sample at any direction other than at right-angles to the beam also produces a Doppler shift which can be measured using the setup shown in Fig. 3.19. Here, use is made of the coherence of the laser light to produce interference between the beam and radiation back-scattered from the sample. This interference produces a beat frequency which can be electronically measured, and directly related to the velocity. Velocities up to 40 m s^{-1} can be measured in the laboratory with commercially available instruments, although higher speeds can be measured at a distance; the technique is, for example, applied in the speed checking equipment used by the police. Even wind velocity can be measured by this technique (Doppler anemometry). Chemical uses of the principle usually involve fluid flow measurement at velocities down to 100 μm s^{-1}; there are also important biological applications for the non-invasive measurement of blood flow.

3.6.5 Lidar

A very different type of laser instrumentation is found in 'lidar', an acronym for *l*ight *d*etection *a*nd *r*anging, where once again light scattering is the principle involved. This is a method primarily used for the remote sensing of trace chemical species in the atmosphere, and is an important tool in pollution control. The essential details of the apparatus are shown in Fig. 3.20a. A pulsed laser emits short pulses of light directed into the atmosphere, and a telescopic detection system usually placed alongside the laser picks up any light which is back-scattered. The time interval between the emission of each pulse and the subsequent detection of scattered light is a direct measure of the distance of the substance responsible for the scattering. Hence a plot of the detected

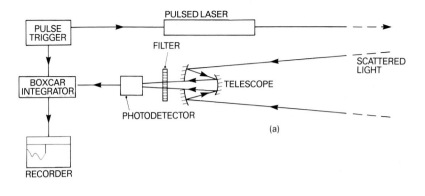

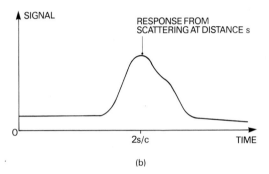

Fig. 3.20 a and b. (a) Essential features of a lidar system, and (b) the response measured from the time of laser pulse emission

intensity against time provides information on the variation of concentration with distance, as illustrated in Fig. 3.20 b.

Various lasers are used for this type of application, usually Q-switched and emitting pulses of the order of 10 ns duration. Since the only other limitation on ranging precision is the response time of the detection electronics, also typically measured in nanoseconds, the distance of the scattering species can easily be measured to an accuracy of a few metres. Where the species of interest is in the form of aerosol particles, use can be made of the fact that such particles scatter light one or two orders of magnitude more effectively than gases. Here, then, a large response from the detector signifies the presence of the pollutant. For this purpose, ruby or Nd-YAG lasers are most often used since their wavelengths in the red and near infra-red result in the most intense scattering. The high degree of collimation of the laser beam means that scattering can be detected over a large range of distances, up to 100 km in the case of clouds.

There are several chemically more specific means of using the principles of lidar for atmospheric analysis. One possibility is the measurement of laser-induced fluorescence, which for good range resolution necessitates population of very short-lived excited states in the species of interest. This technique is therefore only practicable with uv/visible lasers which populate electronic states of nanosecond lifetime. A far more attractive method is to infer the presence of a pollutant from the reduction in scattering intensity when the laser wavelength coincides with one of the compound's absorption bands. Usually, the scattering is monitored at two different laser wavelengths, one resonant with an absorption band of the pollutant and the other not; this technique is known as *differential absorption lidar*. The differential absorption of the resonant wavelength over a particular range of distance thus demonstrates the presence of the pollutant in a particular vicinity, and the concentration can be deduced from the relative reduction in scattering intensity. For this purpose, a tunable diode laser emitting in the infra-red may be employed; a carbon dioxide laser has much greater power and range, but is of more limited use due to its fixed emission wavelengths. Concentrations as low as 0.1 ppm are measurable by this method.

A more sophisticated version of lidar instrumentation for gas detection is based on the Raman effect, described in detail in Chap. 4, and relies on the characterisation of pollutants by their molecular vibrations. The setup requires use of a uv/visible laser source, and placement of a monochromator between the telescope and photodetector. In this way Stokes-shifted Raman wavelengths can be detected, thereby enabling the chemical composition of the pollutant to be identified unambiguously. Indeed, the temperature of the gas may itself be calculated from the intensities of the rotational Raman lines. All such lidar methods have the enormous advantage of enabling range-sensitive chemical analysis of the atmosphere to be performed from ground level. One of the prime areas of application is in the monitoring of stack gases from industrial chimneys. The remote lidar of the atmosphere above volcanoes has also recently been used as a predictor of volcanic activity, based on the known correlation of such activity with the sulphur dioxide content of the volcanic gases.

3.7 Polarimetry

When plane polarised light passes through a chiral substance, it experiences a rotation of its plane of polarisation known as *optical rotation*. In the case of isotropic liquids or solutions, the requirement for optical rotation to occur is that the solvent or solute molecules are in themselves chiral. In other words they must possess no centre of symmetry nor any reflection- or rotation-reflection symmetry. Such molecules are known as *optically active,* and include most large polyatomic compounds. The quantitative measurement of the extent of polarisation rotation is known as *polarimetry,* and can be used to provide information on molecular optical activity, solution concentration, and path length. The parameter in terms of which most optical rotation results are expressed is the specific optical rotatory power, or *specific rotation* [α], defined by

$$[\alpha] = \theta/cl, \tag{3.9}$$

where θ is the angle of rotation, c the concentration and l the path length through the solution. The *molecular rotation* is obtained on multiplying by the molecular weight. Additional information characterising the particular substance involved can be obtained by measuring the dependence of optical rotation on the wavelength of light used for its detection; this technique is known as *optical rotatory dispersion.*

Lasers are ideally suited to polarimetric analysis. A typical setup is illustrated in Fig. 3.21, and involves polarising optics and pulsing techniques discussed earlier in this Chapter. Also involved is a polarisation analyser, which is essentially a variable-orientation polariser. Since the minimum amount of light passes through this polariser

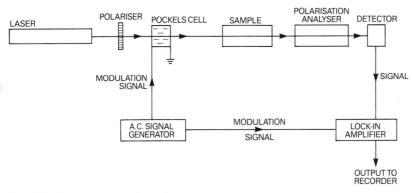

Fig. 3.21. Instrumentation for polarimetry

when it is set at 90° to the plane of polarisation of the incident light, rotation of the analyser enables the orientation of the polarisation plane to be determined.

The instrumentation is easiest to understand if we first consider the setup with no
sample present. The Pockels cell modulates the orientation of the polarisation plane, resulting in modulation of the intensity of light passing through the analyser and onto
the photodetector. However, when the polarisation plane of the analyser lies exactly inbetween the two extreme polarisations emerging from the Pockels cell, the photodetector signal oscillates sinusoidally at exactly twice the modulation frequency. Now with
the sample present, the same condition is only fulfilled if the analyser is rotated
through some small angle; this measures the optical rotation of the sample. Rotations
as small as 0.003 mrad can be routinely detected with this type of apparatus, although
the best polarising optics can enable measurement of rotations almost two orders of
magnitude smaller. The technique is quite amenable to application for very small volumes of solution, and is thus particularly useful in conjunction with liquid chromatography.

Before leaving this subject, it is worth noting that *circular dichroism* measurements
are also very amenable to laser instrumentation, using the same principles of electro-
optic modulation of polarisation. In this case, it is the *differential absorption* of left- and
right-handed circularly polarised light which is used to characterise chiral compounds.
Results are often expressed in terms of Kuhn's *dissymmetry factor,* given by

$$g = (\varepsilon^L - \varepsilon^R)/\bar{\varepsilon}, \tag{3.10}$$

where ε denotes the molar absorption coefficient (see section 4.1), and $\bar{\varepsilon}$ is the mean of
the results ε^L and ε^R for left- and right-circularly polarised light. The value of the dis-
symetry factor for absorptions in the visible/uv region is typically of the order of 10^{-4}.
Since the spectral lineshape associated with circular dichroism bands is often appreciably narrower than the corresponding optical rotatory dispersion curves, there is generally a better spectral resolution, and the circular dichroism technique is therefore often to be preferred.

3.8 Laser Detectors in Chromatography

For the detection of trace quantities of organic compounds in mixtures with other substances, one of the most powerful methods is chromatography. The general principle
behind chromatography is the separation of different components of a mixture by their
different rates of transport through various media. The most fully developed and sensitive type of chromatography, applicable to mixtures which are readily volatilised, is gas
chromatography; the preferred method for relatively involatile liquid samples is high-
performance liquid chromatography (hplc). One of the difficulties with this type of
chromatography has been finding a sufficiently sensitive detector; there is now every
indication that lasers may provide the answer.

The methodology of hplc is essentially similar to conventional liquid chromatography, in which a solution of the sample is carried by a solvent (eluant) under gravity
down a column packed with a fine powder; different components of the solution travel
through the column at different rates depending on their transport properties and affinity for the column material, and hence have different, and characteristic column retention times. The chromatogram is obtained by monitoring a suitable physical property
of the solution eluted from the bottom of the column, and plotting it as a function of
time.

High-performance liquid chromatography involves the same principles, but the column material is composed of much finer particles, typically 5–10 μm in diameter, and
consequently the liquid has to be forced through under a high pressure, typically of the
order of 100 atmospheres. A preparation of hydrated silica particles is the most commonly used material for the separation, in which case a non-polar solvent is employed,
and the most polar solutes tend to have the longest retention times. Alternatively, or-
ganically substituted silica may be used (the OH group most often replaced by
straight-chain $OC_{18}H_{37}$) so that the solvent is more polar, and the tendency is for polar
solutes to be retained the shortest times; this method is known as *reverse-phase* liquid
chromatography.

Various laser techniques can be employed to monitor eluant from an hplc column.
Mostly, these entail spectroscopic principles which are described in detail in the next
Chapter. The narrow beamwidth and high intensity of laser light generally means that
very small detection volumes can be employed; it is usually the flow design which is
the limiting factor, but volumes as small as 10^{-8} l are in principle possible. By the
same virtue, laser methods are also singularly appropriate for 'on-column' detection.
Nonetheless, since very low concentrations are usually involved, the two most obvious
methods of detection, namely absorption and fluorescence, are both applicable only
where the substances of interest display appreciable absorption at the operating wavelength of the laser. Absorption features are generally quite broad in liquids, and the
process of chromatography can itself provide a means for the separation and identification of the components in a mixture. One of the absorption methods which shows most
promise for this purpose is thermal lensing spectroscopy (see Sect. 4.2.3). Where measurements are based on fluorescence, the sensitivity towards species which fluoresce
only weakly can be markedly improved by 'tagging' with a strongly fluorescent label;
this approach is often used for the detection of biomolecules. Despite the fact that fluorescence may occur over a wide range of frequencies, the monochromaticity of a laser
detector is nevertheless important in these measurements since with suitable filters it

facilitates discrimination against Rayleigh and Raman scattering. It has been demonstrated that picogram quantities can be detected with hplc fluorescence instrumentation.

Measurements based on the Raman effect itself, or the related method of CARS (see
Sect.4.4 and 4.5.4) also require highly intense and monochromatic sources, and lasers
are the obvious choice. Here, the price of spectral resolution is a loss in sensitivity;
these methods are therefore more useful for characterisation than for detection. Another option for the detection of optically active species is laser polarimetry. Finally,
the principles of nephelometry can be employed. Since hplc eluants are rarely turbid,
this method is most successful if a precipitating agent is introduced before the detection stage.

3.9 Laser Microprobe Instrumentation

A number of laser applications in chemical analysis entail use of microprobe instrumentation. Such techniques are often used to probe the varying chemical composition
of heterogeneous solid materials such as minerals or corroded metals. The instrumentation basically consists of an optical microscope, through which laser light is focussed
onto the sample. As shown in Fig.3.22, visual inspection through the microscope facilitates movement of the sample so as to probe its response at various sites. Microprobe
methods thus have microsampling capability without the need for any sample preparation. They also offer the advantages of real-time analysis, and in certain cases the ability to obtain depth profiles of chemical concentration.

The principal disadvantage of a laser microprobe is its relatively poor spatial resolution, especially compared to electron microscopic methods. The resolution limit for a
focussed laser is determined by the diffraction-limited beam width, and is of the order
of one wavelength. For a laser operating in the visible range, resolution can therefore
seldom be less than about 500 nm. Various mechanisms are employed for analysis of
the response of the sample to a microprobe laser. Trace element analysis is usually performed by using a powerful laser source to vapourise minute quantities of the sample.
These are subsequently characterised by atomic fluorescence or mass analysis in a
mass spectrometer. An alternative method which is entirely non-destructive involves
measurement of the Raman scattering induced by the laser beam. These methods are
discussed in more detail under the appropriate headings in the next Chapter.

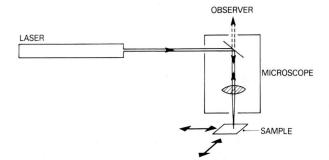

Fig. 3.22. Essential laser
microprobe setup. The de-
tection instrumentation depends on the spectroscopic
principle employed

A very different kind of laser microprobe technique applicable to liquids has recently been developed using optical fibre instrumentation. This is known as *remote fibre fluorimetry,* and is used for the chemical analysis of solutions. The basic principle is the passage of laser light along an optical fibre to a minute lens at its end known as an *optrode,* immersed in the solution to be monitored. Fluorescence from ions or other species excited by the laser light then passes back through the fibre and can be detected and converted to an electrical signal in the usual way. The optrode can be designed to be ion-specific, by using a coating of a suitable chemical reagent to react with the ion and give a fluorescent product. Strongly fluorescent solution species can be monitored with a high degree of sensitivity using an optrode protected by a suitable semipermeable membrane. Detection limits in the parts-per-trillion range have been demonstrated using this fluorimetric method.

3.10 Laser Safety

Safety has in the last decade become more a matter of legislation than individual assessment. Since regulations differ from country to country, it is only appropriate to present a brief overview of laser safety here. In fact, lasers have a remarkably good safety record, despite the fact that beams have in the past often been allowed to propagate freely through laboratory space. In such cases the room is often darkened to prevent stray light entering the detection system, and where visible wavelengths are concerned the high intensity and collimation of the laser beam usually make its path clearly visible through light scattering. There should be little scope for accidents in such situations.

Many of the hazards associated with laser operation are not directly connected with the beam itself. Indeed, the greatest danger often lies in the high-voltage power supply generally required for lasers and associated items of electro-optical equipment, and it appears that the majority of serious accidents so far incurred by laser users have involved electrocution. There are often other supplementary dangers, such as those connected with the cryogenic apparatus used for cooling high-powered cw sources; there is even a noise hazard in the case of a gas-dynamic laser. However, most such hazards can be dealt with by the adoption of obvious and well-established precautions. We shall therefore concentrate on the optical radiation dangers associated with the laser beam.

The greatest cause for concern is the possibility of the collimated laser beam either directly or by reflection entering the eye. Various types of damage can be caused, depending on the wavelength, intensity and exposure time. The precise mechanism for tissue damage in the infra-red and visible regions generally involves thermal effects, or even in some cases photoacoustic shock, whilst in the ultraviolet, damage is initiated by photochemical processes. With most laser radiation in the ultraviolet or far infra-red, there is an appreciable risk of accidental eye injury compounded by the invisibility of the light (although sometimes a visible wavelength is also passed through the system collinearly so as to make the beam pathway visible). Such radiation is not focussed onto the retina, but is absorbed by the cornea and lens, where it causes the damage. If only the outermost layer of the cornea is damaged, the normal process of tissue replacement means that despite the intense pain, complete healing may occur within a couple of days. Much more serious is the possibility of lens damage from the absorption of near ultraviolet or infra-red radiation. This may result in the formation of cataracts many years later.

Although the risk of injury is less likely with visible light, radiation anywhere in the visible/near infra-red region 400–1.400 nm is potentially more damaging since it is focussed onto the retina, and thereby suffers an increase in intensity. A human eye adapted to dark conditions has a focussing power of approximately 5×10^5. Since, for example with ruby laser radiation, energy densities of less than $1\,\text{kJ}\,\text{m}^{-2}$ can cause retinal damage, then allowing a safety factor of 10 the maximum permissible exposure level at the cornea is of the order of $10^{-4}\,\text{J}\,\text{m}^{-2}$. A single pulse from a Q-switched ruby laser will thus cause severe tissue disruption at the retina; of course, this is the basis for some ophthalmic procedures. Diffusely reflected laser light is less of a problem; there, a safety limit of $100\,\text{J}\,\text{m}^{-2}$ should be acceptable. Even moderately low-power cw lasers can cause severe retinal burns. The retina is not normally exposed to *continuous* intensities above $1\,\text{W}\,\text{m}^{-2}$; anything more intense results in aversion response. The Sun, for example, produces an image intensity on the retina of about $10^5\,\text{W}\,\text{m}^{-2}$. Retinal burns will occur before there is time for the natural blink response if the image intensity exceeds approximately $10^6\,\text{W}\,\text{m}^{-2}$. The resultant damage to the rod and cone cells produces an immediate and permanent degradation of both colour and night vision.

Skin on other parts of the body is most susceptible to injury over the so-called 'actinic' range 200–320 nm. However, unless there is prolonged exposure to beams exceeding several tens of kW m^{-2}, damage is unlikely to occur; usually the warmth response enables evasive action to be taken before this stage is reached. Skin hazards are therefore usually only associated with high-power beams. The maximum permissible exposure varies very much according to the wavelength, not least because of the variation in skin reflectivity. For example 40 % of the 1.06 µm radiation from a neodymium laser is reflected from an average skin surface, whilst the corresponding figure for 10.6 µm carbon dioxide radiation is only 4 %. Thus an energy density of $4 \times 10^5\,\text{J}\,\text{m}^{-2}$ may be possible before a neodymium laser causes skin burns, whilst the carbon dioxide laser may cause damage at energy densities a factor of ten less. The exposure limit for radiation which is pulsed on the nanosecond or shorter timescales may be still smaller, because there is less time for heat dissipation. For example the maximum exposure figure for a Q-switched CO_2 laser is $2 \times 10^3\,\text{J}\,\text{m}^{-2}$.

Lasers are generally classified into one of four classes depending on the degree of hazard associated with the beam. Class I lasers, for which it is deemed that no safety precautions are necessary, include only low-power GaAs or similar semiconductor sources, or other lasers enclosed within machine equipment. Class II lasers, such as low-power He-Ne lasers with output below 1 mW, are dangerous only in the unlikely event of prolonged staring into the beam *(intrabeam viewing)*. Class III lasers can be defined as those which can cause eye damage within the time taken for a blink response, which is about 0.2 s. However, such lasers are not hazardous to the skin under normal conditions. The principal consideration with this type of laser is that apparatus should be constructed in such a way that it is impossible to place the eye directly in line with the beam. Finally, we have Class IV lasers which cause serious skin injury, and where even diffuse reflections can damage the eye. For this reason, protective goggles are traditionally worn by the laser operators. The heating effect associated with any such source may also be considered a fire hazard. All cw lasers with output power above 0.5 W fall into this category and call for very comprehensive safety precautions.

4 Chemical Spectroscopy with Lasers

'Why is the grass so cool, fresh, and green?
The sky so deep, and blue?'
Get to your Chemistry.
You dullard, you!
'The Dunce', Walter de la Mare

Spectroscopy is the study of the wavelength- or frequency-dependence of any optical process in which a substance gains or loses energy through interaction with radiation. In the last Chapter, we considered several strictly *non*-spectroscopic chemical techniques, mostly based on interactions with laser light at a *fixed* wavelength. The advantage of studying the wavelength-dependence is the much more detailed information which is made available. Since the exact spectral response is uniquely determined by the chemical composition of a sample, there are two distinct areas of application. Firstly, spectroscopy can be employed with pure substances for the purpose of obtaining more information on their molecular structure and other physicochemical properties; such are the research applications. Secondly, the characteristic nature of spectroscopic response can be utilised for the detection of particular chemical species in samples containing several different chemical components; these are the analytical applications. In both areas, lasers have made a very sizeable impact in recent years.

Although simple absorption and emission of light are the classic processes used for spectroscopic analysis, there are a great many other radiative interactions which are now being used for spectroscopic purposes. Even within the context of a single process such as absorption, there is a wide range of laser techniques which can be used for its detection. Also there are several inherently different classes of absorption spectroscopy, depending on the region of the electromagnetic spectrum being used. For example, molecular absorption spectra in the infra-red region generally result from vibrational transitions in the sample, and hence provide information on the structure of the nuclear framework, whereas absorption spectra in the visible or ultraviolet result from electronic transitions, and so relate to electronic configurations. Whilst high intensity radiation can now be obtained from laser sources (in conjunction with frequency-conversion devices if necessary) at virtually any wavelength longer than 100 nm, the crucial feature to consider for spectroscopic applications is the essential monochromaticity of the source.

In principle, laser monochromaticity can be used to very good effect, since the extremely narrow linewidth which can generally be obtained naturally lends itself to high-resolution spectroscopic techniques. Also, the small beam divergence facilitates the use of long path lengths through samples, which in the case of samples with very weak spectral features can helpfully improve the sensitivity. Ultimately, the resolution in any form of spectroscopy depends on the linewidths both of the radiation and the sample. The linewidth of a laser source is determined by various factors already considered in Sect. 1.5.3, such as natural line-broadening, collision broadening and Doppler broadening, and the same processes can contribute to broadening of spectroscopic features in a sample. However, there are two other line-broadening mechanisms which may come into effect in a sample in a particularly significant way due to the unique properties of laser light. These are *time-of-flight broadening,* and *power- or saturation-broadening.*

The former, applicable only to fluid media, results from the fact that molecules moving across a narrow laser beam with a perpendicular component of velocity v experience the radiation for only a very short period of time given by d/v, where d is the beam diameter. The result is the introduction of a frequency uncertainty of the order v/d. This effect is of course far more significant in gases than in liquids. Saturation broadening, on the other hand, is a reflection of the high intensities often associated with laser light. When the radiation has the appropriate frequency to induce molecular transitions (as of course is necessary for spectroscopic purposes) it can effect a considerable redistribution of population amongst the molecular energy levels. In other words, there is a departure from the conditions of thermal equilibrium under which the Boltzmann distribution applies. The intensity of absorption can thus decrease due to a reduction in the number of molecules left in the ground state; this is what is meant by *saturation.* Since saturation is most effective at the centre of an absorption band, the lineshape associated with the transition again broadens. Neither of these effects, however, alters the fact that with their inherently narrow linewidth, lasers offer. the best spectroscopic resolution available.

Finally, there are certain other practicalities to consider when dealing with a monochromatic source. Many of the general instrumental features such as ratiometric recording have been discussed in the last chapter (Sect. 3.5.1). The primary consideration is that since any individual laser can only be operated over a limited range of the electromagnetic spectrum, the spectroscopic analysis of a particular sample may call for a particular laser system. Again, the monochromatic emission of a laser is such that conventional spectrometers operating with continuum radiation and monochromators are not directly applicable. Moreover, the traditional light sources often have the compensating virtues of greater reliability, simplicity of operation and more constant output. For these reasons, it has often tended to be the newer forms of spectroscopy which have been specifically developed around laser instrumentation. However, lasers have also made substantial inroads into all more traditional areas of spectroscopy by virtue of their high resolution capabilities. We begin with a consideration of some of the methods which are used to record absorption spectra using lasers.

4.1 Absorption Spectroscopy

Absorption spectroscopy is based on the selectivity of the wavelengths of light absorbed by different chemical compounds, and involves monitoring the variation in the intensity of absorption from a beam of light as a function of its wavelength. The selectivity over absorption results from the requirement that photons absorbed have the appropriate energy to produce transitions to states of higher energy in the atoms or molecules of which the samples are composed (in the case of crystalline solids, more delocalised transitions involving the entire lattice can also take place). Each such transition requires the absorption of a single photon, and is governed by spectroscopic selection rules. Before proceeding further, we briefly recap on the basis equations for the absorption of light, for later reference.

To begin, it is clear that the rate of loss of intensity in a beam passing through an absorbing medium is proportional to both the instantaneous intensity and the concentration of the absorbing species, and we thus have the relation,

$$-dI/dl = \alpha IC, \tag{4.1}$$

where α is the constant of proportionality known as the *absorption coefficient,* l represents the path length through the sample, and C the concentration of the absorbing species. The solution to this simple differential equation is the exponential decay function,

$$I = I_0 e^{-\alpha lC} \tag{4.2}$$

This result is generally known as the *Beer- Lambert Law,* and expresses the intensity of light transmitted through the sample in terms of the intensity I_0 incident upon it. In order to facilitate expressions based on logarithms to base 10, Eq. (4.2) may alternatively be written as

$$I = 10^{-\varepsilon lC} I_0, \tag{4.3}$$

where $\varepsilon = \alpha/2.303$ is generally termed the *extinction coefficient,* or the *molar absorption coefficient.* The product εlC is often given the symbol A, and referred to as the *absorbance* of the sample; it is evidently related to the transmittance $T = I/I_0$ by

$$A = -\log_{10}T. \tag{4.4}$$

One final relation which follows from (4.3) is

$$\Delta I/I_0 = 1 - 10^{-A}, \tag{4.5}$$

where $\Delta I = I_0 - I$; the left-hand side of Eq. (4.5) thus represents the fractional loss in intensity at a given wavelength on passage through the sample. This last relation will be useful for considerations of the sensitivity of the various methods for absorption spectroscopy.

Before the advent of the laser, most absorption spectroscopy was performed using broadband or continuum sources, with wavelength scanning carried out by passing the light through a monochromator placed either before or after the sample, as illustrated in Fig. 4.1 a. In such a set up, the spectrum is obtained by detecting the intensity of radiation after passage through the sample, and plotting it as a function of the wavelength allowed through the monochromator. It is still the case that the majority of infra-red and uv/visible spectroscopic equipment in chemical laboratories operates on these principles, using conventional light rather than lasers. For later reference, it is interesting to note that for normal analytical purposes, absorbances of the order of 10^{-2} are considered the minimum which can be measured by conventional spectrophotometric techniques.

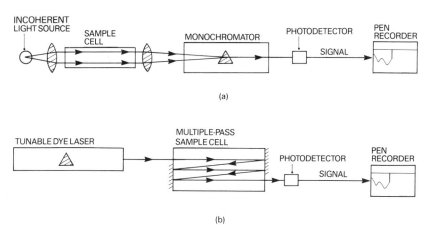

Fig. 4.1a and b. Schematic diagrams of **(a)** a conventional absorption spectrometer, and **(b)** the modified setup used with a tunable laser source. In each case, the signal received by the photodetector is ratioed with that obtained from a reference beam (not shown) derived from the same source, but which does not pass through the sample

Laser-based systems for absorption spectroscopy can be based either on fixed-frequency or tunable laser sources. Fixed-frequency lasers providing emission at only one, or a few discrete wavelengths, are not at all amenable to the usual absorption methods which require scanning over a continuous range of wavelengths, and thus call for specialised techniques. Two such methods, both developed in the late 1960's, are *laser magnetic resonance* and *laser Stark spectroscopy*, which are discussed later in this Chapter. However, the introduction of tunable dye lasers at about the same time made it possible to obtain an absorption spectrum by scanning the source itself across the appropriate wavelength range. This kind of approach has many advantages, and obviates the need for any monochromator: it is also a method which is applicable to other types of laser source, insofar as they can be tuned across the width of the gain profile (see Fig. 1.12). Although we shall mostly concentrate on electronic spectroscopy in the visible range using a dye laser, the general principles discussed here are of much wider application, and include infra-red spectroscopy with diode lasers.

We shall begin, then, by considering a simple setup for absorption spectroscopy using a dye laser, as illustrated in Fig. 4.1b. This arrangement produces a spectrum by monitoring the transmission through the sample as a function of wavelength; the attenuation relative to a reference beam provides a direct measure of the absorption, as in conventional spectrometry. However, absorbances as low as 10^{-5} can be detected in the laser configuration. Since the absorption signal is proportional to the distance travelled through the sample by the radiation, long path lengths through the sample are often used for laser spectroscopy; the Figure illustrates one simple means of obtaining a long path length by multiple passes of the radiation through the sample medium. With a cell 20 cm in length, for example, it is fairly straightforward to generate 50 traversals, resulting in an effective path length of 10 m. In this way, an absorption band producing a decrease in intensity per traversal of as little as one part in 10^4 ultimately produces a drop in intensity of 0.5 %, which is easily measurable. The technique is, of course, particularly well suited to studies of gases, which have very low absorbance; however, it is also of use in obtaining the spectra of components in dilute solutions, in which absorption features due to the solute have to be distinguished from the often very much more intense features due to the solvent. In such a case, the reference beam would be passed through a multiple-pass cell containing pure solvent. An alternative method of obtaining long path lengths is to use long hollow glass or quartz fibres filled with sample fluid.

Another method used to increase the absorption signal from samples is *intracavity enhancement*. As the name indicates, this phenomenon relates to the apparent increase in the intensity of absorption displayed by samples placed within the laser cavity. The detailed mechanism for this effect is somewhat complicated, since the losses resulting from sample absorption become an integral part of the laser dynamics. However, three effects contribute to the enhancement. Firstly, there is simply a multiple-pass effect resulting from the propagation of radiation back and forth within the cavity. Secondly, if the laser is operated close to threshold, then the introduction of extra losses can result in a drop to below threshold, so reducing the intensity of laser emission. The emission intensity is a very sensitive function of the losses close to threshold, so that even a very weak absorption can lead to a dramatic reduction in the output intensity. Finally, if the absorption linewidth of the sample is narrower than the laser gain bandwidth, competition between the various modes within the cavity can result in an effective

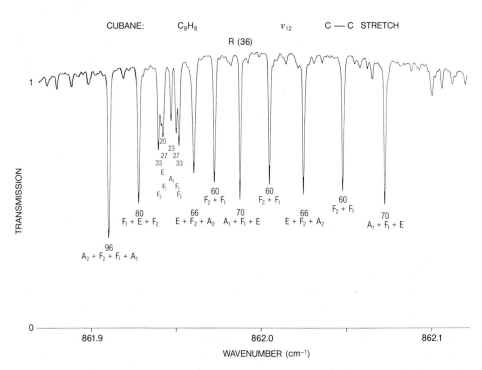

Fig. 4.2. Diode laser absorption spectrum of cubane obtained by Pine et al., showing the rotational structure of the v_{12} C-C stretch. Reprinted by permission from reference 1. Copyright 1985, American Chemical Society

transfer of energy to modes outside the absorption range, again reducing the emission intensity at the absorption wavelength. Hence, comparison of the intensity of laser emission with the sample, and then with a reference material placed within the cavity, provides a highly sensitive method for obtaining an absorption spectrum; in optimum cases, extinctions as low as $10^{-8}\,cm^{-1}$ can be detected. However, it should be noted that the mechanisms for the enhancement are such that there is no simple quantitative relationship between laser output and sample absorption.

Finally, sensitivity can be improved by amplitude- or frequency-modulation of the incident light; here, the absorption signal is obtained by phase-sensitive detection of the oscillation in intensity of the transmitted light; usually the latter type of modulation is more successful. It is worth noting, however, that strikingly good results can be obtained without any of the refinements described above. A very fine example is afforded by the infra-red absorption spectrum of cubane (C_8H_8) obtained by Pine et al.[1] shown in Fig. 4.2. This spectrum, obtained with a lead-salt diode laser, reveals the rotational fine structure in the $C-C$ v_{12} stretching band; notice that the entire scan covers a range of less than $0.2\,cm^{-1}$, and the wavenumber resolution is a remarkable $0.0004\,cm^{-1}$.

4.2 Specialised Absorption Techniques

Thus far, we have considered mostly direct methods of detecting absorption, involving measurement of the intensity of light after passage through the sample. As seen above, a common problem with such transmittance methods is the difficulty of detecting weak absorption features, since the signal ΔI is generally very small compared to I_0. There are, however, several alternative, but highly sensitive measurement techniques particularly suited to laser spectroscopy. These methods are all based on the monitoring of physical processes which take place *subsequent* to the absorption of radiation. Before examining these in detail, however, it is worth emphasising that all the methods to be described in this Section involve precisely the same absorption process in the initial excitation of the sample, and also produce spectra through analysis of the dependence on excitation frequency; hence they may all properly be described as types of absorption spectroscopy. There is one other type of absorption spectroscopy, entailing multiphoton absorption, which does not share the same excitation mechanism; discussion of this method is reserved for Sect. 4.6.

4.2.1 Excitation Spectroscopy

The deactivation of atoms and molecules excited by the absorption of visible or ultraviolet light can often involve the emission of light at some stage. In the case of atomic species, fluorescent emission generally takes place directly from the energy level populated by the excitation. In the case of molecular species, as illustrated in Fig. 2.16, there are usually a number of different decay pathways which can be followed, of which spin-allowed fluorescence from the electronic state initially populated provides the most direct, and usually the most rapid, means of deactivation. However, radiationless decay processes which occur in the vibrational levels prior to fluorescence result in emission over a range of wavelengths, so that even if the initial excitation is at a single fixed wavelength, the spectrum of the emitted light may itself contain a considerable amount of structure which can provide very useful information. This is the basis for fluorescence spectroscopy, discussed later in Sect. 4.3. *Excitation spectroscopy* or *laser-excited fluorescence,* by contrast, is concerned not with the spectral composition of the fluorescence, but with how the overall intensity of emission varies with the wavelength of excitation. Figure 4.3 a illustrates the various transitions giving rise to the net fluorescence from an atomic or simple molecular species with discrete energy levels.

The sensitivity of this method of absorption spectrometry stems from the fact that the signal is detected relative to a zero background; every photon collected by the detection system (usually a photomultiplier tube) has to arise from fluorescence in the sample, and must thus result from an initial absorptive transition. If every photon absorbed results in a fluorescence photon being emitted, in other words if the *quantum yield* is unity, then in principle the excitation spectrum should accurately reflect both the positions and intensities of lines in the conventional absorption spectrum. In practice, collisional processes may lead to non-radiative decay pathways and thus a somewhat lower quantum yield, although this problem may to some extent be overcome in a gaseous sample by reducing the pressure. Also, the imperfect quantum efficiency of the detector has to be taken into consideration, along with the fact that a certain proportion of the fluorescence will not be received by the detector because it is emitted in the wrong direction. Nevertheless, it has been calculated that fractional extinctions $\Delta I/I_0$ as low as 10^{-14} are measurable by commercially available instrumentation using the method of excitation spectroscopy described above.

Amongst other applications, the high sensitivity of this method makes it especially well suited to the detection of short-lived chemical species, and Fig. 4.4 shows part of the spectrum of Xe_2 obtained by this method[2]. The vacuum ultraviolet radiation for the excitation was in this case produced by mixing the frequencies from two dye lasers

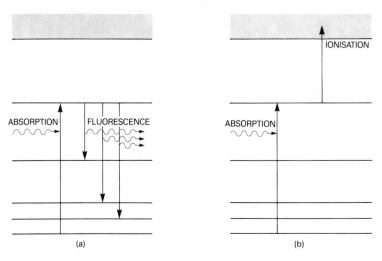

Fig. 4.3 a and b. Energy level diagrams for (**a**) excitation spectroscopy, and (**b**) ionisation spectroscopy of a simple species with discrete energy levels

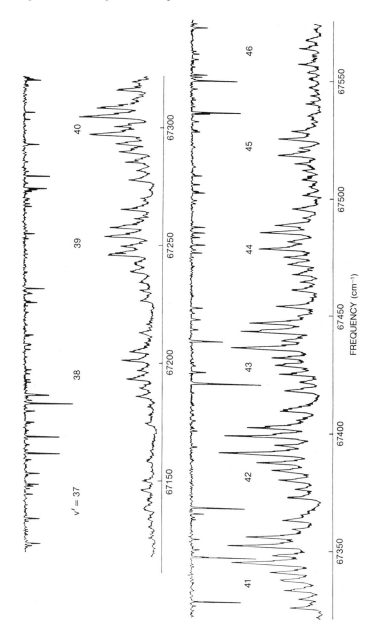

Fig. 4.4. Excitation spectrum of the B O_u^+ (v′) ← X O_g^+ (O) system of Xe_2 obtained by Lipson et al. Reprinted with permission from reference 2

pumped by a XeCl exciplex laser source. Excitation spectroscopy is equally useful for identifying the transient species involved in gas-phase reactions, which represents one of its most important research applications (see Sect. 5.2.6).

4.2.2 Ionisation Spectroscopy

The second specialised technique which is used to monitor absorption in the uv/visible range is ionisation from the electronically excited states populated by the absorption, as illustrated in Fig. 4.3 b. In this case, a spectrum is obtained by monitoring the rate of ion production as a function of the irradiation frequency. Various methods can be used to produce the ionisation, but clearly the process has to be sufficiently selective that ground state species are not ionised. For excited states close in energy to the ionisation limit, ionisation can be induced either by application of an electric field, or by collisions with other atoms or molecules.

Alternatively, a photoionisation technique can be employed, using either the laser, or any other suitable frequency light source to produce photons with enough energy to bridge the gap to the ionisation continuum. The ions or free electrons so produced can be detected by electrical methods, which often have close to 100 % efficiency; ionisation spectroscopy is thus one of the most sensitive methods of detecting absorption, and under ideal conditions virtually every atom or molecule excited by the laser radiation is subsequently ionised and detected. The simplest arrangement, in which laser radiation provides the energy for both the initial excitation and the subsequent ionisation, falls under the general heading of multiphoton absorption, and is discussed further in Sects. 4.6 and 4.7.

4.2.3 Thermal Lensing Spectroscopy

In this form of spectroscopy, and also in photoacoustic spectroscopy discussed in the following Section, the optical absorption from a tunable laser beam is monitored through an effect based on the heating produced by absorption. The initial heating itself results from the radiationless decay of electronically excited states, and may be termed a *photothermal* effect. In the case of thermal lensing spectroscopy, the specific phenomenon utilised is the temperature-dependence of the refractive index, which results in nonuniform refraction around the laser beam as it passes through any absorbing gas or liquid.

The mechanisms at work in this kind of spectroscopy, also known as *thermal blooming* spectroscopy, are in many respects more complicated than in any other absorption method. In the first place, the production of heat as the result of absorption depends on the detailed kinetics of relaxation in the sample molecules. The localised change in temperature which then ensues depends on the bulk heat capacity, and the extent of both conduction and convection in the sample; consequently, liquids are much more amenable to study by this method than gases. The characteristics of the laser beam also play a crucial role. Since the spatial distribution of intensity across the laser beam should normally conform to a well-defined Gaussian profile, as shown in Fig. 4.5, the small volume of sample traversed by the centre of the beam experiences a greater intensity and thus exhibits a greater temperature rise than sample at the outside edge of the beam. The localised refractive index gradient which is consequently established is

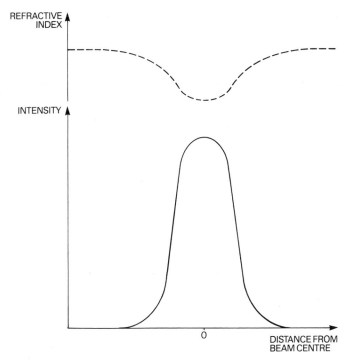

REFRACTIVE
INDEX

INTENSITY

0

DISTANCE FROM
BEAM CENTRE

Fig. 4.5. Diagram showing the Gaussian variation in intensity of a laser beam across its beam (solid curve), and the resulting variation in refractive index in an absorbing medium (dotted curve)

a function of the sample's thermal expansion properties, and leads to a defocussing of the laser beam which is ultimately measured in the setup illustrated in Fig. 4.6. Sensitivity can be increased by chopping the beam, and detecting the modulation in the signal detected.

The instrumentation required for thermal lensing spectrometry places several constraints both on the kind of laser which can be employed, and on the nature of the sample which can be studied. It is a method which is more appropriate for absorption in the visible range than in the infra-red where direct heating effects occur, and it calls for a tunable laser with a good transverse mode structure, so as to produce a consistent spatial intensity cross-section. Since the beam has to be focussed close to the sample

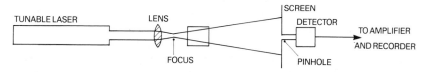

TUNABLE LASER   LENS

SCREEN

DETECTOR

TO AMPLIFIER
AND RECORDER

FOCUS   PINHOLE

Fig. 4.6. Diagram of the apparatus used in thermal lensing spectroscopy. The defocussing which arises as the laser is tuned through an absorption band of the sample results in a broadening of the transmitted beam, and is observed as a decrease of intensity through the pinhole

in order for the defocussing to be observed, only a very small volume of sample is required to generate the spectrum, and the intensity of the signal does not increase in proportion to any increase in the path length.

Samples must be carefully filtered, to overcome false signals due to light scattering impurities; also, the method is basically inappropriate for flowing samples, unless the flow is very slow indeed (less than 1 ml per minute!) The dependence on the thermal properties of the sample make it a far more sensitive method for organic solvents than for water; in the former case, relative absorbances $\Delta I/I_0$ as low as 10^{-7} can be detected. This indicates that a detection limit as low as 10^{-11} M may be achievable for solutions of strongly coloured solutes. Indeed, perhaps the most outstanding feature of thermal lensing spectroscopy is that it represents one of the cheapest means of obtaining high sensitivity in analytical absorption spectroscopy.

4.2.4 Photoacoustic Spectroscopy

As mentioned earlier this form of spectroscopy, also known as *optoacoustic* spectroscopy, is similar to thermal lensing spectroscopy in that both hinge on the heating effect produced by optical absorption. In this case, however, the specific effect which is measured results from the thermally produced pressure increase in the sample, and makes use of the fact that if the laser radiation is modulated at an acoustic frequency, pressure waves of the same frequency are generated in the sample, and can thus be detected by a piezoelectric detector or a microphone. Since the intensity of sound must depend on the amount of heating, it reflects the extent of absorption; hence a spectrum can be obtained by plotting the sound level against the laser frequency or wavelength. Once again, the method produces spectra of very high resolution.

The instrumentation for laser photoacoustic spectroscopy is shown in Fig. 4.7. Visible or infra-red lasers can be used for the source, so that either electronic or vibrational (or in some cases vibration-rotation) spectra can be obtained. The laser light first passes through a chopper which provides the acoustic frequency modulation of the beam incident upon the sample; a modulation signal is also derived at this point from a beam-splitter and optical detector (not shown on the diagram). The acoustic signal generated as the beam passes through the sample cell, or *spectraphone*, is then sent to the phase-sensitive amplifier locked into the modulation signal, and produces an output which is plotted by a pen recorder. Alternatively, a pulsed laser can be used in conjunction with an averaging system.

In contrast to thermal lensing spectrometry, photoacoustic absorption methods are

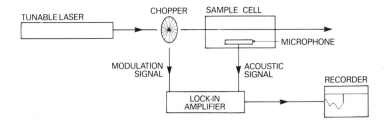

Fig. 4.7. Schematic diagram of the apparatus used in the photoacoustic spectroscopy of a gas

easily applicable to both liquids and gases, and can also be adapted to solid samples; in the latter case, a microphone can be used to pick up an acoustic signal from the gas above the sample. Flowing fluid samples are also relatively easy to study, provided the acoustic noise generated by the flow is electronically filtered out of the microphone signal. In the case of liquids, it is again the case that solutions in organic solvents provide a much greater sensitivity than aqueous solutions, due to the high heat capacity of water. Nonetheless, it is important to choose a solvent which is comparatively transparent over the spectral range being examined, or else there may be little improvement over the spectra obtainable by conventional absorption methods.

Gases are the simplest to work with, and optimum sensitivity is obtained if the sample cell dimensions support resonant oscillations of the acoustic waves, which in practice means tailoring the modulation frequency to the particular sample being studied. Relative absorbances of 10^{-7} are measurable in both liquids and gases, although in gases the detection limit may be as much as two orders of magnitude lower. Sensitivity is in fact comparable to that of a good gas chromatography-mass spectrometry system, and concentrations of gases in the ppb (parts per billion) range can be detected over a wide range of pressures, from several atmospheres down to 10^{-3} atmospheres. The technique is thus particularly well suited to investigations of gaseous pollutants in the atmosphere, and there are few other absorption techniques which match photoacoustic spectroscopy for sensitivity.

4.2.5 Optogalvanic Spectroscopy

The last method which is commonly used to measure absorption spectra with a tunable laser source is based on the *optogalvanic effect*, which is essentially a change in the electrical properties of a gas discharge (or in some cases a flame) when irradiated by certain frequencies of light. In contrast to the other specialised methods discussed so far, this type of spectroscopy provides information on atomic or ionic species, rather than molecules. Also, the optical transitions involved are not only those originating from the ground state, since excited states are also appreciably populated.

The electrical current passing through a gas discharge results in ionisation of the atoms, and the efficiency of this process depends on the energy they already possess, being most effective for atoms in highly excited states near to the ionisation continuum. The current is thus a sensitive function of the various atomic energy level populations. In the presence of a beam of light with suitable frequency, absorption processes result in atomic transitions to states of higher energy, so that the relative populations of the various levels are changed. This effect can be registered as a change in the voltage across the discharge, which may be either positive or negative. Plotting the variation in discharge voltage against the irradiation frequency thus provides a novel kind of absorption spectrum, and completely obviates the need for an optical detection system.

The apparatus used in discharge photogalvanic spectroscopy is shown in Fig. 4.8. In practice, the voltage change across an external resistor is monitored as a dye laser focussed into the discharge in a hollow cathode lamp is tuned across its spectral range. To increase the sensitivity, the laser radiation may again be chopped, and the modulation of the signal at the chopping frequency detected with a lock-in amplifier and recorder. By using a lamp containing uranium or thorium, whose spectra are already

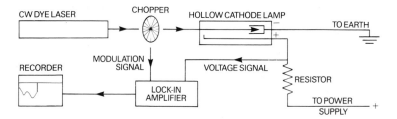

Fig. 4.8. Apparatus used for optogalvanic spectroscopy in a gas discharge

known with high precision, optogalvanic spectroscopy can incidentally be used as a method of wavelength calibration for tunable laser sources.

The primary analytical applications of optogalvanic spectroscopy arise elesewhere however, in flame methods similar to those already widely used in atomic absorption and fluorescence spectrometry. Samples are usually in solution form, and are introduced as a narrow jet into flames of an ethyne (acetylene)/air mixture. However, the optogalvanic instrumentation additionally requires the coupling of the burner head to the positive side of a power supply, and the insertion of high-voltage cathode plates on either side of the flame, as shown in Fig. 4.9. The remainder of the setup is then similar to that shown in Fig. 4.8, with the burner assembly and cathode plates replacing the hollow cathode lamp. One alternative arrangement is to replace the cw dye laser with one which is flashlamp-pumped, using a photodiode to detect each pulse and trigger a signal-averaging amplifier.

In flames, it is generally found that the optogalvanic effect results in an *increase* in the extent of ionisation between the electrodes each time an absorption line is encountered. For this reason, the alternative designation *laser enhanced ionisation* spectroscopy has also gained usage. Compared to the optical detection methods used in conventional flame spectrometry, this method suffers none of the problems usually caused by the flame background and scattering of the laser light. It also has improved sensitivity, both in spectral resolution and in its detection limits for certain elements. The best example of this sensitivity is its use in the analysis for lithium, which has been detected in concentrations as low as $1\,\text{pg/ml}$ $(10^{-12}\,\text{g/ml})$ by this method, representing an improvement over the detection limit of most flame analytical methods by several powers of ten. The optogalvanic spectrum of rubidium is shown in Fig. 4.10.

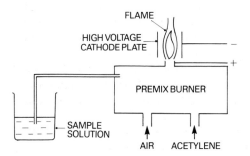

Fig. 4.9. Modified burner used for flame optogalvanic measurements in laser enhanced ionisation spectroscopy

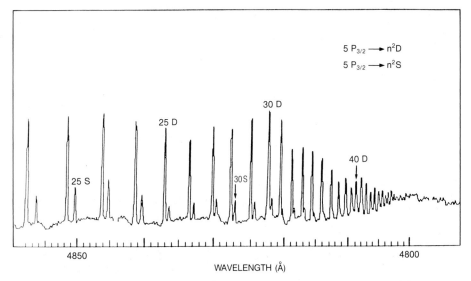

$5\,P_{3/2} \longrightarrow n^2D$

$5\,P_{3/2} \longrightarrow n^2S$

30 D

25 D

40 D

25 S

30 S

4850   4800

WAVELENGTH (Å)

Fig. 4.10. Optogalvanic spectrum of rubidium. Reprinted with permission from Ref. 3

4.2.6 Laser Magnetic Resonance

We now come to the first of the two methods mentioned earlier which enable absorption spectra to be obtained with fixed-frequency laser sources. Both of these, laser magnetic resonance and laser Stark spectroscopy, operate on the principle of tuning the *absorption* frequency of the sample to the frequency of the light source, rather than the more usual converse; the difference between the two is that one method involves tuning using a magnetic field, and the other an electric field. Both methods were contrived before tunable lasers had become commonplace and facilitated the various methods described above; however, they still have a more limited role to play in cases where high intensities are required, and only fixed-frequency laser sources provide sufficient power near to the absorption region of interest. For this reason, these two methods are often used for vibrational analysis in the low-frequency 'fingerprint' region below 1000 cm^{-1}, for example using a carbon dioxide laser, and also for obtaining rotational spectra in the far infra-red, using for example a hydrogen cyanide laser (wavelength 330 μm).

Laser magnetic resonance spectroscopy is based on the phenomenon known as the *Zeeman effect*; indeed, it is sometimes referred to as *laser Zeeman spectroscopy*, and is generally applicable to molecules which possess a permanent magnetic dipole moment. Such a moment may result from various types of angular momentum, i.e. electron spin, electronic orbital angular momentum, nuclear spin, or even molecular rotation. However, the largest effects result from electron spin, and are hence associated with *paramagnetic* species which have one or more unpaired electrons. States with a quantum angular momentum J usually have a $(2J + 1)$-fold degeneracy associated with the range of values of the azimuthal quantum number $M_J = -J$ through zero to $M_J = J$. Application of a static magnetic field results in each of these $(2J + 1)$ levels being shifted in energy to a different extent, according to the equation

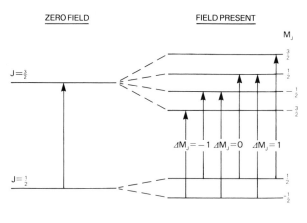

Fig. 4.11. Splitting of the $J = \frac{1}{2}$ and $J = \frac{3}{2}$ energy levels of a paramagnetic species in the presence of a static magnetic field, showing the allowed transitions

$$\Delta E = g\mu_B M_J B, \tag{4.6}$$

where g is the Landé factor determined by the kinds of angular momentum involved, μ_B is the Bohr magneton, and B is the magnetic induction field. Thus, for example, a $J = \frac{1}{2} \rightarrow J = \frac{3}{2}$ transition is split into six components satisfying the selection rule $M_J = 0, \pm 1$, as shown in Fig. 4.11. This leads to six different absorption frequencies (or three, if the Landé g factors of the $J = \frac{1}{2}$ and $J = \frac{3}{2}$ states are equal). Since the extent of energy shift is proportional to the applied magnetic field (Eq. 4.6), it is possible to bring absorption lines into resonance with a fixed frequency laser by varying the magnetic field strength; the resonance can then be detected by absorption of the laser light. A magnet capable of producing a 2 T field, for example, provides a tuning range of the order of $1\,\mathrm{cm}^{-1}$.

The instrumentation for laser magnetic resonance is shown in Fig. 4.12. Because the method is a highly sensitive one, it is often applied to samples in the gas phase, and the lasers used are also generally gas lasers. It is therefore most straightforward to adopt an intracavity configuration, which brings about the advantage of signal enhancement discussed in Sect. 4.1. The spectrum is thus obtained by plotting the laser output as a function of the intensity of the magnetic field applied across the sample. Even greater sensitivity can be obtained by modulating the magnetic field and using phase-sensitive detection methods, as in the related field of electron spin resonance spectroscopy; with such methods, it is estimated that a detection limit of 10^8 molecules cm^{-3} is achievable with current instruments.

Probably the most valuable attribute of laser magnetic resonance is its particular suitability to studies of short-lived paramagnetic species; for this purpose gas may be flowed through the sample cell immediately following the microwave irradiation or

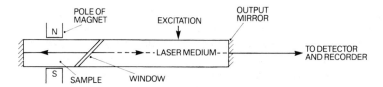

Fig. 4.12. Intracavity measurement of laser magnetic resonance

chemical reaction in which they are produced. The technique is thus very well suited to the monitoring of the free radicals involved in many photochemical reactions; the first high resolution gas phase spectrum of HO_2, for example, was obtained in this way. Recent laser magnetic resonance studies have also made possible the identification of an ethynyl (CCH) radical in the reaction of methane with free fluorine atoms[4]. Ethynyl, which is one of the most ubiquitous molecules in interstellar clouds, had not previously been detected on Earth.

4.2.7 Laser Stark Spectroscopy

In the Zeeman effect described above, a magnetic field removes the $(2J + 1)$-fold degeneracy from spin states with angular momentum J in molecules with a magnetic moment. In much the same way, an electric field lifts the degeneracy from rotational states in molecules possessing a permanent electric dipole moment; this is known as the Stark effect. In any polar compound, there is always a shift in energy which varies with the square of the electric field, and is thus referred to as a *second-order*, or *quadratic* Stark effect. In symmetric rotor species, and in linear molecules in states with non-zero orbital angular momentum, there is an additional shift which is directly proportional to the field stength, and is thus termed a *first-order*, or *linear* Stark effect. Again, the selection rule $\Delta M_J = 0, \pm 1$ applies to transitions between the various components of each rotational state. Thus, an absorption frequency associated with a single rotational transition in the absence of the field splits into several different absorption frequencies when the field is applied, and the exact frequency of each transition can be varied by changing the electric field strength; this is the basis of laser Stark spectroscopy.

The source used for this kind of spectroscopy is usually a carbon monoxide or carbon dioxide laser, with apparatus basically similar to that shown in Fig. 4.12. The only main difference is that the poles of the magnet are removed, and electrode plates are inserted inside the sample cell so as to create a field transverse to the axis of the laser. Since an electric field of 10^4 V cm^{-1} can produce a tuning range of only about 0.3 cm^{-1} even in a sample with a sizeable molecular dipole moment of 2 D, it is important to produce the highest possible field strength. As the magnitude of the field between the plates is inversely proportional to their separation, the optimum setup is where the plates are placed as close together as possible; however, the separation must remain sufficient for laser radiation to propagate back and forth in the cavity without interruption. Thus plate separations of a few millimetres are common, with applied voltages in the kilovolt range producing fields of the order of 10^4–10^5 V cm^{-1}. In cases where this is insufficient, it may be necessary to use more conventional extracavity methods so that the plates can be placed even more closely together. Sensitivity can again be improved by modulation of the applied field, coupled with phase-sensitive detection.

The method of laser Stark spectroscopy is rather more widely useful than laser magnetic resonance, since it is limited only to dipolar molecules. It is also sufficiently quantitative to enable very precise determinations to be made of molecular dipole moments. Its main use, however, lies in providing high resolution molecular rotation spectra in regions normally inaccessible by the traditional microwave methods. Figure 4.13 shows the rotational structure in the laser Stark spectrum of the v_4 band of $CH_3C^{15}N$ obtained by Mito et al.[5].

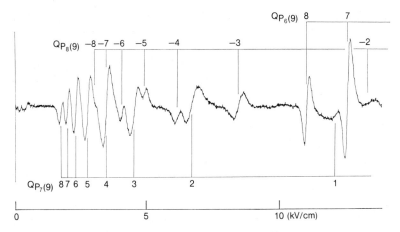

Fig. 4.13. Laser Stark spectrum of the v_4 band of $CH_3C^{15}N$ obtained using the P(25) line from a CO_2 laser. Reprinted with permission from Ref. 5

4.2.8 Other High-Resolution Methods

Finally, we can briefly consider some other high-resolution absorption methods. One technique for obviating the effects of collisional and Doppler broadening is to use a molecular beam, probed by a laser beam intersecting it at right angles (see Sect. 5.2.6). For most samples, however, a more practical method is based on detection of the resonant frequency v_0 at the *centre* of a Doppler-broadened absorption band by *saturation spectroscopy*, for which the underlying principle is as follows.

Consider an absorbing sample strongly pumped by two counterpropagating laser beams of the same frequency v. If the laser frequency exactly coincides with the resonant frequency, then by virtue of the Doppler shift (Sect. 1.5.3) only molecules with axial velocity component $v_k = 0$ can be excited. Consequently, if passage of one beam through the sample is sufficient to appreciably deplete the ground state population of these molecules, in other words to result in *saturation* taking place, then comparatively

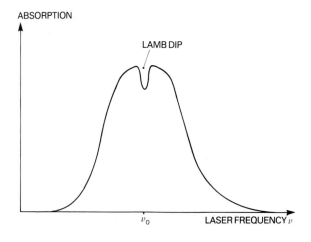

Fig. 4.14. Reduction of absorption intensity at the centre of an optical absorption band observed with counterpropagating beams of the same frequency

little light will be absorbed by the second beam. This effect is often loosely referred to as *bleaching*. However if the laser frequency is off-resonance by an amount Δv, then whilst one beam can excite molecules with axial velocity $v_k = c\Delta v/v$, the beam travelling in the opposite direction can excite other molecules with axial velocity $v_k = -c\Delta v/v$.

Thus the beam intensity is actually diminished *less* by absorption in the former case, when its frequency coincides with the centre of the absorption band. This phenomenon is known as the *Lamb dip*, and is illustrated in Fig. 4.14. In practice, the experimental setup for saturation spectroscopy usually involves two *nearly* collinear counterpropagating beams as shown in Fig. 4.15. The saturation technique is widely employed to eliminate Doppler-broadening, and its sensitivity can be improved by fluorescence detection. One further degree of sophistication can be introduced by separately chopping the two laser beams at different frequencies, and observing fluorescence modulated at the sum frequency.

One final type of absorption spectroscopy based on specialised laser instrumentation is *optical-optical double resonance*, which calls for simultaneous irradiation of the sample by two laser beams of different frequency. There are various schemes for this method based on different combinations of optical transitions; the two which involve a pair of absorption transitions are illustrated in Fig. 4.16. In each case, however, the basic principle is the same; one transition $E_1 \rightarrow E_2$ is strongly driven by a modulated beam from a pump laser, and thus produces a fluctuation in the populations of the two

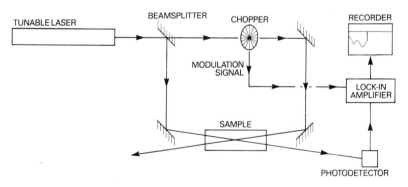

Fig. 4.15. Instrumentation for saturation Lamp-dip spectroscopy

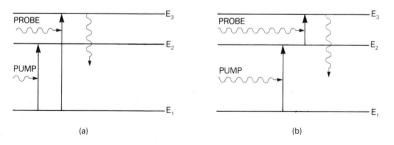

Fig. 4.16a and b. Configuration of absorption transitions for optical-optical double resonance

energy levels it connects. A probe laser beam is then used to induce transitions originating from either level E_1, as in (a), or from E_2, as in (b). The fluctuating population
of the initial state for the second transition thus results in a modulation of the probe
beam intensity at the same frequency as the chopping of the pump laser. Phase-sensitive detection of this effect using a lock-in amplifier thus provides a sensitive means of
monitoring absorption transitions, and is useful for the assignment of high-resolution
spectral lines in small molecules.

4.3 Fluorescence Spectroscopy

In every type of laser spectroscopy examined so far, the essential mechanism for introducing spectral discrimination has been the absorption of radiation by sample atoms
or molecules. We now consider an alternative but equally well established branch of
spectroscopy based on the emission of radiation by the sample. The laser provides a very
selective means of populating excited states, and the study of the spectra of radiation
emitted as these states decay is generally known as *laser-induced fluorescence*. There are
two main areas of application for this technique; one is atomic fluorescence, and the
other molecular fluorescence spectrometry.

4.3.1 Laser-Induced Atomic Fluorescence

This method is now increasingly being used for trace elemental analysis. It involves at-
omising a sample, which is usually in the form of a solution containing the substance
for analysis, in a plasma, a furnace, or else in a flame, and subsequently exciting the
free atoms and ions to states of higher energy using a laser source. Since the
atomisation process populates a wide range of atomic energy levels in the first place,
the emission spectrum is complex, despite the monochromaticity of the laser radiation. The fluorescence frequencies are nonetheless highly characteristic of the elements present, and whilst detection capabilities are typically in the 10^{-11} g/ml range,
concentrations two orders of magnitude smaller have been measured by this technique, corresponding to a molarity of 10^{-19} M. In fact, it has been demonstrated that
laser-induced atomic fluorescence has the ultimate capability of detecting single atoms, although this is in an analytically rather unuseful setup based on a beam of identical atoms.

All sorts of lasers have been utilised for atomic fluorescence measurements. The
most important factor to take into consideration is the provision of a high intensity of
radiation in the range of absorption of the particular species of interest. For this reason, although individual measurements are made with a fixed irradiation wavelength,
it can be helpful to employ a tunable source in order to provide a detection facility for
more than one element. For example use of a frequency-doubled dye laser in conjunction with the traditional air/acetylene flame has been shown to offer dramatically improved detection limits for several precious metals[6]. In common with several other
types of spectroscopy considered previously, sensitivity is often increased by either modulating a cw laser with a chopper, detecting the signal with a phase-sensitive lock-in
amplifier, or else by integrating the signal using a pulsed laser, as shown in Fig. 4.17.
The principles of laser-induced atomic fluorescence may also be employed in a micro-

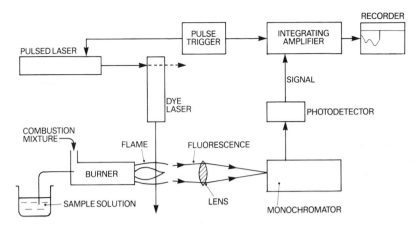

Fig. 4.17. Schematic diagram of the instrumentation for laser-induced atomic fluorescence measurements using a dye laser pumped with a pulsed laser, and with flame atomisation of the sample solution

probe, as discussed in Sect. 3.9. Here, material is vapourised from the surface of a heterogeneous solid sample by a pulsed laser, and its elemental composition is characterised by its atomic fluorescence.

One other variation on this theme is *laser-induced breakdown spectroscopy*. Here, high intensity pulses of laser light, as for example from a Nd:YAG laser, are focussed into the sample and result in the formation of a spark through the process of dielectric breakdown. Once again, elemental analysis can then be carried out by measurement of the resultant atomic fluorescence. This method, which has evolved from the field of electric spark spectroscopy, has the advantage of obviating the need for electrodes, and hence removes any possibility of spectral interference associated with atomisation of the electrode surface. It also has the advantage of speed over many other analytical techniques since it requires little or no sample preparation. This kind of procedure enables ppb concentrations to be detected in optimal cases. Applications to tissue analysis and the identification of trace constituents of blood and sweat have been demonstrated, making the technique an attractive alternative to many of the methods more traditionally used in medical and forensic laboratories.

4.3.2 Laser-induced Molecular Fluorescence

Compared to atomic fluorescence, this method suffers from much poorer sensitivity due to the much broader lines in the emission spectra of all but the smallest polyatomic molecules. These result from the complicated radiationless decay processes involving vibrational energy levels, as discussed in the context of dye lasers in Sect. 2.5. The width of the emission bands can be reduced if the frequency of intermolecular collisions and other line-broadening interactions is diminished; for this purpose samples may for example be analysed in the gas phase, or in the frozen matrix of another compound, as we shall see. However, the effects of reduced concentration associated with all such methods also have to be borne in mind.

In general, then, the high monochromaticity and narrow linewidth of a laser source does not result in a comparable resolution in laser-excited molecular emission spectra. However, it does present some advantages over the more traditionally used black-body radiation sources. Fluorescence spectra are often complicated by features which owe nothing to fluorescence but rather to Raman scattering in the sample, which results in the appearance of frequencies shifted by discrete amounts from the irradiation frequency (see the following Section). The principal advantage of laser-induced fluorescence measurements is that when monochromatic light is used, the Raman features occur at discrete frequencies, and can easily be distinguished from the broad fluorescence bands. It is also worth noting that for certain analytical applications fluorescence measurement may be more selective than absorption spectroscopy, since even when two different compounds absorb at the same frequency, their fluorescence emission frequencies may be quite different.

The energetics of molecular excitation and decay are such that fluorescence spectra can only be collected at wavelengths longer than the irradiation wavelength; subject to this restriction, however, there is comparative freedom over the choice of laser wavelength. Only in the spectroscopy of diatomic and other small molecules does the facility for selectively exciting a particular state play a significant role. However, fluorescence spectroscopy is not solely concerned with chemically stable systems; many important applications concern study of the photolytic processes which can occur in a sample through the interaction with laser light, as will be discussed in more detail in Chapter 5. When the energy absorbed by sample molecules is sufficient, fragmentation processes can occur, resulting in the formation of short-lived transient species which are often strongly fluorescent. For example laser photolysis of cyanogen C_2N_2 produces free CN radicals which can be detected through collection of the fluorescence spectrum.

Another useful feature of laser-induced molecular fluorescence is that with a pulsed source, it is possible to monitor the time-development of the emission process. This provides a further means of discriminating against the Raman scattering which takes place only during the period of irradiation, since the fluorescence is associated with decay processes subsequent to absorption. Different radiative transitions can in fact be discriminated by their different decay constants in some cases. Perhaps more useful in connection with the study of photochemical reactions, however, is the possibility of measuring excited state and transient lifetimes by this method, using either a Q-switched or a mode-locked source. The methods of time-resolved laser measurement are thus particularly pertinent to the study of chemical reaction kinetics.

In general both continuous-wave and pulsed lasers can be used for laser-induced molecular fluorescence measurements. Helium-cadmium and argon lasers are the most popular cw sources; nitrogen lasers and nitrogen-pumped dye lasers are usually adopted for pulsed sources. As usual, the latter are generally operated with a signal-integrating detection system. With such a setup, detection limits for solution samples can be as low as 10^{-13} M, or 10^{-5} ppb. With comparatively complex samples where one is analysing for a particular component in admixture with a large concentration of other species, a detection limit of 1 ppb is more realistic. In air, a detection limit of 10 ppb has recently been established for formaldehyde, in an experiment which provides accurate concentrations within 100 seconds, and without the need for sample collection, water extraction or chemical treatment of any kind[7] (see Fig. 4.18).

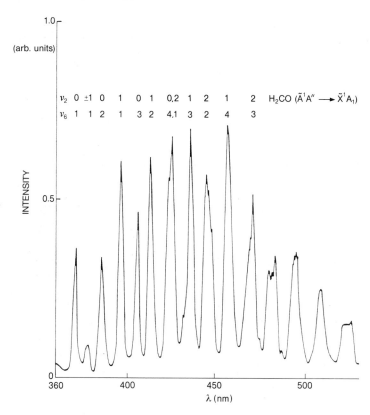

Fig. 4.18. Fluorescence spectrum of 2 mg l^{-1} formaldehyde in air, obtained by G. R. Möhlmann using a frequency-tripled Nd:YAG laser. The numbers denote the vibrational quanta in the electronic ground state, and the $\pm$ sign means that ν_2 is also excited in the electronically excited state. Reprinted with permission from reference 7

Laser-induced fluorescence is increasingly being used for the analysis of eluant from liquid chromatography (Sect. 3.8). It is also possible to 'tag' suitable biological species with fluorescent compounds, so as to facilitate laser fluorimetric detection. Mention should also be made of the novel analytical method which, although not strictly spectroscopic, is nonetheless based on fluorescence and is known as *remote fibre fluorimetry* (see Sect. 3.9). Quite another field of application lies in the remote sensing of oil spills at sea, using an airborne ultraviolet laser source and telescoping detection system to scan the surface of the water for the characteristic fluorescence. Finally, laser fluorescence spectroscopy is widely employed for combustion diagnostics, since it enables the concentrations and temperatures of transient species within a flame to be ascertained with a high degree of spatial resolution, without interfering with the gas flow or chemistry. Intermediates which have been detected by this method range from atomic oxygen and nitrogen to comparatively large fragments such as CH_3O, although OH is most commonly monitored since it provides a relatively straightforward measure of the degree of reaction.

Before leaving this subject, it is worth briefly taking a look at some of the ways in which the resolution of molecular fluorescence spectra can be improved in research on chemically pure samples. Of course the simplest method is to study the substance of interest in the gas phase, but this is rarely useful other than for small molecules with significant vapour pressure at low temperatures. One method makes use of a *supersonic molecular beam*, in which a jet of the sample mixed with helium is forced at high pressure through a small aperture into a vacuum. This process results in a beam of molecules in which there is only a very narrow distribution of translational energies, and only the lowest rotational states are populated. Hence, molecular collisions are comparatively infrequent, and both Doppler and rotational broadening of vibronic transitions are minimised, leading to a very significant improvement in spectral resolution. This kind of technique is particularly useful in the detailed study of reaction kinetics (see Sect. 5.2.6).

Other methods involve isolating individual molecules of the sample from each other in a low-temperature solid phase of some other substance. The solid phase may be constituted by a solvent, in which the sample is dissolved and then frozen to liquid helium temperatures. Most commonly, however, the sample is vapourised and mixed with a large excess of an inert gas (typically 10^4–10^8 atoms per mole of sample), and then frozen onto a cold substrate in a vacuum line, a method known as *matrix-isolation*. In a sample prepared this way, even solute-solvent interactions are eliminated, and normally short-lived chemical transients can be studied since they are trapped in the matrix.

4.4 Raman Spectroscopy

Raman spectroscopy is based on the *Raman effect*, a phenomenon involving the inelastic scattering of light by molecules (or atoms). The term 'inelastic' denotes the fact that the scattering process results in either a gain or loss of energy by the molecules responsible, so that the frequency of the scattered light differs from that incident upon the sample. The energetics of the process are illustrated in Fig. 4.19. The two types of Raman transition, known as Stokes and anti-Stokes, are shown in Fig. 4.19a and b; the former results in an overall transition to a state of higher energy, and the latter a transition to a state of lower energy. The elastic scattering processes repesented by Fig. 4.19c and d, in which the frequencies absorbed and emitted are equal, are known as *Rayleigh scattering*.

The first thing to notice about Raman scattering is that the absorption and emission take place together in one *concerted* process; there is no measurable time delay between the two events. The energy-time uncertainty relation $\Delta E \, \Delta t \geq h/2\pi$ thus allows for the process to take place even if there is no energy level to match the energy of the absorbed photon, since the absorption does not populate a physically meaningful intermediate state. It is therefore quite wrong to regard Raman spectroscopy as a fluorescence method, since fluorescence relates to an emissive transition between two physically identifiable states. In passing, we also note that despite their widespread adoption and utility, transition diagrams like Fig. 4.19 are strictly inappropriate for any such processes involving the *concerted* absorption and/or emission of more than one photon, since for example in this case they incorrectly imply that emission takes place *subsequent* to absorption.

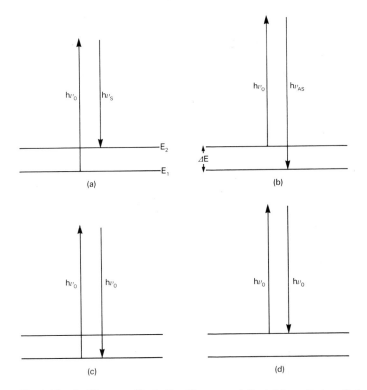

Fig. 4.19 a-d. Diagrams illustrating Raman and Rayleigh scattering. Only the energy levels directly involved are depicted: (**a**) shows a Stokes Raman transition, and (**b**) an anti-Stokes Raman transition; (**c**) and (**d**) show Rayleigh scattering from the two different levels

Raman scattering generally involves transitions amongst energy levels which are separated by much less than the photon energy of the incident light. The two levels denoted by E_1 and E_2 in Fig. 4.19, for example, may be vibrational levels, whilst the energies of both the absorbed and emitted photons may well be in the visible range. Hence the effect provides the facility for obtaining vibrational spectra using visible light, which has very useful implications as we shall see shortly. Raman transitions are, moreover, governed by different selection rules from absorption or fluorescence. Thus whilst in centrosymmetric molecules only *ungerade* vibrations show up in the infra-red absorption spectrum, only *gerade* vibrations appear in the Raman spectrum. This illustrates the so-called *mutual exclusion rule*, applicable to all centrosymmetric molecules, which states that vibrations active in the infra-red spectrum are inactive in the Raman, and vice versa. Even for complex polyatomic molecules lacking much symmetry, the intensities of lines resulting from the same vibrational transition may be very different in the two types of spectrum, so that in general there is a most useful complementarity between the two methods.

Returning to the diagrams of Fig. 4.19, it is clear that the Stokes Raman transition from level E_1 to E_2 results in scattering of a frequency given by

$$\nu_S = \nu_0 - \Delta E/h, \tag{4.7}$$

and the corresponding anti-Stokes transition from E_2 to E_1 produces a frequency

$$\nu_{AS} = \nu_0 + \Delta E/h, \tag{4.8}$$

where $\Delta E = E_2 - E_1$, and ν_0 is the irradiation frequency. For any allowed Raman transition, then, two new frequencies usually appear in the spectrum of scattered light, shifted to the negative and positive sides of the dominant Rayleigh line by the same amount, $\Delta\nu = \Delta E/h$. For this reason Raman spectroscopy is always concerned with measurements of frequency *shifts*, rather than the absolute frequencies. In general, of course, a number of Raman transitions can take place involving various molecular energy levels, and the spectrum of scattered light thus contains a range of frequencies shifted away from the irradiation frequency. In the particular case of vibrational Raman transitions, these shifts can be identified with vibrational frequencies in the same way as absolute absorption frequencies in the infra-red spectrum.

One other point is worth raising before proceeding further. Although the Stokes and anti-Stokes lines in a Raman spectrum are equally separated from the Rayleigh line, they are not of equal intensity, as illustrated in Fig. 4.20. This is because the intensity of each transition is proportional to the population of the energy level from which the transition originates; the ratio of populations is given by the Boltzmann distribution (Eq. 1.13). There is also a fourth-power dependence on the scattering frequency, as the detailed theory shows. Hence the ratio of intensities of the Stokes line and its anti-Stokes partner in a Raman spectrum is given by

$$I_{AS}/I_S = \{ (\nu_0 + \Delta\nu) / (\nu_0 - \Delta\nu) \}^4 (g_2/g_1) \exp(-\Delta E/kT), \tag{4.9}$$

and the anti-Stokes line is invariably weaker in intensity. The dependence of this ratio on the absolute temperature T can be made good use of in certain applications, for example in determining flame temperatures.

The Raman effect is a very weak phenomenon; typically only one incident photon in 10^7 produces a Raman transition, and hence observation of the effect calls for a very intense source of light. Since the effect is made manifest in shifts of frequency away from that of the incident light, it also clearly requires use of a monochromatic source. Not surprisingly, then, the field of Raman spectroscopy was given an enormous boost by the arrival of the laser, so that Raman spectroscopy is today virtually synonymous

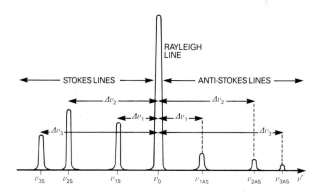

Fig. 4.20. Model Raman spectrum showing the intensity of light scattered with frequency ν'

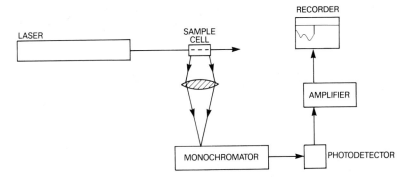

Fig. 4.21. Instrumentation for laser Raman spectroscopy

with *laser* Raman spectroscopy. A typical setup is shown in Fig. 4.21. Light scattered by the sample from a laser source (usually an ion laser) is collected, usually at an angle of 90°, passed through a monochromator, and produces a signal from a photodetector. The Raman spectrum is then obtained by plotting the variation in this signal with the pass-frequency of the monochromator. Even with a fairly powerful laser, Raman spectroscopy requires a very sensitive photodetection system capable of registering single photons to detect the weakest lines in the spectrum.

In addition to providing data on molecular vibration frequencies, information can also be obtained on the symmetry properties of the vibrations themselves. This is accomplished by measurement of the *depolarisation ratios* of the lines in the Raman spectrum. The procedure is very simple; the sample is irradiated with plane polarised laser radiation with polarisation perpendicular to the plane of Fig. 4.21. The Raman spectrum is then collected with a polarising filter inbetween the sample and monochromator, so as to analyse for the two perpendicular polarisation components of the scattered light, as shown in Fig. 4.22. The ratio of intensities in the spectra obtained from the two configurations, defined by

$$\varrho_1 = I(z \rightarrow y) / I(z \rightarrow z), \qquad (4.10)$$

can then be calculated for each Raman line, and denotes the depolarisation ratio for the corresponding vibration. In the case of gases and liquids, ϱ_1 takes the value of $\frac{3}{4}$ for vibrations which lower the molecular symmetry, but less than $\frac{3}{4}$ for vibrations which are totally symmetric.

All phases of matter are amenable to study by the Raman technique, and there are several advantages over infra-red vibrational spectroscopy in the particular case of aqueous solutions. Firstly, because only wavelengths in the visible region are involved, conventional glass optics and cells can be used. Secondly, since water itself produces a rather weak Raman signal, spectra of aqueous solutions are not swamped by the solvent. For these reasons, Raman spectroscopy is especially well suited to biological samples. As with most types of laser spectroscopy the spectrum is derived from a relatively small number of molecules because of the narrow laser beamwidth, and even with a comparatively insensitive spectrometer only a few ml's of sample is sufficient. Raman spectroscopy is thus also a useful technique for the analysis of the products of

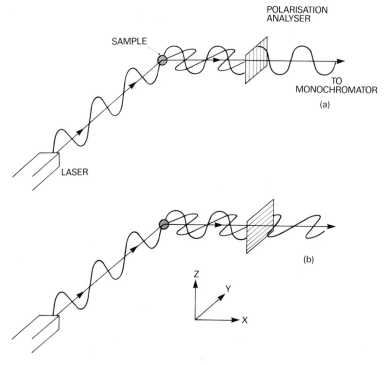

Fig. 4.22. Measurement of Raman depolarisation ratios; (**a**) represents the I(z → z) configuration and (**b**) I(z → y)

reactions with low yield. One further point worth noting is that compared to infra-red spectroscopy, sample heating is much reduced by using visible radiation. The only case where heating does cause a real problem is with strongly coloured solids, where it has become a common practice to spin the sample so that no single spot is continuously irradiated; this is particularly important in the case of *resonance* Raman studies, discussed in Sect. 4.5.1.

One of the major problems with Raman spectroscopy is interference in the spectrum of scattered light from sample fluorescence. There are various ways to overcome this problem; for example use can be made of the fact that the fluorescence signal will not generally exhibit any shift in frequency when the excitation frequency is changed, whereas of course the entire spectrum of Raman signals will shift by the same uniform amount. A more sophisticated method is to use pulsed lasers and time-gating techniques to separate the slower fluorescence from the spontaneous Raman emission. However in some cases much simpler remedies are possible. For example in solids, prolonged exposure to the laser radiation can often reduce the extent of fluorescence from defects and impurities, and in solutions it is sometimes expedient to add a fluorescence quencher with a known spectrum to accomplish the same result.

One type of application that is gaining popularity is known as the *laser Raman microprobe*, and is principally used for heterogeneous solid samples, for example geological specimens. Two different methods have been developed based on the microprobe

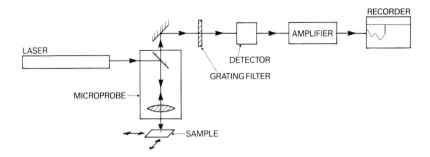

Fig. 4.23. Raman microprobe (point illumination) instrumentation. In the alternative global illumination configuration, a larger surface area is illuminated, and the detection equipment is replaced with an image intensifier phototube and camera

principles discussed in Sect. 3.9. In one method illustrated in Fig. 4.23, various points on the surface of the sample (or indeed if it is transparent various regions within the sample) are irradiated with a laser beam and the Raman scattering is monitored. Since any particular chemical constituent should produce Raman scattering at one or more characteristic wavelengths, it is possible to filter out the emission from the surface at one particular wavelength, and hence produce a map of the surface concentration of the substance of interest. Used in conjunction with a visual display unit and image-processing techniques, this method undoubtedly has enormous potential as a fast and non-destructive means of chemical analysis. A less sensitive, but simpler technique involves global irradiation of the sample. In this way a small area of the surface can be imaged directly, and the entire image can be passed through a filter to a camera system.

4.5 Specialised Raman Techniques

As with absorption spectroscopy, there is a wide range of modifications to the standard methods of Raman spectroscopy described above. To some extent, these are modifications to the Raman process itself, rather than simply different means of detection; nonetheless, all the techniques discussed in this section involve essentially the same types of Raman *transition*. The one method which does not fit into this category, namely *hyper-Raman* spectroscopy, is discussed separately in Sect. 4.6.4.

4.5.1 Resonance Raman Spectroscopy

In Sect. 4.4, it was pointed out that the frequency of radiation used to induce Raman scattering need not in general equal an absorption frequency of the sample. Indeed, it is generally better that it does not, in order to avoid possible problems in accurately recording the Raman spectrum caused by the interference of absorption and subsequent fluorescence. However, there are certain special features which become apparent when an irradiation frequency is chosen close to a broad intense optical absorption band (such as one associated with a charge-transfer transition) which make the technique a

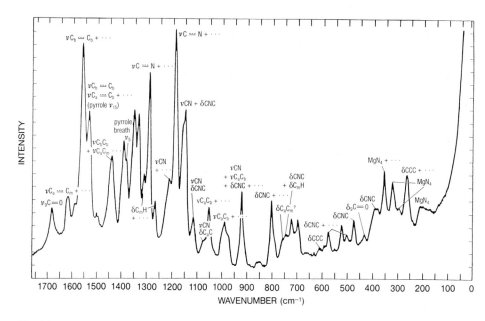

Fig. 4.24. Resonance Raman spectrum of chlorophyll *a* at 30 °K, excited by a He-Cd laser at 441.6 nm (for the structure of chlorophyll *a*, see Fig. 5.6, p. 150). Reprinted from Ref. 8 by permission of John Wiley and Sons Ltd

useful one despite its drawbacks. Quite simply, the closer one approaches the resonance condition, the greater is the intensity of the Raman spectrum. The selection rules also change, so that certain transitions which are normally forbidden become allowed, thus providing extra information in the spectrum. Lastly, in the case of large polyatomic molecules where any electronic absorption band may be due to localised absorption in a particular group called a *chromophore*, the vibrational Raman lines which experience the greatest amplification in intensity are those involving vibrations of nuclei close to the chromophore responsible for the resonance. In principle, this facilitates deriving structural information concerning particular *sites* in large molecules. This feature has been much utilised, for example, in the resonance Raman vibrational spectrosopy of biological compounds containing strongly coloured groups, and is well illustrated by the spectrum of chlorophyll *a* in Fig. 4.24.

Figure 4.25 shows the energetics of Raman scattering under different conditions with increasing irradiation frequency, with the conventional non-resonant process represented by (a). In the situation corresponding to (b), there is a sizeable *pre-resonance* growth in intensity even before the irradiation frequency reaches that of an absorption band. When the incident frequency is coincident with a discrete absorption frequency as in (c), real absorption and emission transitions can take place, and *resonance fluorescence* starts to complicate the picture. Nonetheless, it is when the incident photons have sufficient energy to bridge the gap to a continuum state (d) that the largest Raman intensities are produced, often with an amplification factor of up to a thousand over the normal Raman signal: this phenomenon is known as the *resonance Raman effect*. If the intensity of a given Raman line is plotted as a function of the irradiation

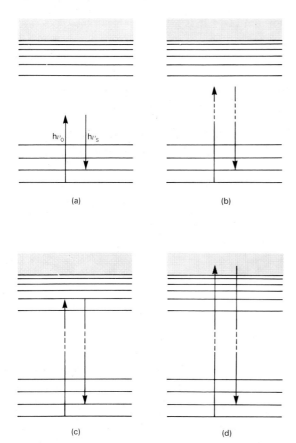

Fig. 4.25a-d. Diagrams showing the same Stokes Raman transition with variation in the irradiation frequency ν_0: (a) conventional Raman effect; (b) pre-resonance Raman effect; (c) resonance fluorescence; (d) resonance Raman effect

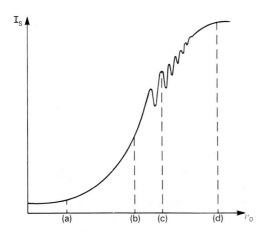

Fig. 4.26. Resonance Raman profile of the transition illustrated in Fig. 4.25

frequency, then a graph like that shown in Fig. 4.26 is produced; this is known as a *resonance Raman excitation profile*.

Resonance Raman spectroscopy can be performed with any laser source provided it emits a wavelength lying within a suitable broad absorption band of the sample. Clearly, for application to a range of samples, a tunable dye laser is to be preferred. Unfortunately, resonance fluorescence occurs with increasing intensity, and over an increasingly broad range of emission wavelengths as the resonance condition is approached. However, this problem may be overcome by use of a mode-locked source providing ultrashort (picosecond) pulses. Since there is no detectable time-delay for the appearance of the Raman signal, whilst resonance fluorescence is associated with a lifetime typically in the nanosecond range, the signal from the photodetector can be electronically sampled at suitable intervals and processed so that only the true Raman emission is recorded.

4.5.2 Stimulated Raman Spectroscopy

As has been stressed before, the Raman effect is essentially a very weak one, and it is in general only under conditions of resonance enhancement that the scattering intensity could be regarded as appreciable. However, there is one other way in which the effect can be enhanced, which is as follows. If a sufficiently intense laser is used to induce Raman scattering, then despite the initially low efficiency of conversion to Stokes frequencies, Stokes photons which are emitted can *stimulate* the emission of further Stokes photons through Raman scattering of laser light by other sample molecules in the beam. This process evidently has the self-amplifying character always associated with stimulated emission, and is therefore most effective for Stokes scattering in approximately the direction of the laser beam through the sample. In fact with a giant

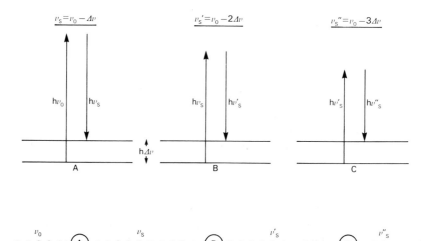

Fig. 4.27. Illustration of the stimulated Raman effect. As the laser photon of frequency ν_0 travels through the sample, it suffers three consecutive frequency conversions through identical Stokes Raman transitions in three different molecules A, B and C

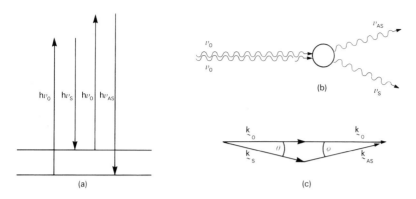

Fig. 4.28 a–c. Four-wave process involved in stimulated Raman scattering and CARS: (a) shows the energetics, (b) depicts the physics and (c) illustrates the wave-vector matching condition

pulse laser this *stimulated Raman scattering* can lead to generation of Stokes frequency radiation in the 'forward' direction with a conversion efficiency of about 50 %.

Once a strong Stokes beam is established in the sample, of course, it can lead to further Raman scattering at frequency

$$v'_S = v_S - \Delta v, \tag{4.11}$$

and this too may be amplified to the point where it produces Raman scattering at frequency

$$v''_S = v'_S - \Delta v, \tag{4.12}$$

and so on. Hence a series of regularly spaced frequencies $(v_0 - m\Delta v)$ appears in the spectrum of the forward-scattered light, as illustrated in Fig. 4.27. Because of the self-amplifying nature of the process, it is usually the case that only the vibration producing the strongest Stokes line in the normal Raman spectrum is involved in stimulated Raman scattering. The effect does not, therefore, have the analytical utility of conventional Raman spectroscopy.

One other phenomenon generally accompanies stimulated Raman scattering, and is not always clearly differentiated from it. This is a *four-wave mixing* process, whose energetics are illustrated in Fig. 4.28. It provides a mechanism for the conversion of two laser photons of frequency v_0 into a Stokes and anti-Stokes pair, with frequencies $v_S = (v_0 - \Delta v)$, and $v_{AS} = (v_0 + \Delta v)$. Since this process returns the molecule in which it takes place to its initial state, there is no relaxation period necessary before two more laser photons can be converted in the same molecule, which of course significantly improves the conversion efficiency. There are other reasons why the process is particularly effective, however. One is the fact that if the photons involved propagate in a suitable direction, four-wave mixing can take place without any transfer of momentum to the sample; this means that the total momentum of the absorbed photons must equal that of the two emitted photons. Photons always carry a momentum $hk/2\pi$ in the direction of propagation, where k is the *wave-vector* of magnitude $k = 2\pi v/c' = 2\pi n/\lambda$, and n

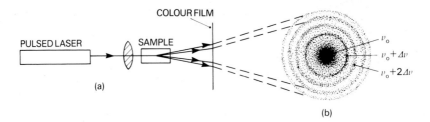

Fig. 4.29. Photographic demonstration of the four-wave stimulated Raman effect: (a) shows the apparatus, and (b) the photograph, assuming that the *Stokes* frequencies $v_0 - m\Delta v$ are outside the range of sensitivity of the photographic emulsion

is the refractive index of the medium. Because the refractive index generally depends on wavelength too, the required *wave-vector matching* is generally produced when there are small angles between the wave-vectors as shown in Fig. 4.28 b and c.

The four-wave process thus returns the molecule to its initial state, and confers no momentum to it. Any mechanism which fulfils these two conditions is known in the field of nonlinear optics as a *coherent parametric process*, and is invariably associated with a substantially increased conversion efficiency. Since in the experimental context described above the Stokes transition involved can also be a stimulated process, this four-wave interaction can therefore take place at a very significant rate, and it plays an important role in the production of further anti-Stokes frequencies ($v_0 + m\Delta v$). The angle-dependence of the emission can be graphically illustrated by placing a colour-sensitive photographic film as shown in Fig. 4.29 a, producing rings of colour as in (b). Most of the converted frequencies are emitted within an angle of about 10° away from

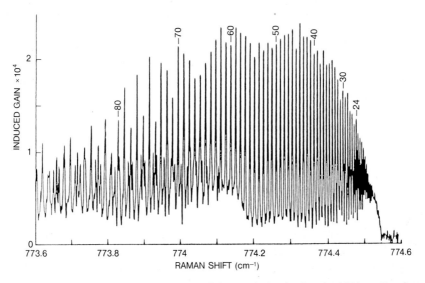

Fig. 4.30. Raman gain spectrum of SF_6 at 3.8 torr, obtained using the 647.1 nm line from a krypton ion laser as probe, showing v_1 fundamental Q(J) transitions and underlying $v_1 + v_6 \leftarrow v_6$ hot band transitions. Reprinted from reference 9 by permission of John Wiley and Sons Ltd

the laser beam direction. The efficiency of this process in producing Stokes and anti-Stokes frequencies in liquids or highly pressurised gases is sufficiently high to make the effect useful as a means of laser frequency conversion (see Sect. 3.2).

Finally, we note that there is more than one other spectroscopic method based on the principle of stimulated Raman scattering. One of the alternatives is to simultaneously irradiate the sample with both a 'pump' laser beam of frequency v_0 and with another tunable 'probe' laser beam. When the frequency of the latter beam is tuned through a Stokes frequency v_S, it stimulates the corresponding Raman transition, and thus experiences a gain in intensity. The Raman spectrum is thus obtained by plotting the intensity of the probe beam after passage through the sample against its frequency; this method is known as *Raman gain spectroscopy*. The beautiful example of a Raman gain spectrum shown in Fig. 4.30 illustrates the rotational fine structure in the fundamental breathing mode of SF_6. The other common method involving stimulated Raman transitions is inverse Raman spectroscopy, which has a rather different methodology as described below.

4.5.3 Inverse Raman Spectroscopy

The stimulated Raman processes discussed in the last Section involve the absorption of a frequency v_0 and the stimulated emission of a Stokes frequency v_S. *Inverse* Raman spectroscopy is based on the converse radiative processes, in other words absorption of a Stokes frequency (or for that matter an anti-Stokes frequency), and stimulated emission of v_0, as shown in Fig. 4.31. In practice, it is usual to employ an intense monochromatic laser source for the pump beam of frequency v_0, and a secondary *continuum* source of lower power to act as probe; inverse Raman transitions are then detected as absorptions from the probe beam at each Stokes or anti-Stokes frequency. Thus although it is possible to use a tunable dye laser pumped by the pump laser for the probe, the broad band fluorescence from the laser dye can in fact be used directly without being monochromatised, as shown in Fig. 4.32 a. Note that the energetics of the inverse Raman effect are such that the *anti-Stokes* transition is normally the more intense since it originates from a state of lower energy, and hence with a higher population than does the Stokes transition. Thus the spectrum is best obtained with

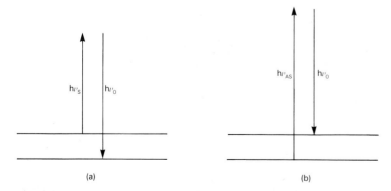

(a)   (b)

Fig. 4.31 a and b. Stokes and anti-Stokes inverse Raman transitions

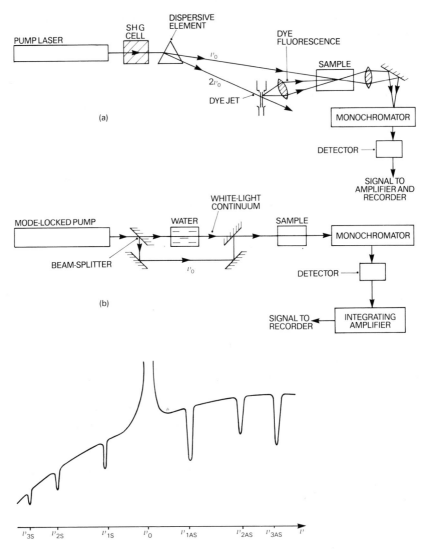

Fig. 4.32 a–c. Measurement of inverse Raman spectra: (**a**) using a laser dye to generate a continuum, and (**b**) using a white light continuum produced by self-focussing of picosecond laser pulses in water; (**c**) illustration of the spectrum (compare Fig. 4.20)

probe frequencies *higher* than the pump beam; this is the reason for the use of a frequency-doubling crystal in Fig. 4.32 a.

An alternative method makes use of the broad-band *ultrafast supercontinuum laser source* produced by focussing intense ultrashort laser pulses into an optically transparent but physically dense material (see Sect. 3.3.3). The physical mechanism for this process is complex, but the effect is delightfully simple to produce; focussing into a beaker of water will often suffice to produce a continuum covering the whole range of the visible spectrum. The setup used to obtain inverse Raman spectra using this source as the probe is shown in Fig. 4.32 b, and c shows the kind of spectrum obtained by either inverse Raman method.

4.5.4 CARS Spectroscopy

Next we consider *coherent anti-Stokes Raman scattering spectroscopy*, usually abbreviated to *CARS*. On the molecular level, the mechanism for this process is precisely the same as the four-wave interaction encountered in Sect. 4.5.2 in connection with the stimulated Raman effect, and all of the illustrations in Fig. 4.28 are directly applicable to CARS. The difference arises in the method used to induce the effect. Whereas in the case of the stimulated Raman effect the Stokes wave is spontaneously generated by conventional Raman scattering from a beam of frequency v_0, in CARS it is produced by a second laser beam directed into the sample, whose frequency v_1 is tuned across the frequency range below v_0. The four-wave interaction (see Fig. 4.33) produces coherent emission at a frequency

$$v' = 2v_0 - v_1, \tag{4.13}$$

so that when v_1 coincides with *any* Stokes Raman frequency v_S, we have

$$v' = 2v_0 - v_S = 2v_0 - (v_0 - \Delta v) = v_0 + \Delta v = v_{AS}, \tag{4.14}$$

corresponding with the anti-Stokes frequency. Since it is once again a coherent parametric process, the CARS emission is produced in a well-defined direction governed by the wave-vector matching condition as in Fig. 4.28 (c).

In contrast to the stimulated Raman effect, emission at the frequency given by Eq. (4.13) can also occur, albeit with reduced intensity, when v_1 does *not* equal a Stokes frequency. The distinction is illustrated in Fig. 4.33: (a) shows the *resonant* CARS mechanism (cf. Fig. 4.28 a); (b) shows the *non-resonant* process which occurs when $v_1 \neq v_S$. Hence although the strongest lines in the CARS spectrum occur when v_1 strikes resonance with a Stokes frequency, this is usually seen against a background emission due

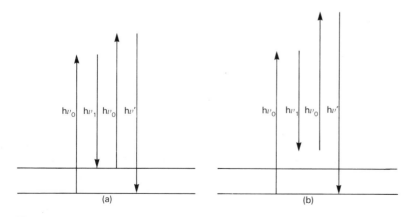

Fig. 4.33 a. Resonant, and (b) non-resonant four-wave interactions in CARS spectroscopy

to the relatively frequency-insensitive non-resonance mechanism. For this reason, CARS spectroscopy is not well suited to applications in trace analysis. However since it does produce a spectrum typically 10^4–10^5 times more intense than the normal Raman effect, but subject to the same selection rules, there are other important areas of application, as we shall see.

Although CARS is a two-beam method basically requiring two laser sources, part of the output from a single pump laser may be frequency-converted in a dye laser cavity to produce the second beam, as shown in Fig. 4.34. The two laser beams are directed into the sample at a small angle θ, and as the dye laser is tuned, CARS emission is detected through a pinprick hole in a direction making an angle φ with the ν_0 beam, as shown. In fact, the two angles θ and φ are nearly equal, since the wave-vector matching triangle of Fig. 4.28 c is almost isosceles. In principle, since the CARS emission frequency is given by Eq. (4.13), it is completely determined by the two incident frequencies, and the spectrum may be obtained by plotting the intensity of light received by the detector against the frequency difference $\nu_0 - \nu_1$.

Occasionally a monochromator may be placed before the detector to cut out stray frequencies resulting from light scattering and fluorescence processes in the sample. However it is often sufficient to use a simple optical filter with a cut-off just above ν_0 for this purpose, making use of the fact that only higher frequencies can result from the CARS process.

An unusual feature of CARS, but one which it holds in common with other coherent parametric processes, is that it depends quadratically on the number of molecules per unit volume. Hence whilst the vibrational spectra of liquids can be obtained with cw lasers, giant pulsed lasers producing megawatt pulses are required in order to obtain the vibration-rotation CARS spectra of gaseous samples; a frequency-doubled Nd:YAG source is a common choice. It is, nonetheless, a most useful technique for samples obtainable only in very small quantities; since the CARS signal is produced only by molecules at the intersection focus of the two applied beams, only μl volumes of liquid are required, or gas pressures of 10^{-6} atmospheres. Two other points are worth noting concerning gaseous samples. Firstly, because the refractive index is close to unity for each frequency involved, the angles θ and φ in the wave-vector matching diagram are sufficiently small that in practice a collinear beam geometry can be used. Secondly, there is almost no Doppler broadening of spectral lines, since the effect is based on the absorption and emission of photons with almost identical wave-vector. Thus CARS can be used to produce spectra of very high resolution, with bandwidths as low as 10^{-4} cm^{-1} in optimum cases.

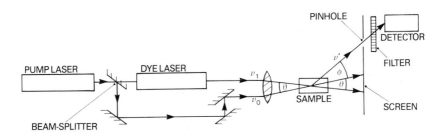

Fig. 4.34. Instrumentation for CARS spectroscopy

One special area in which CARS spectroscopy has found widespread application, by virtue of its particular suitability to highly luminescent samples, is in combustion and other high-temperature reaction diagnostics. The main advantage over traditional Raman spectroscopy is the fact that a coherent high-frequency output beam is produced, which can be much more easily detected against a strong fluorescence background. Here the CARS technique often provides results which usefully complement the data on transients available from laser-induced fluorescence measurements. By making the two applied laser beams intersect within the flame or reaction volume to be studied, the CARS spectrum not only helps to reveal the chemical composition in any small region, but also the temperature can be accurately determined from the relative intensities of rotational lines. This method of temperature determination not only surpasses that of any standard thermocouple device in its degree of accuracy, but also in that measurements are possible well above the usual thermocouple limit of about 2500 K. The technique is moreover non-instrusive, and hence in no way affects the process being studied. A good industrial example of this type of application lies in the analysis of gas streams in coal gasification plants.

4.5.5 Surface-enhanced Raman Spectroscopy

Another special type of Raman spectroscopy involves the study of surface-adsorbed species. In view of the low intensities normally expected in Raman spectra, the technique does not immediately seem a very sensible choice for the study of very low surface concentrations. However, it has been found that for species adsorbed onto suitable metallic substrates (silver is the favourite), the Raman signal from the adsorbate is often many orders of magnitude higher than might be expected. There has been a great deal of debate about the reason for this enhancement and the factors by which it is influenced, and it is likely that more than one mechanism is at work. Surface roughness

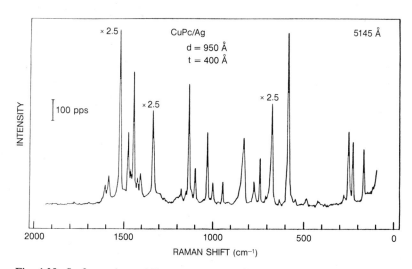

Fig. 4.35. Surface-enhanced Raman spectrum of copper phthalocyanine on silver obtained by Hayashi and Samejima using 514.5 nm radiation from an argon ion laser. Reprinted from reference 10 by permission of North Holland Physics Publishing

is one effect which seems to play an important role. It is also clear that it is the molecules closest to the metal surface whose Raman transitions experience the greatest enhancement. Figure 4.35 shows the surface-enhanced Raman spectrum of a 95 nm layer of copper phthalocyanine on an evaporated silver film substrate. The high luminescence level and low solubility of phthalocyanine compounds usually result in a comparatively poor signal-to-noise ratio in the Raman spectrum, but the enhanced thin film spectrum gives a very good result. The enhancement factor here is approximately 20, but can be as large as 10^4 for monolayer coverage on silver-island films.

There is potentially a great deal of practical utility in the effect, and surface-enhanced Raman spectroscopy (SERS) is rapidly gaining popularity, albeit for certain rather specific applications. One particular area of interest lies in solution studies using electrochemical cells with silver electrodes. Silver electrodes can be very effectively roughened by oxidation-reduction cycles, and solute species can subsequently be studied at the electrode surface by enhanced Raman spectroscopy. Symmetry considerations often permit the orientation of adsorbed molecules to be deduced by this means. Other potential applications include the surface-adsorbed species involved in heterogeneous catalysis.

4.5.6 Raman Optical Activity

The last subject we shall deal with in this Section concerns Raman scattering measurements on optically active compounds, usually in the liquid or solution state, using circularly polarised laser light. Using the same type of electro-optic modulation of polarisation discussed in Sect. 3.1 and 3.7, it is possible to obtain a spectrum showing the *difference* in the Raman intensity $I^R - I^L$ as a function of scattering frequency. In contrast to most conventional measurements of chirality such as optical rotatory dispersion or circular dichroism, which are based on molecular electronic properties, *circular differential Raman spectroscopy* is based on the molecular vibrations. For this reason it is more directly related to the detailed stereochemical structure which is ultimately responsible for any manifestation of chirality. The information contained in a circular differential Raman spectrum is also far more directly useful. In particular, the extent of differential scattering in a region of the spectrum associated with a particular group frequency can be interpreted in terms of the chiral environment of the corresponding functional group. Figure 4.36 shows the Raman circular intensity sum and difference spectra of pure (R)-(+)-3-methylcyclohexanone obtained by Laurence Barron, one of the pioneers of the method.[11] The enormously complicated form of the lower trace illustrates the wealth of structural information which is contained in the differential spectrum.

The only other method which offers data of this kind is vibrational circular dichroism using infra-red radiation, a technique which is very insensitive for low-frequency vibrations. However, circular differential Raman scattering is not only a more sensitive technique across a full 50–4000 cm^{-1} range of vibrations, but it has all the experimental advantages of spectroscopy in the visible/uv region. For example, by making use of resonance conditions, it is possible to study the stereochemistry of specific sites of biological activity in large biomolecules in aqueous solution. The same principles can be applied to optically inactive (achiral) molecules by inducing chirality with a magnetic field.

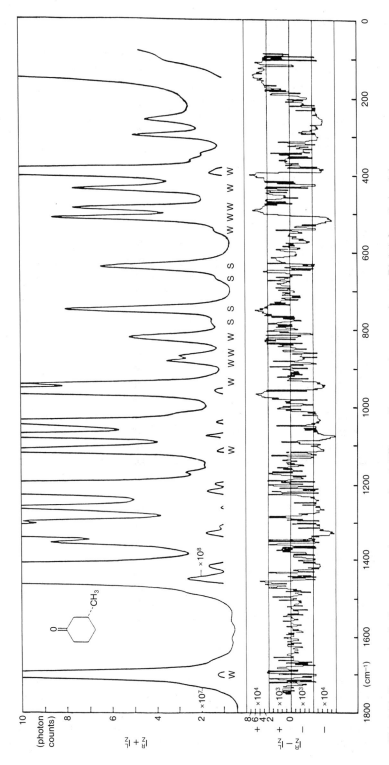

Fig. 4.36. Raman circular intensity sum (upper trace) and difference (lower trace) spectra of pure (R)-(+)-3-methylcyclohexanone

4.6 Multiphoton Spectroscopy

Multiphoton processes involve the concerted interaction of two or more photons with individual atoms or molecules. We have already dealt with some examples; simple Raman scattering is one case, in which one photon is absorbed and one is emitted in each molecular transition. Although the term 'multiphoton' is seldom used to describe this process, it does illustrate the essential concerted nature of the interaction, in that the absorption and emission are inextricably bound up with each other; it is not the same thing as absorption followed by fluorescence. However, the term 'multiphoton' is generally applied to processes involving the concerted absorption of two or more photons. The four-photon interactions connected with stimulated Raman scattering and CARS provide good examples, but the simplest multiphoton processes are those in which only absorption is involved. Figure 4.37 illustrates the essential difference between (a) a process involving two sequential single-photon absorption events, and (b) the concerted process of two-photon absorption. The former takes place only if both photons have suitable energies, and it results in population of both levels E_1 and E_2; in the latter case the only restriction is on the *sum* of the photon energies, and only level E_2 becomes populated.

4.6.1 Single-beam Two-photon Absorption

The very simplest case of two-photon absorption is where there is a single monochromatic beam of light and photons are absorbed pairwise by atoms or molecules of the sample. Several principles of more general application can be understood by considering this case in detail. The first thing to note is that it requires a very intense source of light for observation of the process; in fact the effect was not experimentally demonstrated until the first pulsed lasers arrived on the scene. It is not hard to understand why this is so. Because the photons have to be absorbed in a concerted fashion, it is necessary for two photons to pass essentially *simultaneously* through the region of space occupied by one molecule (or else the space occupied by a chromophore, in the case of localised absorption in a large molecule). The likelihood of this depends on the intensity of light and the molar volume, and can be estimated quite easily.

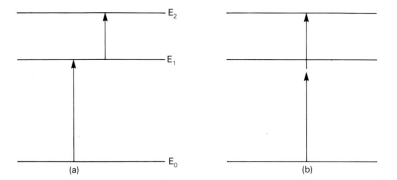

Fig. 4.37a and b. Schematic energetics of (a) two sequential single-photon absorption processes, and (b) two-photon absorption

For comparison purposes, we first note that for an unfocussed cw laser light source, for example an argon laser producing an irradiance of 10^7 W m^{-2} at a wavelength of 488.0 nm, the photon density given by Eq. (1.6) is approximately 7×10^{16} photons per m^3, corresponding to 2×10^{-12} photons per molecule in liquid water. Hence even with this level of intensity, only one molecule in 5×10^{11} is experiencing the transit of a single photon at any instant of time. The probability of *two* photons simultaneously traversing a water molecule can be calculated from equation 1.29, with M = 2×10^{-12} and N = 2; the result is 2×10^{-24}, corresponding to only one in 5×10^{23}, in other words virtually one molecule per mole. We must conclude, then, that even if we had a molecule possessing the correct energy level spacing, there would be no likelihood of ever observing two-photon absorption with this level of intensity. However, if the same laser is mode-locked and focussed, peak intensities of 10^{15} W m^{-2} can easily be produced. Whilst this increase by a factor of 10^8 results in a roughly proportional increase in the probability of finding *one* photon in a molecular volume, the probability of finding *two* photons increases by a factor of approximately 10^{16} to roughly one molecule in 5×10^7. With *pulsed* lasers, then, we may expect to be able to detect two-photon absorption in suitable media.

Two other facets of two-photon absorption are illustrated by these considerations. Firstly, although it is not of much chemical interest, it is worth noting that if the calculations are repeated for thermally produced radiation of the same mean intensity, using Eq. (1.28) rather than (1.29), it turns out that the probability of finding two photons in a molecular volume is twice as large. This illustrates the fact that, unlike conventional absorption, multiphoton absorption is influenced by the photon statistics of the source. Much more important, however, is the fact that the probability, and hence the rate of two-photon absorption, depends *quadratically* on the irradiance. Thus on passage of a beam of light through a medium exhibiting the effect, the rate of loss of intensity is given by

$$-\mathrm{d}I/\mathrm{d}l \propto I^2 C \qquad\qquad\qquad (4.15)$$

in contrast to Eq. (4.1). The solution to this Equation takes the form

$$I = I_0 / (1 + \beta l C) \qquad\qquad\qquad (4.16)$$

where β is a constant depending on the strength of the two-photon transition. Hence the usual Beer-Lambert Law of Eq. (4.2) does not hold for this process, and the concepts of absorbance and molar absorption coefficient do not apply as normally defined.

We now turn to the specifically spectroscopic properties of the two-photon process. Again, there is a substantial difference from conventional absorption in several areas. Foremost is the fact that the selection rules are quite different. As illustrated in Fig. 4.38, although the two-photon transition is subject to the condition

$$2h\nu = \Delta E \qquad\qquad\qquad (4.17)$$

on the laser frequency ν, it is not necessarily the case that the same transitions can be induced by single-photon absorption from a beam of frequency 2ν. In centrosymmetric molecules, for example, whilst the usual absorption selection rules permit only transi-

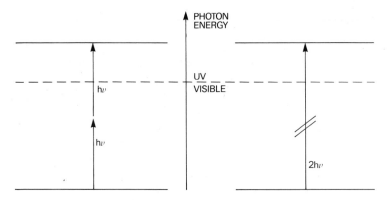

Fig. 4.38. Illustration of a transition which is two-photon allowed but single-photon forbidden

tions between states of *opposite* parity, i.e. gerade—ungerade, the two-photon selection rules permit only transitions between states of the *same* parity, i.e. gerade—gerade or ungerade—ungerade. Indeed, two-photon absorption in the uv/visible range stands in the same relationship of complementarity to conventional uv/visible spectroscopy as Raman spectroscopy does to infra-red. Even for compounds with transitions which *are* both one- and two-photon allowed, two-photon spectroscopy in the visible range usefully provides access to states which would otherwise require uv excitation, as also shown in Fig. 4.38. In such cases, the ultraviolet response can thus be probed without the need for special vacuum techniques, as only *visible* wavelengths are involved.

Another interesting feature is the possibility of *Doppler-free spectroscopy*. If a mirror is placed such that the laser beam traverses the sample in both directions, then as shown in Fig. 4.39 a and b, it is effectively irradiated by two counterpropagating beams of wavevectors k and −k. The two Doppler-shifted frequencies 'seen' by a sample molecule travelling with velocity v are given by the absorption analogue of Eq. (1.24), i.e.

$$v_k' \approx v(1 - v_k/c) \tag{4.18}$$

and hence equally

$$v_{-k}' \approx v(1 + v_k/c) \tag{4.19}$$

Two-photon transitions in which counterpropagating photons are absorbed thus provide an excitation energy in which the two Doppler shifts essentially cancel out;

$$\Delta E = hv_k' + hv_{-k}' = 2hv \tag{4.20}$$

The result is a very sharp and symmetrical Doppler-free absorption line on top of a broader band due to the two-photon absorption of photons travelling in the same direction, for which the usual Doppler shift occurs; a typical result is illustrated in Fig. 4.39 (c).

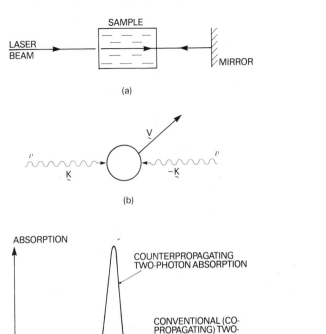

SAMPLE

LASER
BEAM

MIRROR

(a)

(b)

ABSORPTION

COUNTERPROPAGATING
TWO-PHOTON ABSORPTION

CONVENTIONAL (CO-
PROPAGATING) TWO-
PHOTON ABSORPTION

Fig. 4.39 a–c. Doppler-free two-photon spectroscopy

(c)

The last major spectroscopic difference between two-photon absorption and conventional absorption is that even in isotropic samples such as gases or liquids, the two-photon process is very sensitive to the polarisation of the laser beam. In the conventional (single-photon) absorption spectroscopy of fluid samples, such effects are generally very small, and are manifested only by optically active compounds. However, the whole appearance of a two-photon spectrum can be changed dramatically if the beam polarisation is altered, as illustrated in Fig. 4.40. The precise way in which the intensity of each two-photon absorption band changes with laser polarisation provides useful information on the type of transition responsible. For example, in any two-photon transition from the usual totally symmetric ground state to a state which is not totally symmetric (in other words one which lacks the full symmetry properties of the molecule), then the intensity of the associated absorption band increases by a factor of exactly $\frac{3}{2}$ on changing from plane to circular polarisation.

The recording of two-photon excitation spectra can most simply be accomplished either by directly monitoring the absorption from the laser beam, or else by detecting the fluorescence due to subsequent relaxation of excited molecules in the sample. The former method is not easy, since the fractional absorption is generally very small indeed; it is simpler to detect the fluorescence which is relative to an essentially zero background. The apparatus for fluorescence-detected two-photon absorption spectroscopy is illustrated in Fig. 4.41, and is similar in principle to the apparatus used for the

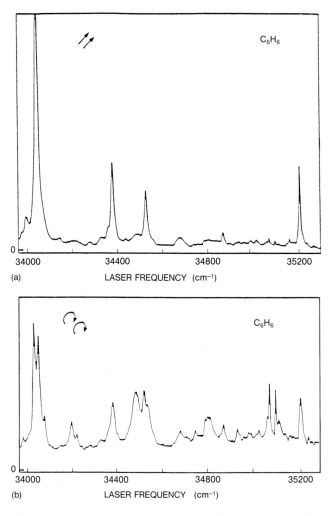

Fig. 4.40a and b. The two-photon spectrum of benzene obtained by Whetten et al. (a) with plane polarised light, and (b) with circularly polarised light from a frequency-doubled dye laser. Reprinted from reference 12 by permission of the American Institute of Physics

first demonstration of two-photon absorption by Kaiser and Garrett in 1961. A filter is used to cut out any fluorescence near to and below the irradiation frequency; any fluorescence observed at appreciably higher frequencies has to result from multiphoton excitation. An alternative detection method based on ionisation of the two-photon excited state is discussed in Sect. 4.6.3.

4.6.2 Double-beam Two-photon Absorption

At the molecular level, the types of transition involved in double-beam two-photon absorption are very similar to those described above. The energetics are such that

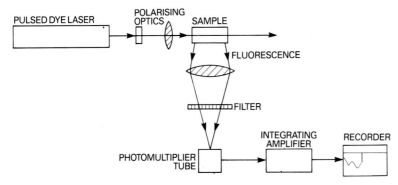

Fig. 4.41. Instrumentation for fluorescence-detected single-beam two-photon absorption studies

$$h\nu_1 + h\nu_2 = \Delta E \qquad (4.21)$$

and the rate of absorption is proportional to the *product* of the intensities of the two beams. In contrast to the single-beam method, however, the frequencies, directions and polarisations of the two absorbed photons can be independently varied. Typically, one beam could be derived directly from a primary laser source, and the other from a dye laser cavity pumped by the same primary laser, with various polarising and reflective optics as shown in Fig. 4.42; the fluorescence is detected by a photomultiplier tube and the signal passed to a recorder as before. There are two main advantages to be had from use of double-beam methods.

The first advantage is the possibility of resonance enhancement, similar to that which we discussed in connection with the Raman effect (Sect. 4.5.1). If one of the beams has a frequency somewhere close to an optical absorption band of the sample, then the rate of two-photon excitation is greatly increased, and a stronger signal is obtained. The second advantage is the facility to study the variation in the two-photon spectrum as both the polarisations of the two beams and the angle between them is varied. A very clever scheme devised by McClain[13] involving measurement of the spec-

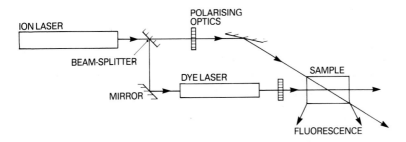

Fig. 4.42. Modification of the instrumentation in Fig. 4.41 for double-beam excitation. The angle between the two beams incident upon the sample will usually be smaller than shown here, to increase the intersection volume

trum under three different sets of conditions provides the most complete information on the symmetry of the two-photon excited states, and is a powerful tool for the determination of molecular electronic structure.

The types of substance in which two-photon absorption has been detected in the uv/ visible range now extends from inert gases to human chromosomes. The new atomic and molecular excited states to which the effect provides access may have photochemical significance, although this is still a largely unexplored area. As a spectroscopic technique, two-photon absorption is very sensitive, providing the facility for single-atom detection in optimum cases. Perhaps the most intriguing technical application lies in the possibility of cutting three-dimensional patterns in a block of clear plastic by focussing two laser beams inside it, and moving the point of intersection around. Providing the frequencies are correctly chosen, two-photon absorption can only occur where the beams cross, initiating a photochemical reaction to soften or harden the plastic. This idea is currently being evaluated for commercial viability.

4.6.3 Multiphoton Absorption Spectroscopy

Multiphoton studies where more than two photons are absorbed are generally based on a single beam of laser light, and transitions are subject to the condition

$$m h \nu = \Delta E \qquad\qquad\qquad (4.22)$$

where m is an integer. Once again, the selection rules differ from conventional absorption, and indeed are distinctive for each value of m. The intensity of absorption, in the absence of resonance enhancement, depends on I^m, where I is the laser beam irradiance. Arguments similar to those at the beginning of Sect. 4.6.1 quickly reveal that the rate of multiphoton absorption decreases very rapidly with any increase in the number of photons involved. In fact very few spectra involving more than three photons in the excitation process have been recorded, and these have generally required special techniques as discussed below. However, one distinction should be made before proceeding further. Whilst three and four-photon absorption generally represent the limit for uv/visible *spectroscopic* purposes, transitions involving the absorption of many more photons have certainly been observed in the *infra-red* laser excitation of polyatomic molecules. We shall return to this subject in Chapter 5; suffice it for the present to note that resonance enhancement is strongly involved.

The resonance aspect of multiphoton absorption lies at the heart of probably the most successful method for its detection, *multiphoton ionisation spectroscopy*. We saw in Sect. 4.2.2 that the detection of ions is a highly sensitive method in absorption spectrometry, and this sensitivity is especially well suited to the small signals expected from multiphoton studies. Since the method is applicable to gaseous samples, spectral resolution is also good. The principle involved is illustrated in Fig. 4.43. As the laser frequency is increased over its tuning range, the number of photons required to promote molecules from their ground state to the ionisation continuum drops from four to three, and the ion current increases accordingly. However, by far the largest signals are obtained under resonance conditions when the energy of one, two or three photons coincides with that of a bound state, as shown. For example at frequency ν_c, although three photons are responsible for the ionisation process itself, it is the two-photon res-

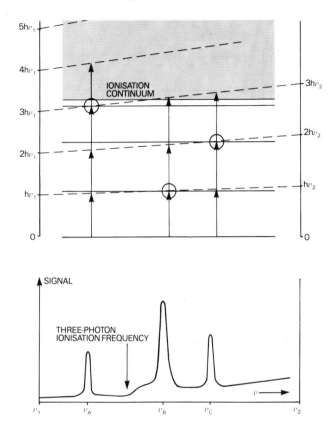

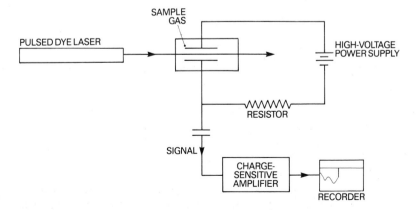

Fig. 4.43. Resonance-enhanced multiphoton ionisation spectroscopy. As the laser frequency is tuned from ν_1 to ν_2, multiphoton resonances occur at ν_A (three-photon), ν_B (one-photon) and ν_C (two-photon). With each resonance, the ionisation signal increases as shown in the lower half of the diagram

Fig. 4.44. Schematic apparatus for multiphoton ionisation spectroscopy

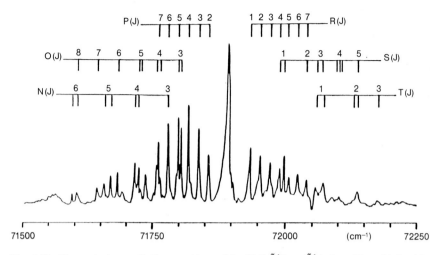

Fig. 4.45. Three-photon excitation spectrum of the H_2S $\tilde{H}^1B_1 \leftarrow \tilde{X}^1A_1$ transition obtained by resonant four-photon ionisation. Reprinted from Ref. 14 by permission of North Holland Physics Publishing

onance which results in a peak in the spectrum. A typical apparatus used to obtain the spectrum is shown in Fig. 4.44, and Fig. 4.45 shows the three-photon excitation spectrum of H_2S obtained by Dixon et al.[14] using this method. The principle of multiphoton ionisation is also made use of in a laser mass spectrometer, which is discussed in Sect. 4.7.

4.6.4 Hyper-Raman Spectroscopy

The hyper-Raman effect is in many senses a hybrid between multiphoton absorption and Raman scattering. The process takes place in samples irradiated by an intense laser beam, and involves three-photon transitions in which two photons are absorbed and one is emitted. Since two photons are absorbed from the laser beam in each transition, the hyper-Raman effect has a quadratic dependence on intensity, and can only be detected using pulsed lasers; even with mode-locked laser pulse intensities in the region of 10^{15} W m^{-2}, the intensity of hyper-Raman scattering is approximately 10^{-6} times weaker than that of Raman scattering.

The molecules involved in hyper-Raman scattering may undergo transition either to a state of higher energy, or to one of lower energy, as illustrated in Fig. 4.46: the two cases correspond to 'Stokes' and 'anti-Stokes' hyper-Raman scattering. Because of the weakness of the effect, it is usually only the relatively more intense Stokes transitions which are observed; these result in the production of frequencies ν' given by

$$h\nu' = 2h\nu - \Delta E \tag{4.23}$$

Hence the energy uptake is measured in frequency terms by a shift away from the *harmonic* frequency 2ν. The selection rules for hyper-Raman scattering are very different from those which apply to conventional Raman scattering; for example in centrosym-

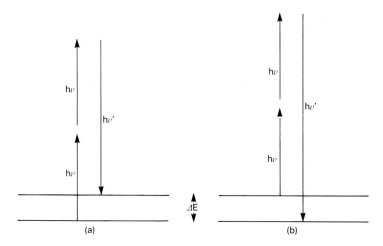

Fig. 4.46a and b. The hyper-Raman effect: (**a**) 'Stokes' transition, (**b**) 'anti-Stokes' transition

metric molecules, only *ungerade* vibrations show up in the spectrum. However, many more transitions are allowed than in single-photon absorption spectroscopy, and indeed one of the principal advantages of the hyper-Raman effect is its use for the direct observation of fundamental vibrations which are otherwise made manifest only in overtone or combination bands.

4.7 Laser Mass Spectrometry

The last topic to be discussed in this chapter is one which is quietly bringing about a revolution in the well-established field of mass spectrometry. This technique involves sample ionisation using a tunable laser rather than the traditional electrical methods. Since ionisation of most molecules requires energies in excess of that which can be supplied by a single photon unless very short wavelengths below 150 nm are used, laser ionisation generally has to involve a multiphoton excitation process; hence the term *multiphoton ionisation mass spectrometry* is also applied to this method. The advantage of using a mass spectrometer is that it enables the different types of ion produced by multiphoton absorption and subsequent fragmentation to be distinguished.

A typical laser mass spectrometer is based on a pulsed dye laser pumped by a nitrogen laser. Many details of the ionisation cell are determined by the nature of the sample. Samples are often introduced into the system in the gas phase, at vapour pressures usually not less than 10^{-6} atmospheres but in some cases as high as 50 atmospheres. Supersonic molecular beams are also amenable to laser ionisation analysis, and even liquid samples can be studied directly. Compared to a conventional instrument, the ionisation efficiency may be several orders of magnitude better, approaching 10 % for the ground state molecules. However, the ionisation volume is much smaller, again by several orders of magnitude, in a laser setup. The ionisation volume can of course be increased by using an unfocussed laser, but this has to be played off against

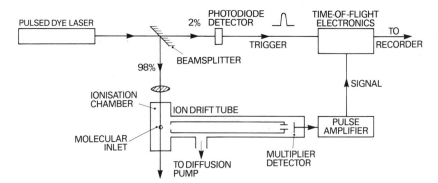

Fig. 4.47. Instrumentation for time-of-flight laser mass spectrometry. The molecular inlet is normal to the laser beam and is at its focus

the resultant reduction in multiphoton ionisation efficiency. Detection methods also vary, as in other mass spectrometers, but one of the most effective is time-of-flight (TOF) analysis, as illustrated in Fig. 4.47. Here, the various ion fragments are electrically accelerated, and pass along a tube about a metre long. The time taken by each ion to reach the detector is determined by its mass, and is typically measured in microseconds. Since laser pulses of only nanosecond duration are used to produce the ionisation, the mass spectrum can be obtained by monitoring the ion current as a function of the time elapsed since the laser pulse.

One of the most important aspects of laser mass spectrometry is the fact that the mass spectrum changes with the laser wavelength. As discussed earlier and illustrated in Fig. 4.43, the multiphoton ionisation process is most efficient when a resonance condition applies. A different mass fragmentation pattern can thus be recorded at each resonance, so providing a great deal more information on the molecular structure of the sample. Indeed, the mass spectrum can be represented by a three-dimensional plot, where ion current is plotted against mass in one direction and laser wavelength in the other. This technique is known as *resonance-enhanced multiphoton ionisation mass spectrometry.* Since multiphoton absorption generally produces molecular fragmentation patterns which are quite different from those of the normal mass spectrum, this method additionally provides a useful insight into the dynamics of multiphoton excitation. Finally, there is the possibility of obtaining still further information by varying laser polarisation. Whilst these methods can be used directly on chemically complex samples, it is also possible to treat samples by conventional 'wet chemistry' techniques to produce solutions for resonance ionisation mass spectrometry. This kind of approach is particularly expedient in analysis for inorganic elements; for example, Fe atoms can be detected by their resonant two-photon ionisation at a wavelength of 291 nm. By varying the irradiation wavelength, different elements can be selectively ionised and quantitatively measured in the mass spectrometer.

One other type of laser mass spectrometry more directly aimed at analytical applications should be mentioned before we leave the subject. This is known both as *laser microprobe mass analysis* (LAMMA), or *laser-induced mass analysis* (LIMA), and is based on the same microprobe concept discussed earlier. As shown in Fig. 4.48, high-power

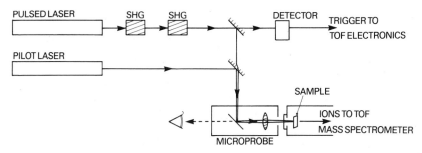

Fig. 4.48. Laser microprobe mass analysis apparatus for transmission sampling. The details of the time-of-flight mass spectrometer are as in Fig. 4.47

pulses of frequency-quadrupled light from a Q-switched Nd-YAG laser are focussed onto small areas of the sample, typically vapourising volumes of about 1 μm³ per shot. A pilot He-Ne laser follows a collinear optical path onto the sample surface so as to facilitate visual location of the focus point. Ions released from the sample are subsequently analysed in a time-of-flight mass spectrometer; the method illustrated is appropriate for transmission samples, but laser desorption in a reflection mode is also now practicable. The negative ion LAMMA spectrum of 2-hydroxynaphthalene-6-sulphonic acid is shown in Fig. 4.49. The reflection method is very well suited to the characterisation of different areas of inhomogeneous surfaces; it can also be applied to the subsurface analysis of inhomogeneous solids, since successive laser pulses vapourise and release ions from progressively deeper layers within the sample. Thus it has already established extensive applications ranging from geology to microelectronics.

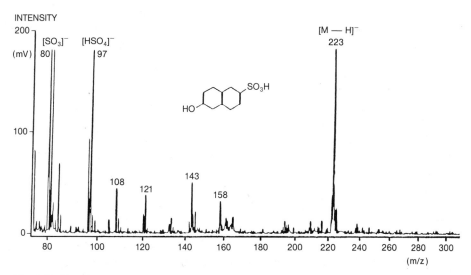

Fig. 4.49. Negative ion LAMMA spectrum of 2-hydroxynaphthalene-6-sulphonic acid, obtained by Holm et al. under normal atmospheric conditions. Reprinted with permission from Ref. 15. Copyright 1984, American Chemical Society

This method is particularly well suited to the study of fibres, environmental particles and metal corrosion processes, where it frequently out-performs conventional microanalytical techniques. Apart from the microsampling capability, the main advantages of laser mass analysis are speed of analysis (a complete mass spectrum can be obtained in a matter of microseconds), an elemental sensitivity typically on the ppm scale, and a mass range of 1-1,000 amu. It can thus also offer a powerful method for the determination of organic and biological residues or contaminants. A laser mass analysis system based on these principles is now in commercial production.

4.8 References

1. Pine, A.S., Maki, A.G., Robiette, A.G., Krohn, B.J., Watson, J.K.G. and Urbanek, Th.: J. Am. Chem. Soc. *106*, 891 (1984)
2. Lipson, R.H., LaRocque, P.E. and Stoicheff, B.P.: J. Chem. Phys. *82*, 4470 (1985)
3. Collins, C.B.: J. de Phys. Colloq. *C7*, 395 (1985)
4. Saykally, R.J., Veseth, L. and Evenson, K.M.: J. Chem. Phys. *80*, 2247 (1984)
5. Mito, A., Sakai, J. and Katayama, M.: J. Mol. Spec. *103*, 26 (1984)
6. Kachin, S.V., Smith, B.W. and Winefordner, J.D.: Appl. Spec. *39*, 587 (1985)
7. Möhlmann, G.R.: Appl. Spec. *39*, 98 (1985)
8. Lutz, M.: in 'Advances in Infrared and Raman Spectroscopy', ed. Clark, R.J.H. and Hester, R.E. *11*, 211 (Wiley, Chichester, 1984)
9. Esherick, P. and Owyoung, A.: in 'Advances in Infrared and Raman Spectroscopy', ed. Clark, R.J.H. and Hester, R.E. *9*, 130 (Wiley, Chichester, 1982)
10. Hayashi, S. and Samejima, M.: Surf. Sci. *137*, 442 (1984)
11. Barron, L.D. and Vrbancich, J.: in 'Topics in Current Chemistry' *123*, 151 (Springer, Berlin, 1984)
12. Whetten, R.L., Grubb, S.G., Otis, C.E., Albrecht, A.C. and Grant, E.R.: J. Chem. Phys. *82*, 1115 (1985)
13. McClain, W.M.: J. Chem. Phys. *55*, 2789 (1971).
14. Dixon, R.N., Bayley, J.M. and Ashfold, M.N.R.: Chem. Phys. *84*, 21 (1984)
15. Holm, R., Kämpf, G., Kirchner, D., Heinen, H.J. and Meier, S.: Anal. Chem. *56*, 690 (1984)

5. Laser-induced Chemistry

May not bodies receive much of their activity from the particles
of light which enter into their composition?
'Opticks', Isaac Newton

In preceding Chapters, we have looked at a wide range of applications in which the laser is used as a probe for systems of chemical interest. Although the application of laser spectroscopic techniques in particular may result in short-lived changes in molecular energy level populations, the laser does not generally induce any *chemical* change in the sample; in that sense it is used as a static, rather than a dynamic tool. Quite distinct from this is the field of applications in which laser excitation is used specifically to promote chemical reaction. Although this is a less well-developed area, it is one which is growing at a very rapid rate, and includes perhaps some of the most exciting research topics in the whole field of lasers in chemistry, as we shall see. There are, for example, indications that laser-induced chemical synthesis may ultimately prove the best and most economically sound method of producing some of the more expensive pharmaceutical compounds. To introduce the subject, we begin with a general overview of the major principles appertaining to laser-induced chemistry.

5.1 Principles of Laser-induced Chemistry

Chemistry induced by optical excitation is by definition *photochemistry*, and the whole of laser-induced chemistry can thus be regarded as one part of this much wider field. Although lasers can replace other light sources in any conventional photochemistry, many laser-induced reactions are not practicable with conventional light sources. In this Chapter the emphasis is firmly placed on these more distinctive applications specifically requiring laser stimulation. First, then, we need to identify the particular characteristics of laser light which mark out laser photochemistry as a separate identifiable discipline. The two most important qualities are undoubtedly the monochromaticity and high intensity of a laser source. The monochromaticity naturally lends itself to applications requiring selective excitation of specific chemical species, whilst the intensity is important for increasing excitation efficiency. Also significant and connected with the intensity is the possibility of producing pulses of very short duration, which is highly useful for inducing and monitoring ultrafast chemical reactions. On the other side of the coin, the narrow beamwidth of a laser is of course a disadvantage in that without divergent optics, and the associated loss of intensity, only very small volumes of material can normally be irradiated. Let us look at the photochemical implications of these properties in more detail.

5.1.1 General Considerations

To start with, the obvious advantage of using an essentially monochromatic source of light for photochemistry is that only selected optical transitions are induced in the sample (although of course the subsequent photochemical reactions may involve states populated by decay processes rather than those directly accessed by optical absorption). This contrasts, for example, with the traditional use of a broadband flashlamp, which usually populates a number of different energy levels to an extent depending on the strength of each transition. These various excited states may lead to different types of chemistry and interfere with any particular photochemical reaction of interest. The same principle applies with even more force to chemical mixtures; here, a laser may be employed to selectively excite one specific component. This is especially useful in isotopically selective photochemistry, as we shall see in Sect. 5.3. The high intensity of monochromatic light available from lasers generally carries the advantage of being able to induce selective photochemistry with far greater efficiency than any filtered conventional source.

The powerful intensity of a laser is also significant in photochemical applications for another reason. Much photochemistry proceeds as a result of an initial excitation to electronically excited states, since it is electronic energies which are involved in the formation and rupture of chemical bonds. Whilst this excitation normally necessitates irradiation with the comparatively energetic photons of uv/visible light, we have seen at the end of the last Chapter that by a process of multiphoton absorption, it is possible for individual molecules to acquire more energy than is associated with a single photon. Hence with a sufficiently intense laser source, even infra-red radiation can produce multiphoton excitation to electronically excited states and so lead to useful photochemistry. This represents a highly distinctive branch of photochemistry in which lasers are the only practicable sources. Again, there are important isotopically selective reactions which hinge on this type of laser photochemistry.

Lastly, the ability to pulse laser radiation, quite apart from the high intensities which then ensue, provides the means for both inducing and monitoring ultrafast photochemical reactions. With the instrumentation described in Chap. 3, it is possible to identify short-lived transient species formed as intermediates during reaction. It is even possible to trace the course of processes which occur on a picosecond timescale, such as the molecular rotations, torsional motions and electron transfer processes which play crucial roles in many biochemical reactions. Indeed, there are no other physical methods of measuring events which occur over such short times. An additional benefit accruing from the use of short pulses is that some of the slower processes which usually enter into the reaction kinetics, such as relaxation and diffusion, can effectively be suppressed; we shall pursue these subjects in more detail below.

Before proceeding further, let us briefly consider the various types of chemical process which can be initiated by the absorption of laser radiation. In polyatomic molecules, the initial photo-induced transition to an electronically excited state is almost invariably followed by some degree of intramolecular relaxation before any real chemistry takes place. Such unimolecular relaxation processes generally involve redistribution of energy amongst vibrational states and take place typically over nanosecond or sub-nanosecond timescales, as will be discussed in Sect. 5.2.5. The state directly populated by photon absorption may therefore have little *chemical* significance. Relaxation may lead to ionisation, isomerisation or dissociation, as illustrated for a polyatomic molecule ABC in the following scheme;

$$\text{Photoabsorption:} \quad ABC + nh\nu \rightarrow ABC^* \quad (n \geq 1) \tag{5.1}$$

Autoionisation: $ABC^* \rightarrow ABC^+ + e$ (5.2)

Isomerisation: $ABC^* \rightarrow ACB^*$ (5.3)

Dissociation: $ABC^* \rightarrow AB^+ + C^+$ (5.4)

The last of these processes results in bond fission and produces the fragments AB and C, which are generally both in vibrationally excited states, and in some cases are also electronically excited. The term 'photolysis', also occasionally applied to photoisomerisation, is most often applied to this kind of process in which dissociation follows absorption, in other words where molecular fragmentation is induced by the absorption of light. As such, it represents the simplest kind of photochemical reaction, and it also provides the mechanism for the first step in many more complex types of photochemistry. For example, if AB and C are radical species, laser photolysis can provide the initiation step for a chain reaction. We shall return to a more detailed consideration of specific unimolecular processes in Sect. 5.2.1.

Whilst the above considerations apply to unimolecular reactions, lasers can also be used to induce bimolecular reactions in which either one or both of the reactants are initially excited by the absorption of laser light. Even if photoabsorption only takes place in one of the reactants before the collisional reaction, the laser excitation can still play a crucial role in *state preparation*, since the chemistry of excited molecules often differs from their ground-state counterparts. In principle, a wide range of reaction conditions can be obtained by promoting each reactant to various energy levels. Often, infra-red lasers are used for this purpose, promoting reactants to relatively low-lying vibrationally excited states. The problem with electronic excitation is that after the absorption stage, processes (5.2) to (5.4) may occur in either reactant before there is time for any reactive collision. Of course further laser excitation or fragmentation may also be involved at each intermediate stage. For these reasons the kinetics of laser-induced bimolecular reactions are frequently highly complex, as we shall discover in looking at some specific examples in Sect. 5.2.2.

Before leaving these general considerations, it is worth noting that there is one other intriguing possibility for laser-induced bimolecular reaction which does not fit into the scheme outlined above[1]. This is based on the fact that the products of bimolecular reactions are usually formed from reactants X and Y through a transient activated complex XY*, which is energetically unstable. Here the reaction pathway through from reactants to products may be associated with a vibration of the complex; for example the reaction $H + D_2 \rightarrow HD + D$ essentially proceeds via an antisymmetric stretch of H..D..D. In such a case, laser irradiation at the appropriate frequency of the *complex* can enhance the reaction rate. This is a significantly different approach from most other laser-induced chemistry because the irradiation frequency does not correspond to an absorption band of the starting material; indeed, in the example given, the reactants are themselves infra-red inactive. However, relatively few studies appear to have been based on this notion.

5.1.2 Multiphoton Infra-red Excitation

A very distinctive kind of laser photochemistry can be induced by powerful infra-red sources, the carbon dioxide laser being by far the most widely used. As described in

Chap. 4, the multiphoton processes which can be induced by intense radiation become particularly efficient if one or more resonance condition can be satisfied by the molecular energy levels. Where a single frequency of uv/visible radiation is involved, the unequal spacing of most *electronic* levels means that it is rare to obtain even one intermediate state resonance. However, *vibrational* energy levels are more or less equally spaced, at least for the lowest levels of excitation. Hence with infra-red radiation of the appropriate wavelength, multiphoton absorption can become highly significant.

To consider this in more detail, we first take the simple case of a diatomic molecule, where there is only one vibrational frequency. The appropriate energy levels are shown in Fig. 5.1, together with arrows representing the absorption of infra-red photons with the same frequency. The first thing to note is that as we move up the ladder of vibrational states, although the spacing between adjacent levels starts off fairly constant, it diminishes at an increasing rate. Of course it also has to be borne in mind that each vibrational level has its own manifold of much more closely spaced rotational levels. Because each of these levels has an associated linewidth, there comes a point at which we effectively have a quasi-continuum of states, as represented by the shaded area in the

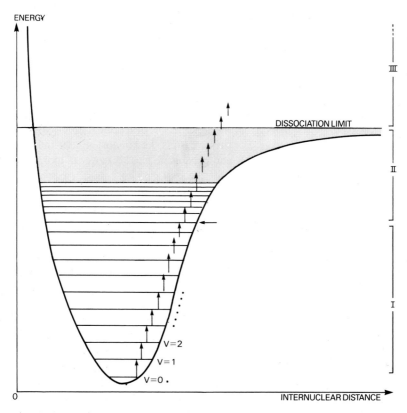

Fig. 5.1. Energy levels and transitions involved in the multiphoton infra-red dissociation of a diatomic molecule. Note that each vibrational level has rotational fine structure (not shown), and also the energies are not discrete but have finite linewidths; both these features are important in the excitation process

diagram. Eventually, an asymptotic limit is reached, at which point there is no longer any restoring force as the two atoms move apart, and dissociation occurs. The question of linewidth is one of the crucial considerations in multiphoton infra-red absorption, both in connection with the radiation and the molecular transitions. As far as the radiation is concerned, linewidths vary according to the nature of the source. The emission from a diode laser can have a linewidth as small as 100 kHz (3×10^{-6} cm^{-1}); at the other extreme the linewidth of a high-pressure CO_2 laser may be anything up to 10^6 times larger. Whilst in energy terms this is still very much smaller than the gap between the lowest-lying vibrational levels, this is not the case for vibrational levels approaching the dissociation limit. Regarding the linewidths of molecular transitions, the usual line-broadening mechanisms apply to an extent which largely depends on the state of the sample. However, power-broadening represents an additional factor which is especially important under conditions of high-intensity irradiation. With laser irradiances of 10^{10}–10^{12} W m^{-2}, power-broadened linewidths typically lie in the 1–10 cm^{-1} region. Again, this contributes to the creation of a quasi-continuum near the dissociation limit.

The process of multiphoton absorption displays different characteristics over different regions of the energy scale, and it has become common to speak in terms of regions I, II and III, illustrated in Fig. 5.1. In region I, vibrational levels are quite widely spaced, and the spacing is greater than the overall absorption bandwidth. Because the spacing is non-uniform, however, the photon energy soon gets out of step, and multiphoton processes occur. In the diagram, for example, the transitions v = 0 $\rightarrow$ 1, 1 $\rightarrow$ 2, 2 $\rightarrow$ 3, 3 $\rightarrow$ 4 and 4 $\rightarrow$ 5 all require energies close to that of a single photon, and lying within the overall bandwidth. These transitions therefore all take place by the process of single-photon absorption. The energy required for the 5 $\rightarrow$ 6 transition, however, is sufficiently different that it lies outside the bandwidth and cannot take place by absorption of one photon. Nonetheless, excitation can proceed up to the v = 10 level, as indicated, by a direct 5 $\rightarrow$ 10 transition involving four-photon absorption. It is the main characteristic of region I that such concerted multiphoton processes take place on the way up the vibrational ladder. As we saw in Sect. 4.6, this necessitates a fairly intense flux of photons, and hence a powerful source of radiation.

Region II is characterised by quasi-continuum behaviour resulting from the fact that vibrational energy level spacing has become *less than* the bandwidth. Here successive photons can be absorbed in a series of energetically allowed single-photon transitions. Since energy conservation is satisfied at every step, the molecule can at each point exist for a finite lifetime before absorbing the next photon; hence excitation through this region does not necessitate the enormously large photon flux which might at first appear necessary. Finally, once the level of excitation has reached the dissociation threshold, a true energy level continuum is encountered, and further photons can be absorbed in the short time before the atoms separate; this is known as region III behaviour. Note that in sharp contrast to region I, the process of excitation through regions II and III is relatively insensitive to the precise irradiation frequency. The entire process leading to dissociation may involve the absorption of 30–40 infra-red photons.

In passing, we note that the term 'coherent excitation' is frequently applied to the multiphoton absorption processes which occur in region I. This refers to the fact that the interval between successive transitions is too short for relaxation to occur. This represents a distinction from the term 'incoherent excitation' applied to the sequential single-photon absorption processes occurring in region II. These are perhaps unfortunate expressions in that 'coherent' is already a rather over-used word with other very different meanings in connection with lasers. Quite apart from signifying the special phase properties of laser light discussed in Sect. 1.6.3, it has a very specific meaning in connection with multiphoton processes in general, as discussed in Sect. 4.5.2. Neither of these senses applies to either region I or region II multiphoton absorption; such processes can certainly take place with incoherent thermal light of sufficient intensity, and they are in the sense of wave-vector matching explicitly *incoherent* by nature. It is therefore preferable to adopt the term *concerted excitation* for non-resonant multiphoton absorption such as takes place in region I, and *sequential excitation* for a series of single-photon absorptions as in region II.

When we turn to the case of *polyatomic* compounds, many of the same principles apply, although as we shall see there are so many additional factors to consider that the detailed theory becomes very complex indeed. In polyatomic molecules containing N atoms, there are generally (3N-6) modes of vibration, each with its own ladder of vibrational and associated rotation-vibration levels. It is therefore possible to induce multiphoton excitation at a number of different wavelengths, corresponding to each of these various vibrational modes. There may also be several different processes of dissociation, involving the fission of different chemical bonds and hence leading to different types of fragmentation. Moreover, further multiphoton excitation may occur in any fragment with suitable vibrational levels, in some sense representing a fourth stage in the overall process of multiphoton decomposition. Nonetheless, in many cases it is not necessary to reach the dissociation level in the parent molecule before useful photochemistry can occur.

One of the most important and distinctive features of multiphoton infra-red absorption by polyatomics concerns the nature of the excitation in the quasi-continuum region II. The multiplicity of energy levels means that even in the gas phase, power-broadening results in formation of an effective continuum far below the dissociation limit, and region II behaviour thus commences at quite low energies. In comparatively small polyatomics, region I may span only a few vibrational quanta; in larger molecules it may be that only one quantum is absorbed before region II behaviour applies. For example in the case of SF_6, region II appears only a few thousand cm^{-1} above the zero-point energy; for molecules with twenty or more atoms it may appear as low as 500 cm^{-1}. In the liquid or solid state, inhomogeneous line-broadening also comes into effect (see Sect. 1.5.3) and contributes to quasi-continuum formation. Hence the onset of continuum behaviour generally takes place much lower down the energy scale than is the case for a diatomic molecule. Another complication is the fact that many polyatomic molecules have electronically excited states lying below the dissociation limit of the electronic ground state. High levels of excitation thus result in some population of these electronic excited states, and may for example lead to fluorescence decay (as in OsO_4) or further vibrational excitation.

Because there are a number of different vibrational modes in polyatomic molecules, there is a considerable amount of overlapping amongst the various energy levels in the quasi-continuum. Since the bandwidth of these levels is generally large compared to the spacing between adjacent levels, this results in a relative inselectivity over the vibrational mode into which the energy of each absorbed photon passes. Also, vibrational energy may be re-distributed into different modes or combinations of modes as a result of anharmonic interactions; the time taken for this relaxation process is typically 10^{-11} s. Collision-induced intermolecular energy transfer also takes place, but over a longer timescale determined by the pressure (typically 10^{-8} s at 1 atmosphere).

It is now generally recognised that under intense infra-red irradiation, the vibrations of the nuclear framework of a polyatomic molecule are in fact most realistically represented by an essentially random, or *stochastic*, mixture of the normal modes of vibration, between which there is completely free energy flow. The result has important implications for the energetics of the multiphoton excitation process. For example, in a certain molecule, absorption at one particular frequency (corresponding to a fundamental vibration) may in principle require only 30 photons to accomplish dissociation. Nevertheless, absorption of 30 such photons will in practice result in various amounts of energy being deposited in each of the vibrational modes, and the level of excitation will therefore fall far short of the dissociation limit. Consequently it is commonly found that the mean number of photons absorbed per molecule far exceeds the energetic minimum, and can in some instances run into hundreds.

Much more significant, however, is the fact that these vibrational relaxation processes greatly diminish the prospects for selectively populating *specific* vibrational modes. In many cases, the results of powerfully pumping a molecule with infra-red radiation can be virtually unchanged even when frequencies corresponding to two completely different bond stretches are employed; it is still *generally* the case that the weakest bond is the first to break. For example the multiphoton dissociation of SF_5Cl produces SF_5 and Cl radicals even if the laser is tuned to the frequency of a sulphur-fluorine vibration. Thus the very appealing notion of mode-selective chemistry, in which a laser could be used to excite and dissociate any chosen bond in a polyatomic molecule according to the exact wavelength employed, is not anywhere near as realistic as was at first hoped. It was, indeed, the expectation behind this largely unfulfilled promise that led to much of the initial funding of research in this area.

With *ultrashort* laser pulses, mode-selective excitation should be more feasible since vibrational relaxation would not have time to occur to any large extent, and the conditions *immediately* subsequent to each pulse might be very far removed from thermal equilibrium; regrettably, it has not yet been possible to achieve the optimal picosecond and sub-picosecond pulse lengths in the infra-red. Despite these problems, there are nonetheless some laser-induced reactions which do take place before complete randomisation of vibrational energy occurs, and these can have unique and highly useful applications, as will become evident during the course of this Chapter.

5.1.3 Reaction Rates and Yields

We have seen that mode-selective chemistry is seldom possible due to the stochastic nature of multiphoton vibrational excitation in polyatomic molecules. This is exactly in accordance with the highly successful *RRKM (Rice-Ramsperger-Kassel-Marcus) theory* of reaction kinetics, in which it is assumed that energy can flow completely freely between different vibrational modes. The energy efficiency of a laser-induced chemical process can thus be rather poor (although no poorer than the corresponding thermally-induced process) since generally more photons are required to bring about reaction than would be expected from the activation energy. There are exceptions to this rule, however, as we shall see. It is often possible to exercise considerable control over the course of reactions induced by multiphoton excitation. The major experimentally controllable factors listed below influence not only reaction rate and yield, but also the relative yields in cases where more than one product can be formed.

1. Laser wavelength. As discussed above, whilst the monochromaticity of laser radiation in principle offers the possibility of delivering energy to specific bonds or vibrational modes in a polyatomic molecule, relaxation processes rapidly redistribute this energy, and so greatly diminish the localisation or selectivity of excitation. Nonetheless, even where substantial relaxation takes place before reaction, there is always some dependence on the wavelength of radiation applied to the system.

2. Fluence (energy density). Most laser chemistry is carried out with pulsed, rather than cw lasers, and the number of pulses applied obviously affects the extent of reaction. Although the sample irradiance during each pulse is determined by fluence, pulse duration and temporal profile, there is good evidence that in the case of multiphoton excitation, it is fluence which exerts the major influence on the course of reaction, particularly in region II. A typical case is represented by Fig. 5.2, which shows the yield of a single-pathway reaction as a function of fluence. The yield generally increases with fluence, until saturation takes place at high energy densities. Not all reactions display the same kind of dependence, however. Other parameters such as pulse duration and repetition frequency generally influence the extent of excitation in a fairly obvious way, but are not always amenable to variation.

3. Pressure. Most laser-induced reactions take place in the gas phase at low pressure. The effect of increasing pressure, whilst often highly significant, is difficult to generalise since it influences the reaction in a number of different ways. The simplest effect is of course an increase in the number of molecules in the laser beam. The effect of pressure-broadening can also change the rate of absorption by increasing the absorption linewidth. Often more important, however, is the increase of collision-induced relaxation with pressure. The resultant increase in the rate of exchange of vibrational energy between excited and unexcited molecules can play a crucial role in populating the quasi-continuum leading to reaction. For example the dissociation yield of silane, SiH_4, increases very appreciably with pressure. In some reactions however, such as the dissociation of SF_6, the yield is relatively insensitive to pressure. The mean number of

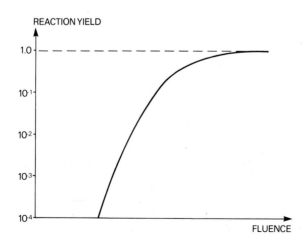

Fig. 5.2. Typical dependence of multiphoton dissociation on laser fluence

photons absorbed by each molecule can also vary markedly with pressure. This is because as pressure increases, collisions can supply the extra energy required to reach the dissociation threshold.

4. Buffer gases. Lastly there is the effect of admixing monatomic or relatively simple polyatomic buffer gases such as argon or nitrogen with the reactant. If relatively few reactant molecules undergo direct laser excitation, transfer of vibrational energy by collision with buffer molecules can increase the extent of reaction in much the same way as an increase in pressure. However, as the buffer gas pressure increases, deactivation of the reactant will ultimately reduce the mean level of excitation and so decrease the reaction yield, as shown in Fig. 5.3a. If the majority of sample molecules are directly excited into the quasi-continuum by the laser, buffer gases can only lead to a decrease in yield, as shown in (b).

A classic piece of work by Zitter and Koster illustrates just how significant some of these factors can be. In a study on laser-induced elimination of HCl from CH_3CF_2Cl it was found that the mean number of photons absorbed per molecular dissociation decreased dramatically both with increasing pressure and cw laser power. Using 966 cm^{-1} CO_2 laser radiation, the number of photons required was shown to vary between almost 500 at 100 torr pressure and 6 W laser power to just over 5 photons at 600 torr pressure and 28 W power[2].

Many of the conventional kinetic and thermodynamic rules governing chemical reactions cannot be directly applied to laser-induced chemistry. For example another very significant result of the study by Zitter and Koster was the observation that under suitable conditions the mean energy requirement can be much *less* than the activation energy (in this case equivalent to 22 photons), and approach the value of the enthalpy change ΔH (equivalent to 5.2 photons). This is only possible because energy released as the products are formed from the activated complex can contribute towards the activation of other reactant molecules. The concept of temperature is also rather ill-defined for a laser-induced reaction since the entire system departs very markedly from thermodynamic equilibrium. It is principally for this reason that equilibrium constants can no longer be directly calculated from the usual relationship K = exp (−ΔG/RT).

It is rare to apply heat in the case of a laser-induced reaction, since it results in non-

REACTION YIELD

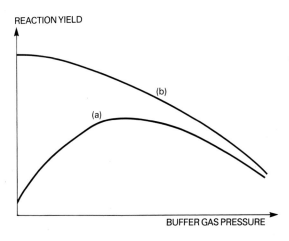

(b)

(a)

BUFFER GAS PRESSURE

Fig. 5.3a and b. Variation of multiphoton dissociation yield with pressure of buffer gas; (a) where there is relatively little direct laser excitation of the reactant into the quasi-continuum region, and (b) where most reactant molecules are directly excited by the laser

specific thermal excitation of vibrational levels, and thus reduces the prospects of observing any novel chemistry. In fact, it is a distinctive feature of most laser-induced reactions that they take place 'in the cold'. In the case of reactions which can be induced either by heat or by infra-red laser radiation, it is generally found that side-reactions take place to a much lesser extent where the laser is employed. This is largely a reflection of the fact that the reaction vessel does not have the hot walls at which reactive intermediates normally form.

5.2 Laser Photochemical Processes

Having discussed the basic principles of laser-induced chemistry, we now move on to a consideration of the various types of reaction which can occur. Although some examples will be given of reactions stimulated by uv/visible radiation, there are many other photochemical reactions which can in principle be laser-induced but which can equally be induced by other less intense, incoherent sources of light. Most of the more novel chemistry induced by lasers involves infra-red multiphoton excitation as described in the last Section, and the following account accordingly places most emphasis on this area.

5.2.1 Unimolecular Laser-induced Reactions

By far the largest number of laser-induced chemical reactions fall into this category, and the carbon dioxide laser, producing powerful emission at numerous discrete wavelengths around 9.6 μm and 10.6 μm (see Table 2.1, p. 33), is the most commonly applied source. The simplest type of unimolecular reaction is isomerisation, and several studies have shown how laser-induced *photoisomerisation* can modify the relative proportions of different isomers in a mixture. Before considering examples of this laser-induced process, however, it is first worth noting the factors which normally determine these relative proportions.

It is often the case that an organic synthesis produces more than one isomer, and unless the reaction is kinetically controlled, the yield of each is determined by the position of equilibrium for the interconversion reaction. As mentioned earlier, the corresponding equilibrium constant is, in turn, usually related to the temperature and the relative thermodynamic stability of the isomers through the well-known relation $K = \exp(-\Delta G/RT)$. Except in cases where the value of ΔG is exceptionally small, the result is that only at very high temperatures can the less stable isomer be present in more than minute proportions. At such high temperatures, decomposition would in any case generally occur.

However the selective laser excitation of one isomer, using a wavelength which no other isomer appreciably absorbs, can substantially modify the relative proportions either towards, or indeed in some cases away from equilibrium. An example of the former case is afforded by 1,2-dichloroethene, where the *cis-* isomer is more stable than the *trans-* isomer by approximately 2 kJ mol^{-1}. Pulsed irradiation of a mixture containing an excess of the *trans-* compound at a frequency of 980.9 cm^{-1} results in conversion to a mixture in which the *cis-* isomer predominates[3]. Pulsed irradiation of hexafluorocyclobutene at 949.5 cm^{-1}, however, results in up to 60% conversion to its isomer hexafluoro-1,3-butadiene[4], which is thermodynamically *less* stable by 50 kJ mol^{-1};

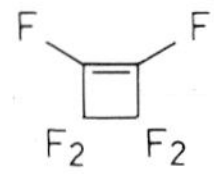

$$\longrightarrow CF_2{=}CF{-}CF{=}CF_2 \tag{5.5}$$

This is simply a reflection of the fact that the cyclic compound absorbs radiation of this frequency much more strongly than its isomer. It is interesting to note that with sufficiently high intensities, further laser-induced reactions take place and lead to the formation of decomposition products and low molecular-weight polymers[5].

A classic case of laser-induced chemistry involves the conversion of 7-dehydrocholesterol (I) (Fig. 5.4) to previtamin D$_3$ (II), which is once again an isomerisation reaction. The usual photolytic method by which this conversion is accomplished involves a series of fractional distillations, and produces relatively low yield. However by using two-step laser photolysis with KrF (248 nm) and N$_2$ (337 nm) laser radiation, the competing side-reactions can largely be eliminated, and the conversion effected with 90% yield[6]. Since the product (II) is reversibly convertible to vitamin D$_3$ (III), the overall

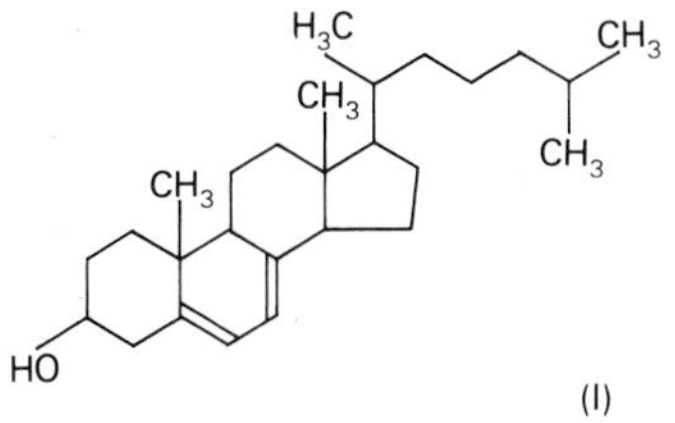

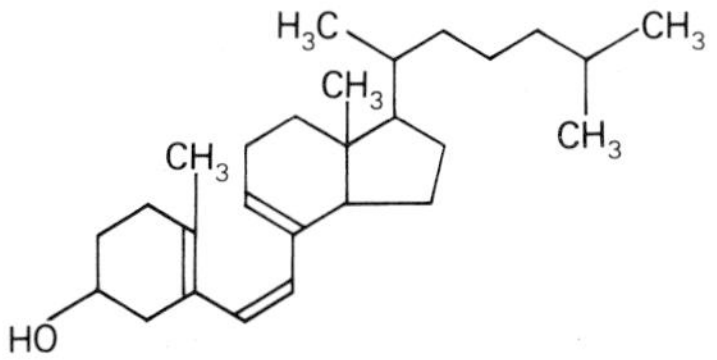

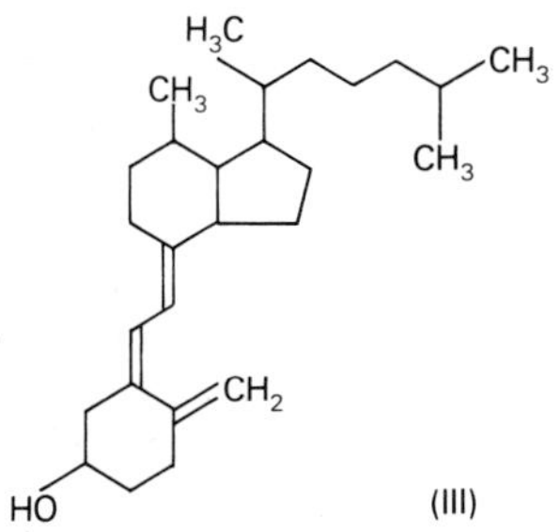

Fig. 5.4. Molecular structures of (I) 7-dehydrocholesterol; (II) previtamin D$_3$; (III) vitamin D$_3$

process repesents a useful and substantially improved method for synthesis of the vitamin.

Most unimolecular laser-induced reactions involve multiphoton infra-red *dissociation*, and several organic elimination reactions come under this heading. These reactions can often produce high yield since the molecular products are formed directly, rather than through secondary reactions of radicals or other reactive intermediates. Some good examples are provided by elimination reactions involving esters, which proceed as follows;

$$
\underset{\substack{\| \\ R-C-OCH_2CH_2R'}}{O} \longrightarrow \left[\begin{array}{c} O---H \\ R-C \qquad CHR' \\ O---CH_2 \end{array} \right]^{\ddagger} \longrightarrow \underset{\substack{\| \\ R-C-OH}}{O} + CH_2{=}CHR' \quad (5.6)
$$

Such reactions can be very effectively induced by laser irradiation at a frequency of around 1050 cm^{-1}, which produces excitation of the stretching mode of the O—CH$_2$ bond, and ultimately results in its fission.

As mentioned above, mode-selective chemistry is not generally as feasible as might at first be expected due to intramolecular relaxation processes. Nonetheless there are certain cases, especially in comparatively small molecules, where irradiation at different laser frequencies genuinely results in different products. An example is afforded by cyclopropane, where it is found that multiphoton excitation at around 3000 cm^{-1} corresponding to the CH stretch frequency results in isomerisation to propene. However irradiation at around 1000 cm^{-1}, corresponding to the CH$_2$ wag, produces both isomerisation and fragmentation in roughly equal amounts.

Other reactions which have been induced by multiphoton dissociation are far too numerous to mention. Often there is more than one end-product, and the product ratios are frequently very different from those obtained in the corresponding thermal pyrolysis experiments. The fact that more than one fragmentation product appears in the laser-induced reactions is of course clear evidence that energy is well distributed amongst various vibrational modes. In such cases the activation energy generally appears to be the major factor in determining the product ratios, although laser pulse power and energy can also exert a remarkably large influence. For example in the following reactions[7], (5.7) overwhelmingly predominates when using pulses of 10^4 J m^{-2}, but with pulse energies four times larger, (5.8) becomes the principal reaction;

$$ (5.7) $$

$$ (5.8) $$

Whilst the precise explanation for this not uncommon type of behaviour is still the subject of debate, it is likely that the competition between alternative relaxation channels with differing kinetics is responsible.

5.2.2 Bimolecular Laser-enhanced Reactions

When laser radiation induces a *unimolecular* reaction, the products may of course participate in further chemistry by a *bimolecular* reaction with another reagent. This principle can be put to good effect, especially where unimolecular photodecomposition results in the formation of highly reactive radicals or other short-lived species. Quite apart from this type of process, however, there exists the possibility of exciting one reactant without chemical change to a state in which it is more reactive towards the other reagent. Mostly this is accomplished by irradiation at an infra-red frequency corresponding to a molecular vibration of one of the reactants, and the method is thus known as *vibrationally enhanced reaction*. Several attempts to induce such reactions have failed, most likely because for these reactions there is insufficient translational energy for reactive collisions without the application of heat (one way around this problem is to use molecular beams; see Sect. 5.2.6). However there is also a large catalogue of successes, and in optimum cases the vibrational excitation of even a single quantum in one of the reactants can lead to an increase in the reaction rate constant by several powers of ten. Most of the successfully demonstrated reactions of this type involve the collisional transfer of single atoms between comparatively small molecules.

As a first example, we can take the reaction between ozone and nitric oxide;

$$O_3 + NO \rightarrow NO_2 + O_2 \tag{5.9}$$

The product nitrogen dioxide can be formed in either its electronic ground state ($^2A_1$) or an electronically excited state ($^2B_2$), and separate rate constants, k_1 and k_2 respectively, are associated with the corresponding rates of formation. Since the reaction involves fission of one of the O_3 bonds, the best way of enhancing the reaction by vibrational excitation is to pump the antisymmetric stretch. It so happens that the corresponding frequency coincides with the P(30) line (1037.4 cm^{-1}) in the 9.6 μm emission of the carbon dioxide laser, and experiment has shown that when this radiation is applied to the system, the reaction rates at 300 K increase by a factor of 4 for k_1 and 16 for k_2 [8].

Another example is the radical reaction between methyl fluoride and bromine;

$$CH_3F + Br \rightarrow CH_2F + HBr \tag{5.10}$$

$$CH_2F + Br_2 \rightarrow CH_2FBr + Br \tag{5.11}$$

Once again, vibrational excitation of the methyl fluoride can be accomplished with CO_2 laser 9.6 μm radiation, in this case the P(20) line at 1046.9 cm^{-1}. The result is an increase in the reaction rate by a factor of approximately 30 at 300 K [9].

Even neater experiments are possible if one of the reactants is a molecule which is itself the active medium in a molecular laser. This obviates the problem of searching for accidental coincidences between the emission lines of a laser source and the absorption features of the reactants. For example, the rate of reaction at 300 K between free (^3P) oxygen atoms and hydrogen chloride,

$$O + HCl \rightarrow OH + Cl \tag{5.12}$$

can be enhanced by a factor of approximately 100 by irradiating the system with pulses from an HCl laser, and so populating the v = 1 level of the reactant HCl [10]

5.2.3 Laser-sensitised Reactions

Quite a different area of laser-enhanced chemistry involves the sensitisation of reactions by the excitation of a species which does not itself undergo chemical change; this can be regarded as a form of *laser-assisted homogeneous catalysis*, although the term *laser-sensitised reaction* is more often used to describe it. This kind of reaction proceeds as a result of the collisional transfer of vibrational energy, often referred to as *V-V transfer*, from molecules of the laser-excited species (the sensitiser) to reactant molecules. The major advantage of this method becomes apparent if the reactants do not themselves strongly absorb in the emission region of the usual CO_2 laser. By choosing a strongly absorbing sensitiser to initiate the reaction, the rate of reaction induced by laser stimulation can be greatly increased. Both sulphur hexafluoride and silicon tetrafluoride have been widely employed as sensitisers; the latter is nonetheless generally preferred since SF_6 decomposes somewhat too readily under even fairly low-power irradiation. In the presence of SiF_4, various types of sensitised gas-phase reaction have been observed. As illustrated below, these include isomerisation (5.13), condensation (5.14), and retro-Diels-Alder reactions (5.15), (5.16);

$$CH_2{=}C{=}CH_2 \rightarrow CH_3C{\equiv}CH \tag{5.13}$$

$$2\,CHClF_2 \rightarrow CF_2{=}CF_2 + 2\,HCl \tag{5.14}$$

$$\tag{5.15}$$

$$\tag{5.16}$$

Many such reactions which are normally carried out at high temperatures, or even with cw laser heating, produce chemically cleaner products if they are induced indirectly by laser sensitisation since the reaction vessel remains cold. In addition to the factors listed in Sect. 5.1.3, such reactions may also be strongly influenced by the choice of sensitiser and the pressure ratio of sensitiser to reagent.

Because the reactants in a sensitised reaction do not need to possess absorption bands in any particular infra-red region, then with a good sensitiser like SiF_4, the range of gas-phase reactions which can be laser-induced is almost limitless. However, there is not a great deal of selectivity in the mechanism for the initial transfer of vibrational energy from the sensitiser to other molecules. This is particularly the case if the reactants are large polyatomics with quasi-continuous vibrational energy levels; here, we cannot in general expect to find isotopic, or even isomeric specificity.

One other related topic is *laser-catalysed* reaction, a term which is a very definite misnomer but is applied to a reaction in which the catalyst is itself produced by laser chemistry. For example the laser pyrolysis of OCS using 248 nm radiation from a KrF laser produces ground state S_2 molecules, which can catalyse the isomerisation of *cis* 2-butene to *trans* 2-butene with an effective quantum yield of about 200[11].

5.2.4 Laser Surface Chemistry

Surface chemistry is an increasingly significant discipline in which lasers are employed. A large number of the chemical applications of lasers to surfaces concern either spectroscopic methods dealt with earlier (see for example Sect. 4.5.5) or other surface-enhanced optical processes. From the point of view of laser-induced chemistry, many of the most important topics in this field concern the treatment of semiconductor surfaces, and therein hold enormous potential for application in the manufacture of microelectronic devices. It is worth noting that exciplex lasers in particular produce emission in a very useful wavelength range, where photon energies are sufficient to break chemical bonds in a variety of compounds involving the Group IV elements. Because a high level of attenuation is associated with the corresponding absorption, surface treatment is a particularly obvious application for these lasers, and the high power levels they produce are such that the rate of processing can be viable for production purposes.

   In this area it is quite hard to draw a clear line between the chemical and physical applications of lasers, although in this section we shall concentrate specifically on laser-induced chemical *reactions*. In passing, therefore, we note that laser machining, annealing, recrystallisation and other essentially physical processes all contribute to the wide-ranging potential for semiconductor processing. In many of these applications, as well as those to be considered below, the simplest alternative to the use of a laser is heat treatment. One of the principal advantages of using a laser, however, is that its tightly focussed beam allows the treatment of very small areas of surface down to 1 μm in diameter without affecting the surrounding material.

   As a first example of laser-induced surface chemistry, we can consider multiphoton dissociation reactions in the gas phase, where the surface plays the role of a heterogeneous catalyst. Here, the course of reaction is influenced by other factors in addition to the four listed in Sect. 5.1.3. For example in the dissociation of propan-2-ol over CuO using 1070.5 cm^{-1} radiation from a CO_2 laser[12], there are two competing reaction pathways leading to different products;

$$\text{CH}_3\text{CH(OH)CH}_3 \quad\overset{\nearrow\quad \overset{\text{O}}{\overset{\|}{\text{CH}_3\text{CCH}_3}} + \text{H}_2}{\searrow\quad \text{CH}_2{=}\text{CHCH}_3 + \text{H}_2\text{O}} \tag{5.17}$$

Here it is found that the product ratio of propanone/propene can be varied from 0.02 to 6, depending on the orientation of the catalytic surface relative to the laser beam.

   Much laser-induced surface chemistry involves the principle of depositing a thin film covering onto a substrate surface by decomposition of a gas. This method is known in the jargon of the laser field as *laser chemical vapour deposition*, and represents an alternative to the high-temperature methods more usually employed for the construction of microelectronic devices. It has, in fact, already been proved possible to fabricate viable integrated circuits using this principle. In the case of laser deposition, although the chemistry of interest actually concerns the vapour, reaction takes place principally at or near to the point at which the substrate is irradiated. The principle involved in the process of deposition may be either pyrolytic or photolytic by nature. For both types of deposition, laser irradiances are typically of the order 10^{12} W m^{-2}, and the partial vapour pressure of the vapour in the range $10^{-3} - 1$ atmosphere. Under these conditions, rates of deposition with a scanning laser beam are typically between 0.1 and 100 µm s^{-1}.

Pyrolytic deposition involves thermal reaction, and is in general an indirect result of the surface heating produced by the laser radiation. For this purpose, it is obviously necessary for the substrate to absorb in the appropriate wavelength region. For example, amorphous films of silicon can be pyrolytically deposited from SiH$_4$ vapour onto quartz or various other surfaces irradiated by 10.59 µm radiation from a carbon dioxide laser, using apparatus such as that shown in Fig. 5.5. In this particular case, the silane can itself absorb the radiation, and it is considered likely that multiphoton dissociation also plays a part in the dehydrogenation process by causing molecular fragmentation before surface deposition occurs. However, even when the chemical mechanism involved is purely a thermal one, the laser method still has several advantages over the traditional heat-induced deposition process, since it leads to higher rates of deposition and much more precisely controlled localisation of the surface coverage.

Photolytic deposition (*photodeposition*) by contrast results directly from the absorption of laser light by molecules of the vapour. This is a technique offering a great deal more control and selectivity, since different compounds will absorb at different wavelengths. Once more, the localisation of reaction is an additionally attractive and significant feature of the laser-induced process. Very often an exciplex laser, or else the frequency-doubled light from an ion laser provides the ultraviolet photon energies necessary to produce dissociation. There have been numerous studies concerning deposition on semiconductor surfaces, and a good example is afforded by the photolysis of dimethyl cadmium, Cd(CH$_3$)$_2$. This compound decomposes to form a surface coating of pure cadmium metal, and represents one of an increasingly large number of organometallic compounds which can be used for the deposition of metals. It is also important to notice that laser methods can be used for *doping* semiconductors, for example by the photolysis of BCl$_3$ or PCl$_3$; here the laser beam has the additional role

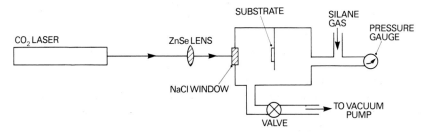

Fig. 5.5. Apparatus for the laser chemical vapour deposition of silicon by the pyrolytic dissociation of silane

of melting the substrate so that the dopant boron or phosphorus atoms can be incorporated into the surface by liquid diffusion.

By contrast, quite a different kind of laser-induced reaction takes place on semiconductor surfaces under bromomethane, since the laser photolysis of CH_3Br releases Br atoms which can subsequently etch the surface. Etching of this kind can be achieved with various organohalogen compounds in the gas phase, and can also be accomplished in solution by electrochemical means. Laser-induced chemical etching is in itself an area which shows enormous potential for the microelectronics industry, since in contrast to the more usual plasma etching methods, it does away with the need for any kind of resist mask, and the rate of etching can be faster by a factor of 50 or more.

Another neat example of laser surface chemistry is afforded by recent work demonstrating the possibility of laying down an InP layer by co-deposition of indium and phosphorus from a mixture of $(CH_3)_3InP(CH_3)_3$ and $P(CH_3)_3$[13]. In this case, using 193 nm radiation from an ArF excimer laser, the photodecomposition reactions are

$$(CH_3)_3InP(CH_3)_3 + 2h\nu \rightarrow In^* + 3CH_3 + P(CH_3)_3 \qquad (5.18)$$

$$P(CH_3)_3 + 3h\nu \rightarrow P^* + 3CH_3 \qquad (5.19)$$

Other examples of laser-induced surface reaction include the formation of oxide layers by the photo-oxidation of metals and semiconductors. A widely studied example is the formation of SiO_2 on the surface of silicon by photo-oxidation in an oxygen atmosphere. The enhancement of the rate of oxidation resulting from laser irradiation of the surface is in this case thought to be a result of the breakage of Si-Si bonds through the absorption of the laser radiation. Metal silicides can also be formed by a direct *thermally* induced reaction between a surface film of metal and a silicon substrate. Finally, it is worth noting that some studies have been made of the photoreactions of adsorbed species, such as the laser-induced polymerisation of methyl methacrylate. In such reactions the substrate can influence the reaction through its physical properties even though it is not necessarily involved in the chemistry.

5.2.5 Ultrafast Reactions

We now turn to a quite distinct branch of laser-induced chemistry concerned with processes induced by ultrashort (picosecond and sub-picosecond) laser pulses. To produce pulses of this duration requires the adoption of mode-locking techniques (Sect. 3.3.3), and indeed the era of picosecond measurements dates back to the mid-1960's, when the first mode-locked solid-state laser was made operational. The shortest pulses can be obtained with radiation at the uv/visible end of the spectrum, where electronic transitions take place. The excitation processes produced by such pulses are in some senses no different from those which could be induced by longer pulses of the same wavelength and power. However, when combined with suitable monitoring instrumentation (see Fig. 3.17, p. 66), the short pulse duration makes it possible to *observe* what goes on at very short times immediately following the absorption of light—the so-called *primary processes*. This has become a very active area of research, particularly in connection with the photochemistry of large aromatic and pseudo-aromatic molecules in which these primary processes occur over sub-nanosecond times. Many of the compounds which have been studied are in fact biochemicals.

Before moving on to consider the specifically chemical processes which can occur, we should first take note of the other physical mechanisms which operate over picosecond and sub-picosecond times. Any cyclic process which occurs at a frequency greater than 10^{12} Hz can of course be complete within a picosecond interval; this includes molecular vibrations and the rotations of some small molecules. Rotations, and *rotational diffusion*, are of particular interest in studies of the liquid state, since they constitute the mechanisms for the most rapid fluctuations in local microscopic structure. The process of rotational *relaxation* following absorption by dyes and other polyatomic molecules usually occurs over tens to hundreds of picoseconds. One means for determining the precise timescale for this type of interaction is to probe the absorption of the sample at short intervals of time after excitation with a picosecond laser pulse. One commonly used technique employs plane polarised light for the excitation, so that only molecules in a suitable orientation can absorb the radiation and proceed to an excited state. Thus until relaxation processes occur there are disproportionately few ground-state molecules in the correct orientation to absorb from a subsequent probe pulse. This anisotropy can be detected by probing with two laser pulses having orthogonal planes of polarisation, and the timescale for anisotropic behaviour to disappear reflects the timescale for relaxation.

Vibrations come into play in other relaxation phenomena, for example those involving *internal conversion* (e.g. $S_1 \rightarrow S_0$) or *intersystem crossing* (e.g. $S_1 \rightarrow T_1$; see Fig. 2.16, p. 42). Some hundreds of cycles of most molecular vibrations can be completed within a picosecond interval, however, and the timescale of an ultrashort laser pulse is therefore sufficient to allow for an appreciable redistribution of vibrational energy over any polyatomic molecule. Such molecular relaxation processes are thought to play a crucial role in determining the outcome of many photochemical reactions; this is certainly true in many gas phase reactions. *Intermolecular energy transfer* by a direct energy exchange mechanism is another highly significant feature of ultrafast chemistry in the condensed phase. Energy can be fairly efficiently transferred by this mechanism over short distances of 1-2 nm in sub-nanosecond time intervals; an exponential dependence on the molecular separation R makes this mechanism of little significance for larger distances. A quite distinct coupling mechanism named after Förster, which operates between polar molecules, has an R^{-6} dependence and hence applies over a longer range, but is generally associated with longer times.

In condensed phase matter, the bulk vibrations known as optical phonons can also dissipate energy over a similar period; here, another mechanism associated with *dephasing* of the vibrations of different molecules also contributes to picosecond relaxation. The time constant τ for vibrational dephasing is generally somewhat smaller than τ', the time constant for energy dissipation, so that it is usually the dephasing which is largely responsible for the relaxation of bulk vibrations. To take a specific example, excitation of the CH stretch of ethanol at 2928 cm^{-1} is associated with an energy dissipation time of approximately 20 ps, compared to a dephasing time of approximately 0.25 ps[14]. This kind of information, obtained from stimulated Raman studies, is obviously only achievable using picosecond laser sources. In passing it is worth noting that for liquid nitrogen, the difference between the timescales of the two effects is even more dramatic; at sufficiently low temperatures the vibrational dephasing time is a few picoseconds whilst the dissipation time is measured in seconds!

Finally, fluorescent decay may also in certain rare cases occur on a picosecond timescale, as for example in cyanine dyes, phthalocyanines and metalloporphyrins with transition metal centres, and also triphenyl methane dyes. Fluorescence can, however, also be limited to picosecond times if there is a non-radiative process competing with it; it thus provides a convenient and widely-used means for monitoring the time-development of other ultrafast chemical processes.

Genuine chemistry on the picosecond timescale tends to be limited to relatively simple unimolecular reaction mechanisms. The reason is simply that on the molecular level, the range of movement over such short times is very limited, and bimolecular reactions which depend on the translational motion and collision of large molecules or molecular fragments therefore become largely unimportant. Protons or other single atoms may however be photodetached by a unimolecular mechanism in sub-nanosecond times, as for example in the reaction

(5.20)

which is associated with a time constant of around 10^{11} s^{-1}. *Electron transfer* reactions in solution may also be accomplished over picosecond times; for a 1 M solution, 100 ps would be typical. Certain types of isomerisation not involving the breaking of chemical bonds, such as those involving cis-trans conversion, can also take place over this kind of timescale. Here an example is afforded by the equilibrium between *cis*- and *trans*-stilbene;

(5.21)

for which the forward reaction in solution or the gas phase also takes place in approximately 100 ps.

As mentioned earlier, some of the most interesting work in the area of ultrafast chemistry concerns elucidation of the primary processes in photobiological phenomena, such as the intricate mechanisms of photosynthesis and vision. The various instrumental techniques used for monitoring these ultrafast processes, which are all based on mode-locking and pulse-selection methods described earlier, are discussed in more detail in Sect. 5.2.6. Some of the key results established by use of ultrashort pulsed laser experiments are described below. Although only results relating to photosynthesis in *plants* are mentioned here, and space does not permit a thorough discussion, it should be pointed out that the much simpler processes in photosynthetic bacteria have also been very extensively studied using picosecond laser instrumentation, and are now even better understood.

In green plants, there are two distinct reaction sequences known as *photosystems* initiated by the absorption of light. Together, these generate the redox potential for the crucial photosynthetic step in which molecules of water are 'split'. Both photosystems involve chlorophyll molecules, the most abundant form of which is chlorophyll *a*, illus-

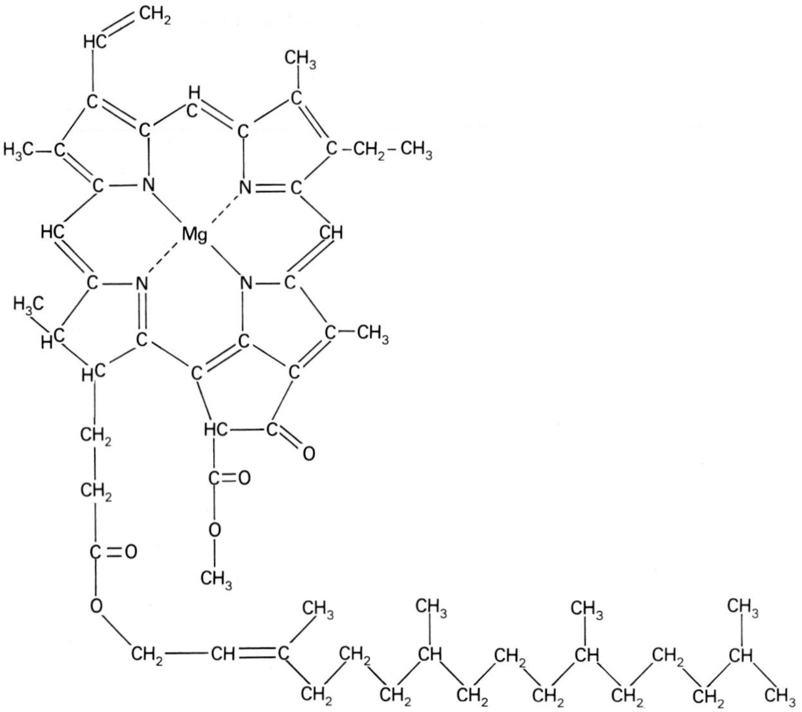

Fig. 5.6. Molecular structure of chlorophyll a; the porphyrin ring at the top of the diagram is responsible for the main spectroscopic properties of the molecule

trated in Fig. 5.6. Amongst other distinctions, the two photosystems can be differentiated on the basis of the different absorption wavelengths of the chlorophyll pigments. In both cases there is absorption in the red, with broad bands of half-width around 30 nm *in vivo*. However, the absorption is centred at around 700 nm in photosystem I, and around 680 nm in photosystem II; the pigments are accordingly referred to as P700 and P680, respectively. In photosystem PS I, the major primary processes can be summarised as follows;

$$\text{Chl } a + \text{hν} \quad \rightarrow \text{ Chl } a^* \tag{5.22}$$

$$\text{Chl } a^* + \text{Chl } a \rightarrow \text{ Chl } a + \text{Chl } a^* \tag{5.23}$$

$$\text{Chl } a^* + \text{P700} \rightarrow \text{ Chl } a + \text{P700}^* \tag{5.24}$$

$$(\text{P700}^*.\text{A}) \quad \rightarrow \text{ (P700}^+.\text{A}^-) \tag{5.25}$$

The photoabsorption step (5.22) represents the initial excitation of the so-called 'antenna pigment' chlorophyll in the chloroplast, and the subsequent equation represents a rapid transfer of the excitation over a series of chlorophyll molecules. Energy is eventually transferred to the reaction centre (Eq. 5.24), resulting in electron transfer to an electron acceptor labelled A (Eq. 5.25). By monitoring the associated changes in spectral characteristics, all of these primary processes have been shown to occur on a picosecond timescale.

The result of the sequence of reactions (5.22) to (5.25) is formation of a strong reducing agent A^- and a weak oxidant $P700^+$. A similar scheme based on P680 chlorophyll takes place in photosystem II, and results in formation of a strong oxidant $P680^+$ and a weak reductant Y^-. The subsequent chemistry may be summarised by the Equations;

$$2\,H_2O \rightarrow 4\,H^+ + O_2 + 4\,e \tag{5.26}$$

$$CO_2 + 4\,H^+ + 4\,e \rightarrow (CH_2O) + H_2O \tag{5.27}$$

Since the oxidation of water in Eq. (5.26) and the reduction of carbon dioxide in (5.27) involve four electrons in each of the two photosystems, the production of a single carbohydrate unit (CH_2O) evidently requires the absorption of eight photons. The entire photosynthetic process generally takes place in units containing about 2500 chlorophyll molecules, and the high speed and efficiency of the primary energy and electron transfer processes involving these molecules is clearly of the utmost importance. Much research in this area has been motivated by the desire to create solar energy cells which can mimic this high efficiency in a non-biological system.

Ultrafast reactions also play an important role in the mechanism of vision. Most research work in this area has involved the photochemistry of rhodopsin, a visual pigment composed of over 300 amino acids[15]. It is the absorption of light by rhodopsin molecules in discs contained in the rod cells of the retina which initiates the basic response of the human eye to light; there are 10^9 such molecules in each cell, and over 10^8 rod cells covering the retina. The absorption of light takes place in a chromophore group known as retinal, an aldehyde derivative of vitamin A displaying broad-band absorption over most of the visible range centred on 498 nm. Retinal (R498) is present as an 11-*cis* isomer with the remaining linkages thought to be *trans*, as shown in Fig. 5.7 a.

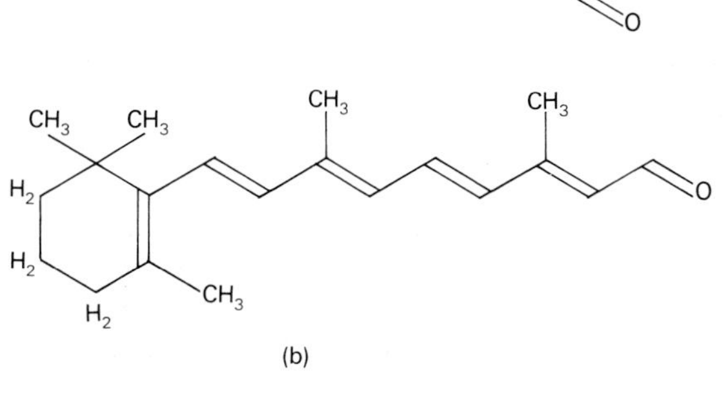

(a)

(b)

Fig. 5.7 a and b. Molecular structure of free retinal, the rhodopsin chromophore. In (a), the retinal is in its 11-*cis* configuration, and in (b) the all-*trans* configuration produced by photoabsorption. In rhodopsin itself, the retinal is linked to the terminal lysine group of a protein via an aldimine bond

Although the detailed mechanism is still not known for certain, it has been established that the primary result of photoabsorption is isomerisation of the retinal to the all-*trans* configuration shown in Fig. 5.7b. The first evidence of chemical change is in fact a shift in the peak of the absorption band to 548 nm, and this has now been shown using mode-locked laser instrumentation to occur within a picosecond of absorption, and with a quantum efficiency of about 0.7. In contrast to photosynthesis, the energy of the absorbed photon is not transferred between pigment molecules; the primary visual processes all take place within the rhodopsin molecule. Subsequently, the activated rhodopsin migrates to the cell membrane, where it results in a change in permeability to sodium ions, and thereby initiates the nerve response.

5.2.6 Laser Reaction Diagnostics

To conclude this section, it is appropriate to recap on some of the major diagnostic methods used for studying the kinetics of laser-induced reactions. Spectroscopic methods based on pulsed lasers are of course particularly appropriate for this purpose, and the relevant principles and instrumentation have mostly been discussed at length in earlier Chapters. Many of these techniques were first developed for the study of reaction rates by *flash photolysis* using pulses of about 10^{-4} s duration from conventional flashlamp sources. Laser flash photolysis has largely superceded these earlier methods because it offers the twin advantages of higher intensity and much shorter pulse length. This of course not only leads to greater precision, but also offers the possibility of investigating much faster reactions such as those discussed in the previous Section.

The high sensitivity of many laser-based methods thus makes them ideal for monitoring the concentrations not only of reactant or product species, but also short-lived transient reaction intermediates. For example many ultrafast reactions can be studied by splitting a single pulse from a mode-locked laser into two parts, one of which is used as a pump to initiate the photochemical reaction, and the other of which is passed through a variable time-delay as shown in Fig. 3.17, p. 66. The delayed part of the pulse can be then be used to generate a probe supercontinuum (see Sect. 3.3.3), and the absorption of different wavelengths by the sample can be used to monitor the appearance and disappearance of various species with different absorption characteris-

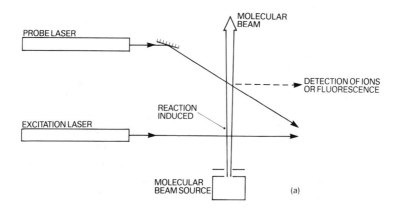

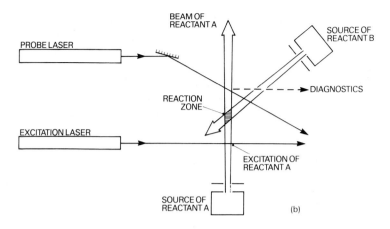

Fig. 5.8 a and b. Laser reaction diagnostics using molecular beams; (a) for the study of a unimolecular reaction, and (b) a bimolecular reaction

tics. Monitoring the variation in spectral response with delay time thus provides very comprehensive kinetic data.

For the majority of laser-induced reactions which occur in the gas phase and over somewhat longer timescales, other methods are more appropriate. For example the build-up and decay of free radicals is commonly detected by measurement of laser-induced fluorescence using boxcar methods, or by laser magnetic resonance. Examples of the species detected in this way are radicals such as CH, CH_2, C_2 and C_3 which play an important role in many organic reactions, particularly those involving combustion. Species such as C_2, however, have also been shown to appear as transients in the multiphoton dissociation of many organic compounds such as ethane, ethene, and their monosubstituted derivatives. In the case of bimolecular reactions, the activated complex (transition state) species can often be monitored in a similar manner. It is not only the reaction kinetics which can be determined by such methods, however; the richness of information in the spectra can also be used to provide structural information on the nature of the transition state, which thereby facilitates elucidation of the reaction mechanism.

One particular technique for studying laser-induced reactions in the gas phase is worth describing in a little more detail, and that involves the use of supersonic molecular beams. As described earlier (Sect. 4.3.2), a molecular beam apparatus produces a beam of molecules with a very narrow velocity distribution, in which collisional processes are minimised. Such a beam typically has an effective temperature of only a few degrees Kelvin, and usually only the molecular rotational states of lowest energy are populated. The great advantage from the point of view of reaction diagnostics is that by crossing a beam of reactant molecules with a laser beam of the appropriate wavelength, any chosen energy levels can be selectively excited in the reactant. For example, the reactant may be excited to energies above a unimolecular reaction threshold, or even into the ionisation continuum. Any subsequent reaction can then be monitored by interception of the molecular beam by a second, probe laser beam, as shown in Fig. 5.8 a. Since the molecular beam travels with a uniform translational speed (typically around

500 m s^{-1}) determined by the source temperature and a velocity selector, the reaction kinetics can be studied by varying the distance between the points of intersection of the molecular beam with the two laser beams.

By inclusion of a second molecular beam of a different species, as shown in Fig. 5.8 b, the kinetics of *bimolecular* laser-induced reactions can be studied. This provides the facility for selectively exciting specific vibration or rotation-vibration levels of either reactant prior to reaction. (Usually the molecules involved are comparatively small diatomic or triatomic species, so that intramolecular relaxation processes do not interfere to the extent that they would in larger polyatomics.) The course of reaction is generally monitored using a fluorescence or CARS technique, or else by ion detection in a photofragment mass spectrometer. The results of such studies provide enormously detailed information and facilitate the determination of rate constants for the elementary reaction steps involved in a complex chemical reaction.

5.3 Isotope Separation

Having examined the principles of laser-induced chemistry, we now move on to discuss one of the major areas of application in isotope separation. The separation of isotopes using lasers is possible by a large number of different methods. Many of these involve photochemical principles examined earlier in the chapter, but there are others such as photodeflection based on different physical principles. It is possible to broadly classify laser schemes for isotope separation into four classes which involve selective ionisation, dissociation, reaction and deflection, each to be discussed below. The common factor in all cases is the selective response of isotopically different compounds *(isotomers)* to laser radiation, based on the monochromaticity of the radiation, and the isotope-dependence of absorption frequencies. This whole area is one in which there is a great deal of interest and enthusiasm, particularly from the nuclear and associated industries. Whilst with certain compounds this connection is not overt, it is fascinating to observe how many studies have been made of the chemically rather uninteresting molecule SF_6. Of course this may be related to the fact that uranium hexafluoride has precisely the same structure.

Isotope separation schemes seldom lead to complete separation, but generally effect an enrichment in the relative abundance of a particular isotope. The essential features of any workable isotope separation scheme are a well-resolved isotopic shift in absorption frequency for the starting material, laser radiation which has a linewidth smaller than the extent of this shift, and an efficient extraction stage in which isotopic selectivity is retained. The starting material and the end-product of the process may be chemically quite different, but once an acceptable isotopic abundance has been achieved, further conventional chemical processing can be carried out. The most useful quantitative measure of the efficiency of a separation scheme is the *enrichment factor*, also known as the coefficient of separation selectivity. This is defined in terms of a ratio of the fractional content of the desired isotope in the enriched mixture to the content in the starting material, and is usually given the symbol β.

Suppose for example we have a scheme in which the starting material is a compound R, and the end-product is P. Jf the original material consists of a mixture of two isotomers R_1 and R_2, and the processing leads to formation of corresponding products P_1 and P_2, then β is given by

$$\beta = \frac{N(P_1)/N(P_2)}{N(R_1)/N(R_2)} = \frac{X(P_1)/(1 - X(P_1))}{X(R_1)/(1 - X(R_1))} \tag{5.28}$$

where N represents the number of moles of each species, and X the corresponding mole fraction. It is assumed that the scheme is designed for enrichment of the isotope contained in R_1 and P_1; hence values of $\beta > 1$ are required for successful enrichment. Typical values for laser separation lie between 1 and 10^4. As a concrete example, we can consider the condition for use of a laser enrichment process to produce nuclear fuel from naturally occurring uranium. The mole fraction of U^{235} in natural uranium is approximately 7.1×10^{-3}, and for reactor grade product a mole fraction of 3.0×10^{-2} is required: it is readily calculated using Eq. (5.28) that an enrichment factor of 4.3 is sufficient for this purpose.

5.3.1 Photoionisation

One of the simplest means for achieving laser isotope separation is atomic photoionisation using ultraviolet radiation, as for example from an excimer laser. In the conventional process of photoionisation involving single photon absorption, a wavelength usually less than 150 nm has to be employed. The ion and free electron thus formed separate with a kinetic energy determined by the excess of the photon energy over that required to reach the ionisation threshold; see figure 5.9a. To avoid recombination of the ion and electron, it is therefore best to use frequencies which provide for transition well up into the ionisation continuum. This, of course, makes the phenomenon completely inselective towards different isotopes. In fact, the wavelength selectivity required for effective isotope separation is almost invariably associated with a transition between states with *discrete energies.*

One way over this problem is to produce ionisation by a two-step absorption process, as illustrated in Fig. 5.9b. Here, the primary absorption of a photon with frequency v_1 produces a transition between two discrete (bound) electronic energy levels, and the isotopic shift is such that only one isotope has the correct energy level spacing to undergo this transition. Subsequent absorption of a second photon of frequency v_2 produces the required ionisation only in the selected isotope. Both laser frequencies must of course be low enough to make direct single-photon ionisation impossible, so that we require

$$hv_1, \; hv_2 < I < hv_1 + hv_2 \tag{5.29}$$

where I is the ionisation energy. Since the frequency of the photon absorbed in the first step is precisely determined by the energy level spacing in the appropriate isotope, a tunable laser is often employed to produce the v_1 radiation. There is, however, relative freedom over the choice of frequency of the secondary, ionising transition. It is therefore often possible to adopt the simplest solution, and arrange for both of the absorbed photons to have the same frequency. In this way, only a single laser beam need be involved.

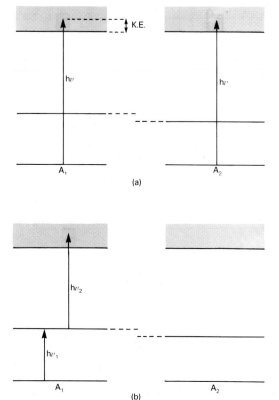

Fig. 5.9a and b. Atomic photoionisation processes. In (a), single-photon absorption results in ionisation of both isotopes A_1 and A_2. In (b), two sequential absorptions result in the selective ionisation of A_1

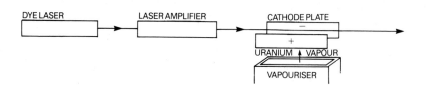

Fig. 5.10. Schematic apparatus for atomic vapour laser isotope separation based on the principle of photoionisation. The uranium vapour contains both ^{235}U and ^{238}U, but only the ^{235}U atoms are ionised by the tuned dye laser beam, and the ions thus formed are collected electrically on a flat cathode plate. The ^{238}U passes on to a separate collector

This method of isotope separation is generally practised by the laser irradiation of a beam of neutral atoms, with removal of the ions so formed by a strong electric field of the order of $10^5\,V\,m^{-1}$. Using this type of scheme, enrichment factors in the range 10^2–10^3 have been successfully achieved. Following a very large research effort ino this method of isotope separation, it has now been selected by the US Department of Energy for the next generation of uranium enrichment plants. The specific principle involved in this application is the selective ionisation of ^{235}U atoms using a dye laser pumped by a powerful 150 W copper vapour laser, in apparatus of the kind illustrated in Fig. 5.10. It is possible to obtain a similar separation effect in isotomeric *molecules*, although in this case photodissociation is more commonly used. There are also several variations on the two-step sequential absorption theme; for example the second transition may be collision-induced, or it may populate a bound state which subsequently autoionises, or else more than two steps may be involved.

5.3.2 Photodissociation

There are a number of ways in which isotopes can be separated by the selective photodissociation of isotomers. Generally, these take advantage of the comparatively large isotopic shifts associated with vibrational energy levels. As with photoionisation, direct single-photon absorption of uv/visible light is by no means ideal for the purpose of isotope separation, since it does not couple two states with discrete energy. Exceptions to this arise in molecules exhibiting *predissociation*, as illustrated in Fig. 5.11. Here, disso-

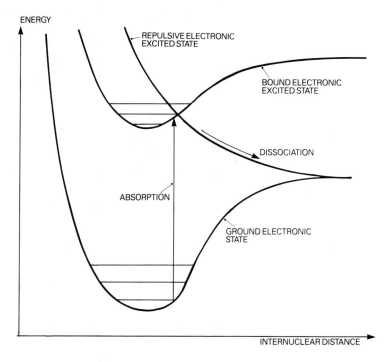

Fig. 5.11. Single-photon predissociation

ciation from a *repulsive* electronic excited state can be accomplished following a single-photon transition which involves only *discrete* vibrational levels, i. e. those belonging to the electronic ground state and a *bound* electronic excited state. A good example is provided by formaldehyde, whose first excited singlet state exhibits predissociation at wavelengths below 354.8 nm. In a mixture of formaldehyde isotomers containing hydrogen and deuterium, absorption of 325.0 nm radiation from a He-Cd laser preferentially produces predissociation of the HDCO;

$$HDCO \rightarrow \left\{ \begin{array}{l} HD + CO \\ H \;\; + DCO \\ D \;\; + HCO \end{array} \right\} \qquad (5.30)$$

The HD product can then be thermally reacted with oxygen to form HDO, and subsequent fractional distillation yields D_2O.

Generally speaking, however, isotopically selective photodissociation necessitates absorption of more than one photon, and provides us with several alternatives. Two-step absorption is an obvious option, where the first step involves an isotope-selective transition between discrete states, and dissociation is induced by absorption of a second photon. The primary absorption may produce vibrational excitation within the ground electronic state, or it may populate a vibrational level of a bound electronic excited state. The former arrangement as illustrated in Fig. 5.12 is more common, and generally requires irradiation with both infra-red and ultra-violet laser light. One condition for satisfactory operation of this method is to minimise *thermal* population of the vibrational levels. The first experiments on ammonia using this method well illustrate its potential[16]. In a mixture of $^{14}NH_3$ and $^{15}NH_3$, irradiation with the P(16) 10.6 μm line (947.7 cm^{-1}) from a TEA carbon dioxide laser produces selective vibra-

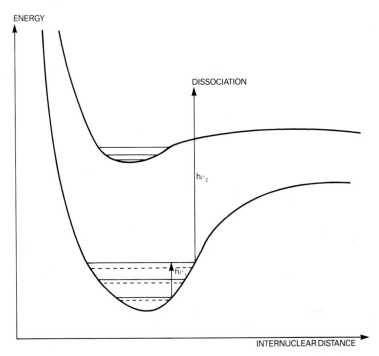

ENERGY

DISSOCIATION

$h\nu_2$

$h\nu_1$

INTERNUCLEAR DISTANCE

Fig. 5.12. Two-step photodissociation involving one infra-red photon ($h\nu_1$) and one ultra-violet photon ($h\nu_2$). The vibrational levels of another isotomer of the same compound are indicated by the dotted lines

tional excitation of $^{15}NH_3$, and subsequent dissociation can be induced by any suitably intense ultraviolet source. The sequence of reactions is as follows;

$$^{15}NH_3 + hv_1 \quad \rightarrow \ ^{15}NH_3^* \tag{5.31}$$

$$^{15}NH_3^* + hv_2 \quad \rightarrow \ ^{15}NH_2 + H \tag{5.32}$$

$$^{15}NH_2 + {}^{15}NH_2 \quad \rightarrow \ ^{15}N_2H_4 \tag{5.33}$$

$$^{15}N_2H_4 + H \quad \rightarrow \ ^{15}N_2H_3 + H_2 \tag{5.34}$$

$$^{15}N_2H_3 + {}^{15}N_2H_3 \rightarrow 2\ ^{15}NH_3 + {}^{15}N_2 \tag{5.35}$$

Since none of the elementary steps following absorption involve molecules of ammonia itself, the $^{14}NH_3$ does not enter into the reaction scheme, and so isotopic selectivity is retained; an enrichment factor of about 4 is typical here.

The most widely studied method of laser isotope separation is undoubtedly multi-photon infra-red dissociation, based on the principles discussed in Sect. 5.1.2 and illustrated by Fig. 5.1. Here, 30 or 40 photons may be involved in the process of excitation, but isotopic selectivity applies only over energy region I where the first few photons are absorbed. Once more, this selectivity results from the relatively large differences between corresponding vibrational energy levels of isomers. The enrichment factor usually reaches its highest values at low temperatures and fairly low pressures in the mbar range where collision broadening is minimised. In large molecules, even if there are discrete vibrational frequencies with suitably large isotopic shifts, the redistribution of vibrational energy amongst the various modes following absorption reduces the isotopic selectivity. Hence, most successful results based on this principle have involved fairly small molecules.

Here, there are a great many case studies, the majority making use of the intense infra-red radiation from a carbon dioxide laser. A recent example involves dissociation of a CF_3CTBrF/CF_3CHBrF mixture with 934.9 cm^{-1} radiation, providing very high selectivity for the tritium isotomer[17]. The most widely studied case is the separation of the sulphur isotopes ^{32}S and ^{34}S, based on selective multiphoton dissociation of SF_6. For example, irradiation at around 10.59 μm selectively excites the v_3 fundamental vibration of $^{32}SF_6$, and leads to its selective dissociation. The magnitude of the enrichment factor in such studies depends on several factors including pressure and laser intensity, and can be either positively or negatively influenced by the presence of other gases. Unfortunately, the prime candidate for laser isotope separation, uranium hexafluoride, does not absorb in this wavelength region, and much effort has been directed into finding a suitable laser source operating in the appropriate 16 μm region. Tunable diode lasers which operate at this wavelength have insufficient power for this purpose, but can nonetheless be used to measure the concentration of fissionable U^{235}.

5.3.3 Photochemical Reaction

We now turn to a consideration of potentially isotope-selective photochemical reactions in which the initial absorption of light does not, in contrast to the processes examined above, directly lead to ionisation or fragmentation. In general, the result of photoabsorption in such cases is the production of an excited state which subsequently undergoes a chemical reaction.

The simplest unimolecular reaction which can take place is isomerisation. Provided this leads to formation of a *geometrical* isomer, in which there are changes in the chemical bonding, the chemical and physical properties of the product and the starting material will generally be quite different, and their separation is a simple matter. The requirement is then for an isotopically selective process of *photoisomerisation*. Once again, excitation of vibrational levels provides the best means of obtaining isotopic selectivity, and the carbon dioxide laser is the favourite infra-red source. An example is afforded by methyl isocyanide, CH_3NC, which isomerises to methyl cyanide, CH_3CN, by strongly pumping at various CO_2 laser wavelengths; the associated enrichment factors for ^{13}C and ^{15}N are in the neighbourhood of 1.2.

Laser-induced reactions involving more than one reactant usually consist of the photoexcitation of one reactant, followed by its reaction with a second species;

$$X + h\nu \rightarrow X^* \tag{5.36}$$

$$X^* + Y \rightarrow Z \tag{5.37}$$

where the species X^* may in fact be either electronically or vibrationally excited. If the absorption process (5.36) is the isotope-selective step, an obvious requirement for retention of specificity is that Y reacts preferentially with X^* rather than X. A useful example is the reaction between iodine monochloride and bromobenzene. Irradiation of the mixture at 605.3 nm selectively excites $I^{37}Cl$, which subsequently undergoes the reaction

$$I^{37}Cl^* + C_6H_5Br \rightarrow I + C_6H_5^{37}Cl + Br \tag{5.38}$$

The chlorobenzene product of the reaction has a ^{37}Cl content well above the level of natural abundance. The isotopic selectivity of the reaction is mainly limited by competing processes such as the collisional transfer of energy from $I^{37}Cl$ to $I^{35}Cl$. It is important to note that the main reaction is *not* one of ICl dissociation followed by radical reaction with C_6H_5Br; the photon energy initially absorbed by the ICl is only sufficient to provide transition to a *bound* excited state ($A^3\pi_1$).

5.3.4 Photodeflection

The last of the principal laser processes used to obtain isotope separation involves photodeflection, and is based on the concept of *radiation pressure*. Every photon carries a momentum of magnitude h/λ which is imparted to any atom or molecule by which it is absorbed. A photon of visible light, for example, has an associated momentum of the order of 10^{-27} kg m s^{-1}, and its absorption by a free atom thus typically increases the atomic velocity by approximately 1 cm s^{-1}. In an atomic beam containing more than one isotope, laser irradiation at a wavelength corresponding to absorption in one particular isotope results in its selective excitation. Thus, if the laser beam and the atomic beam intersect perpendicularly, this produces a deflection of the appropriate atoms away from the beam direction.

Since a typical atomic beam velocity is $500 \, \text{m s}^{-1}$, the angular deflection of $(1 \, \text{cm s}^{-1}/500 \, \text{m s}^{-1}) = 2 \times 10^{-5}$ radians is minimal. However, since the lifetime of electronic excited states is typically measured in nanoseconds, the rapid decay back to the ground state enables a number of photons to be successively absorbed by each atom during the time it takes to traverse the laser beam. Each intervening radiative decay also involves momentum transfer, but this does not cause any net effect since emission is in random directions. Nonetheless, to produce an angular deflection of 10^{-3} radians, which even then corresponds to a displacement of only 1 mm over a 1 m distance, requires each atom to be excited something like fifty times as it traverses the laser beam. In energy terms, this is rather inefficient, and corresponds to an input of about $10 \, \text{MJ mol}^{-1}$. Successful experiments based on this principle have, nonetheless, been carried out with barium, which in its natural occurrence contains isotopes of mass number ranging from 134 to 138. In this case a dye laser tuned to 553.5 nm, for example, can be used to selectively excite and separate the ^{138}Ba isotope.

Before leaving this topic, it is worth drawing attention to another aspect of the same principle of resonant absorption. This concerns the subject of *laser cooling and trapping*. It has been shown[18, 19] that by irradiating a beam of sodium atoms with a counterpropagating beam from a ring dye laser tuned to one of the hyperfine components of the 3s $^2S_{1/2}$–3p $^2P_{3/2}$ transition, (the sodium D-line), each atom can effectively be stopped in its tracks by the successive absorption and emission of about 20,000 photons, reducing the kinetic temperature of the beam to well below 0.1 K. In this case, the laser frequency has to be continually varied to compensate for the gradual removal of the Doppler shift in absorption frequency as the atomic beam is progressively slowed down. This technique represents a new technology which should open some very interesting research possibilities for the study of isolated atoms.

5.4 Miscellaneous Applications

5.4.1 Purification of Materials

Closely related to isotope separation is a more general area of application for lasers, that of material purification. Once again the underlying principle is the specific excitation of a single chemical component in a mixture, in this case usually the impurity. In principle any of the methods discussed in Sect. 5.3 can be applied for this purpose. However, for practical application on an industrial scale, methods based on separation of atoms are clearly inappropriate and photochemical reaction is the only realistic option. An illustration is afforded by the removal of contaminants from silane, SiH_4. This gas is used in the manufacture of silicon-based semiconductor devices, and its purity is crucial for device performance. Using an ArF laser operating at 193 nm, it has been shown that impurities of arsine AsH_3, phosphine PH_3, and diborane B_2H_6 can all be photolysed and so removed from silane gas very effectively; for example 99 % of the arsine can be destroyed at the loss of only 1 % of the silane[20].

Another example based on the argon fluoride laser is the removal of H_2S from synthesis gas (a CO/H_2 mixture obtained by coal gasification). This is particularly significant since H_2S readily poisons the catalysts used for hydrocarbon synthesis. One final case which is again of relevance to the microelectronics industry is the removal from

BCl$_3$ of carbonyl chloride, COCl$_2$, which is often a fairly troublesome contaminant, using the CO$_2$ laser. Applications of this type may prove to be economically viable where there is no cheap alternative based on conventional chemical or physical methods.

5.4.2 Production of Ceramic Powders

In the last few years, much interest has been shown in the development of new methods for the chemical synthesis of ceramic powders. These are inorganic substances with a high degree of thermal stability, which show much promise for applications in mechanical, electronic and chemical engineering. A typical example is silicon nitride, Si$_3$N$_4$, which amongst other commercial methods can be formed by the gas-phase reaction of silane and ammonia;

$$3 \, SiH_4 + 4 \, NH_3 \; \rightarrow \; Si_3N_4 + 12 \, H_2 \tag{5.39}$$

This reaction requires a large input of energy, and is therefore usually carried out in a furnace, or else in the presence of an electric arc. For many practical applications, however, the distribution of particle sizes in the end-product is a crucial factor, and one for which the conventional methods of production fail to be completely satisfactory.

This problem can now be overcome by the use of radiation from a carbon dioxide laser to produce a vibrationally enhanced bimolecular reaction. Both silane and ammonia have several absorption features in the 10.6 µm wavelength region, and both can therefore undergo vibrational excitation. With beam intensities approaching $10^7 \, W \, m^{-2}$, the reaction yield is close to 100 %. In fact, this particular reaction is especially interesting since it represents one of the first cases of laser synthesis where the yield is higher and production cost lower than in the traditional chemical method. In one of the possible configurations, a mixture of the reactant gases traverses a laser beam of a few millimetres diameter, and the particles of silicon nitride powder so produced are carried by the gas to be collected at a filter.

The very short and well-controlled exposure time associated with this method generally results in the formation of a very fine powder with an exceptionally narrow distribution of particle sizes, typically less than a micron across. These characteristics maximize the mechanical strength of articles made by compacting the powder, and they also result in a high surface/volume ratio, which is highly significant for catalytic applications. In particular, laser pyrolysis provides the means for productions of a range of non-oxide based catalysts not hitherto available with such large surface area. Colloidal suspensions of such fine powders can also be prepared; these can be made indefinitely stable, and as such may prove useful for doping semiconductors.

5.4.3 Photoradiation Therapy

We conclude by looking at one last chemical application of lasers which, at least for certain individuals, may prove to have by far the most profound significance. This concerns a recently developed treatment for cancer, based on the unusual biological and photochemical properties of a compound known as haematoporphyrin derivative, usually abbreviated to HpD. Clinical trials have shown that when this dye is introduced into the body of a cancer patient, it accumulates in malignant tumours up to ten times the concentration found in normal tissue, normally within a period of 72 hours. Moreover, powerful laser radiation directed onto the cancerous tissue at wavelengths at the red end of the spectrum around 630 nm is strongly absorbed by the HpD. The result is a dissociation process which involves the formation of short-lived and highly reactive singlet oxygen, which attacks and destroys cells containing the HpD without damage to the surrounding tissue. Whilst the exact mechanisms are not yet established, this method of treatment has been successfully proven and the comparatively new gold laser, emitting several watts at 628 nm, may prove to be the ideal source for this type of treatment.

5.5 References

1. Orel, A.E. and Miller, W.H.: Chem. Phys. Letts 57, 362 (1978)
2. Zitter, R.N. and Koster, D.F.: J. Am. Chem. Soc. 100, 2265 (1978)
3. Ambartzumian, R.V., Chekalin, N.V., Doljikov, V.S., Letokhov, V.S. and Lokhman, V.N.: Opt. Commun. 18, 220 (1976)
4. Yogev, A. and Loewenstein-Benmair, R.M.J.: J. Am. Chem. Soc. 95, 8487 (1973)
5. Yogev, A. and Benmair, R.M.J.: Chem. Phys. Letts 46, 290 (1977)
6. Malatesta, V., Willis, C. and Hackett, P.A.: J. Am. Chem. Soc. 103, 6781 (1981)
7. Danen, W.C. and Jang, J.C.: in 'Laser-Induced Chemical Processes' ed. Steinfeld, J.I. (Plenum, New York, 1981)
8. Kurylo, M.J., Braun, W. and Xuan, C.N.: J. Chem. Phys. 62, 2065 (1975)
9. Krasnopyorov, L.N., Chesnokov, E.N. and Panfilov, V.N.: Chem. Phys. 42, 345 (1979)
10. Arnoldi, D. and Wolfrum, J.: Chem. Phys. Letts 24, 234 (1974)
11. Clark, J.H., Leung, K.M., Loree, T.R. and Harding, L.B.: in 'Advances in Laser Chemistry' ed. Zewail, A.H. (Springer-Verlag, Heidelberg, 1978)
12. Farneth, W.D., Zimmerman, P.G., Hogenkamp, D.J. and Kennedy, S.D.: J. Am. Chem. Soc. 105, 1126 (1983)
13. Donnelly, V.M., Geva, M., Long, J. and Karlicek, R.F.: Mat. Res. Soc. Symp. Proc. 29, 73 (1984)
14. Laubereau, A., von der Linde, D. and Kaiser, W.: Phys. Rev. Letts 28, 1162 (1972)
15. Hargrave, P.A.: Prog. Retinal Res. 1, 1 (1982)
16. Ambartzumian, R.V., Letokhov, V.S., Makarov, G.N. and Puvetskii, A.A.: Sov. Phys. JETP Letts 15, 501 (1973)
17. Takeuchi, K., Kurihara, O., Mahide, Y., Midorikawa, K. and Tashiro, H.: Appl. Phys. B 37, 67 (1985)
18. Prodan, J., Migdall, A., Phillips, W.D., So, I., Metcalf, H. and Dalibord, J.: Phys. Rev. Letts 54, 992 (1985)
19. Ertmer, W., Blatt, R., Hall, J.L. and Zhu, M.: Phys. Rev. Letts 54, 996 (1985)
20. Clark, J.H. and Anderson, R.C.: Appl. Phys. Letts 32, 46 (1978)

6 Appendix 1: Listing of Output Wavelengths from Commercial Lasers

The table below lists in order of increasing wavelength λ the emission lines of the most commonly available discrete-wavelength lasers over the range 100 nm–10 µm. Although continuously tunable lasers are not included, the molecular lasers which can be tuned to a large number of closely spaced but discrete wavelengths are listed at the end of the table. Harmonics are indicated by × 2, × 3 etc. Other parameters such as intensity and linewidth vary enormously from model to model, and no meaningful representative figure can be given. However the annually updated 'Laser Focus Buyers' Guide' and 'Lasers and Applications Designers' Handbook' both have comprehensive data on all commercially available lasers, together with manufacturers' addresses.

λ/nm	Laser	λ/nm	Laser	λ/nm	Laser
157	Fluorine	441.6	Helium-cadmium	534	Manganese
173.6	Ruby × 4	454.5	Argon	539.5	Xenon
193	Argon fluoride	457.7	Krypton	543.5	Helium-neon
222	Krypton chloride	457.9	Argon	568.2	Krypton
231.4	Ruby × 3	461.9	Krypton	578.2	Copper
248	Krypton fluoride	463.4	Krypton	595.6	Xenon
266	Neodymium × 4	465.8	Argon	628	Gold
308	Xenon chloride	468.0	Krypton	632.8	Helium-neon
325	Helium-cadmium	472.7	Argon	647.1	Krypton
333.6	Argon	476.2	Krypton	657.0	Krypton
337.1	Nitrogen	476.5	Argon	676.5	Krypton
347.2	Ruby × 2	476.6	Krypton	687.1	Krypton
350.7	Krypton	482.5	Krypton	694.3	Ruby
351	Xenon fluoride	484.7	Krypton	722	Lead
351.1	Argon	488.0	Argon	752.5	Krypton
353	Xenon fluoride	495.6	Xenon	799.3	Krypton
355	Neodymium × 3	496.5	Argon	904	Gallium arsenide
356.4	Krypton	501.7	Argon	1060	Nd:glass
363.8	Argon	510.5	Copper	1064	Nd:YAG
406.7	Krypton	514.5	Argon	1092.3	Argon
413.1	Krypton	520.8	Krypton	1152.3	Helium-neon
415.4	Krypton	528.7	Argon	1315	Iodine
428	Nitrogen	530.9	Krypton	1319	Nd:YAG
437.1	Argon	532	Neodymium × 2	3391	Helium-neon
				3508	Helium-xenon

2608–3093 (21 lines)	Hydrogen fluoride	a 9160–9840 (45 lines)	Carbon dioxide
3493–4100 (37 lines)	Deuterium fluoride	10400–11040 (58 lines)	Nitrous oxide
5090–6130 (14 lines)	Carbon monoxide	a 10070–11020 (50 lines)	Carbon dioxide

[a] See Table 2.1, p. 33 for a detailed wavelength listing

7 Appendix 2: Directory of Acronyms and Abbreviations

Bearing in mind the origin of the word 'laser' itself, it is perhaps inevitable that the field of laser applications is associated with a plethora of acronyms and abbreviations. The use of these has, as a matter of deliberate policy, been largely avoided in this book. Nonetheless, abbreviations are common in the current laser literature, and the following list has been selected to assist the reader.

AAS	Atomic absorption spectroscopy
ADP	Ammonium dihydrogen phosphate
AFS	Atomic fluorescence spectroscopy
AM	Amplitude modulation
AO	Acoustic-optic
AVLIS	Atomic vapour laser isotope separation
BW	Bandwidth
CARS	Coherent anti-Stokes Raman scattering
CCD	Charge-coupled device
CD	Circular dichroism
CDR	Circular differential Raman (spectroscopy)
CPF	Conversion (of reactant) per flash
CPM	Colliding-pulse mode-locked (laser)
CRT	Cathode-ray tube
CSRS	Coherent Stokes Raman scattering
CTD	Charge-transfer device
CVD	Chemical vapour deposition
CVL	Copper vapour laser
CW	Continuous-wave
DIAL	Differential absorption lidar
DLS	Dynamic light scattering
EL	Electroluminescent: exposure limit
EMR	Electromagnetic radiation
EO	Electro-optic
FEL	Free-electron laser
FFT	Fast Fourier transform
FM	Frequency modulation
FTIR	Fourier transform infra-red
FWHM	Full width at half-maximum
GC	Gas chromatography
GDL	Gas dynamic laser
GTL	Gas transport laser

HFS	Hyperfine structure
HOE	Holographic optical elements
HPLC	High-performance liquid chromatography
IC	Integrated circuit
ICF	Inertial confinement fusion
IM	Intensity modulation
IPL	Iodine photodissociation laser
IR	Infra-red
IRED	Infra-red emitting diode
IRIS	Infra-red interferometric spectrometer
KDP	Potassium dihydrogen phosphate
LADAR	Laser detection and ranging
LAMMA	Laser microprobe mass analysis
LAMS	Laser mass spectrometer
LAS	Laser absorption spectrometer
LASER	Light amplification by the stimulated emission of radiation
LC	Liquid chromatography
LCVD	Laser chemical vapour deposition
LDA	Laser Doppler anemometry
LDV	Laser Doppler velocimeter
LEAFS	Laser-excited atomic fluorescence spectroscopy
LED	Light-emitting diode
LEF	Laser-excited fluorescence
LFBR	Laser fusion breeder reactor
LIA	Lock-in amplifier
LIBS	Laser-induced breakdown spectroscopy
LIDAR	Light detection and ranging
LIFS	Laser-induced fluorescence spectroscopy
LIMA	Laser-induced mass analysis
LIMS	Laboratory information management system
LIR	Laser-induced reaction
LIS	Laser isotope separation
LMR	Laser magnetic resonance
LOG	Laser optogalvanic (spectroscopy)
LPS	Laser photoacoustic spectroscopy
MASER	Microwave amplification by the stimulated emission of radiation
MIS	Metal insulator semiconductor
MOS	Metal oxide semiconductor
MPD	Multiphoton dissociation
MPI	Multiphoton ionisation
MPRI	Multiphoton resonance ionisation
MVL	Metal vapour laser
NLO	Nonlinear optics
OA	Optical activity
OCR	Optical character recognition
OEM	Original equipment manufacturer
OMA	Optical multichannel analyser

OODR	Optical-optical double resonance
OPD	Optical path difference
ORD	Optical rotatory dispersion
PAS	Photoacoustic spectroscopy
PC	Photocathode
PCM	Pulse code modulation
PCS	Photon correlation spectroscopy
PDS	Photodischarge spectroscopy
PDT	Photodynamic therapy
PM	Polarisation (or phase) modulation
PMT	Photomultiplier tube
PRF	Pulse repetition frequency
PSD	Phase-sensitive detector
PWM	Pulse width modulation
PZT	Piezoelectric transducer
QE	Quantum efficiency
QED	Quantum electrodynamics
QLS	Quasi-elastic light scattering
REMPI	Resonance-enhanced multiphoton ionisation
RGH	Rare gas halide
RIKES	Raman-induced Kerr effect spectroscopy
ROA	Raman optical activity
RRE	Resonance Raman effect
RRS	Resonance Raman scattering
SALI	Surface analysis by laser ionisation
SERS	Surface-enhanced Raman scattering
SHG	Second harmonic generation
SIRS	Spectroscopy by inverse Raman scattering
SLAM	Scanning laser acoustical microscope
SNR	Signal-to-noise ratio
SRS	Stimulated Raman scattering
SSL	Solid-state laser
TDL	Tunable diode laser
TEA	Transversely excited atmospheric (pressure)
TEM	Transverse electromagnetic mode
TIR	Total internal reflection
TLV	Threshold limit value
TOF	Time of flight
TPA	Two-photon absorption
TPD	Two-photon dissociation
TPF	Two-photon fluorescence
USLS	Ultrafast supercontinuum laser source
UV	Ultraviolet
VDU	Visual display unit
VUV	Vacuum ultraviolet
XUV	Extreme ultraviolet
YAG	Yttrium aluminium garnet

8 Appendix 3: Selected Bibliography

General References

There are very few textbooks dealing comprehensively with lasers in chemistry; most books dealing with the subject are multi-authored conference proceedings. Many of these provide a useful insight into highly topical fields of research, but are not particularly enlightening without a specialised background knowledge; for this reason they have been excluded from this biliography. The following texts can be recommended for a more general readership;

Ben-Shaul, A., Haas, Y., Kompa, K.L. and Levine, R.D.: 'Lasers and Chemical Change' (Springer-Verlag, Heidelberg, 1981).

Duley, W.W.: 'Laser Processing and Analysis of Materials' (Plenum, New York, 1983).

Hecht, J.: The Laser Guidebook' (McGraw-Hill, New York, 1985).

The book by Ben-Shaul et al. is particularly recommended for its very thorough treatment of theory. Although not comprehensive in its coverage, the 'Laser Handbook' series published by North-Holland is also recommended for its detailed and thorough treatment of many topics in this area.

Laser theory and laser instrumentation

Wilson, J. and Hawkes, J.F.B.: 'Optoelectronics: An Introduction' (Prentice-Hall, London, 1983).

Thyagarajan, K. and Ghatak, A.K.: 'Lasers Theory and Applications' (Plenum, New York, 1981).

Haken, H.: 'Light' Volume 2 (North-Holland, Amsterdam, 1985).

For detailed and up-to-date information on commercially available lasers and laser instrumentation, the reader is referred to the 'Buyers' Guide' produced by the journal 'Laser Focus', and the 'Designers' Handbook and Product Directory' published in conjunction with 'Lasers and Applications', both of which are published annually and are invaluable guides to laser purchasers.

Laser spectroscopy

Demtroder, W.: 'Laser Spectroscopy' (Springer-Verlag, Heidelberg, 1982).

Hollas, J.M.: 'High Resolution Spectroscopy' (Butterworths, London, 1982).

Kliger, D.S. (ed.): 'Ultrasensitive Laser Spectroscopy' (Academic, New York, 1983).

Hieftje, G.M., Travis, J.C. and Lytle, F.E. (ed.s): 'Lasers in Chemical Analysis' (Humana, Clifton NJ, 1981).

Laser-induced chemistry

Letokhov, V.S.: 'Nonlinear Laser Chemistry' (Springer-Verlag, Heidelberg, 1983).

Steinfeld, J.I. (ed.): 'Laser-Induced Chemical Processes' (Plenum, New York, 1981).

One of the best books on the subject is now sadly out of print;

Grunwald, E., Dever, D.F. and Keehn, P.M.: 'Megawatt Infrared Laser Chemistry' (Wiley, New York, 1978).

Laser safety

Many laser textbooks deal with safety matters in brief, but the following provides a very thorough coverage of all aspects of laser safety, and cannot be recommended too highly;

Sliney, D. and Wolbarsht, M.: 'Safety with Lasers and Other Optical Sources' (Plenum, New York, 1980).

Every laser laboratory should hold a copy.

9 Subject Index

Bold type indicates the main references. Terms associated with acronyms (see Appendix 2) are listed under their full titles, and compounds are listed by name.